SCIENCE
AND
SCEPTICISM

SCIENCE
AND
SCEPTICISM

●●●●●●●●●●●●●●●●●●●●●●●●●●●●●●●●●●●

John Watkins

PRINCETON UNIVERSITY PRESS
PRINCETON, NEW JERSEY

All Rights Reserved
Library of Congress Cataloging in Publication Data will be
found on the last printed page of this book

ISBN cloth 0-691-07294-9
pbk 0-691-10171-x

This book has been composed in Times Roman

Clothbound editions of Princeton University Press books
are printed on acid-free paper, and binding materials are
chosen for strength and durability
Paperbacks, although satisfactory for personal collections,
are not usually suitable for library rebinding

Printed in the United States of America
by Princeton University Press
Princeton, New Jersey

The following questions must burningly interest me.
What goal will and can be reached by the science
to which I am dedicating myself?
To what extent are its general results 'true'?

Albert Einstein

We have to try to adopt a highly critical attitude
towards those theories which we admire most.

Karl Popper

CONTENTS

. . . .

CONTENTS

Part Two

CONTENTS

CONTENTS

Immodestly stated, my purpose is to succeed where Descartes failed: to submit our knowledge of the external world to an ordeal by scepticism and then, with the help of the little that survives, to explain how scientific rationality is still possible. No less immodestly stated, it is to find an answer to Hume, but one that accepts the validity of, and is not vulnerable to, his central negative thesis; an answer that resorts to nothing illicit, fishy, or fuzzy; no postulates "proved" by transcendental arguments, no theology in the guise of assumptions about the simplicity of nature or the preattunement of our minds to nature, no attempt to coax out of the probability calculus something it cannot give, and no relaxation of the deductivist idea of a valid inference to accommodate invalid inferences.

More modestly stated, my purpose is to provide a neo-Popperian account of human knowledge, and especially scientific knowledge. I say 'neo-Popperian' because it differs in some important ways from Popper's classical treatment. In the spirit of his exhortation that I have taken as a motto, I have been sharply critical of what I take to be some lingering traces of inductivism in his philosophy; the vexed notion of verisimilitude plays no part in what I propose as the optimum aim for science; and I am in sharp disagreement with his account of the empirical basis for science. But I should add, as will in any case be obvious, that I have arrived at my present positions only after being immersed in his profound and far-reaching system of ideas.

The ordeal by scepticism occupies Part One, or the first three chapters. I needed to make this ordeal as severe as I could, in the hope that the positive account of the rationality of science that comes afterwards would not be vulnerable to some valid sceptical argument that I had failed to take into account. In this part, I am mostly marshalling arguments that are already well known. But there is, I believe, a tincture of novelty in § 2.4 where I present an argument that I have not seen elsewhere and which seems to me to deliver a deathblow to probabilist answers to Hume.

Part Two is constructive. In chapter 4 I seek out that aim for science that is as comprehensive and ambitious as it can be without becoming incoherent or infeasible or failing other obvious adequacy requirements. In chapter 5 I try to elucidate this optimum aim for science with some precision and in a systematic way. This calls for comparative measures for testable content, explanatory depth, and theoretical unity. In chapter 6 I ask whether the statistical theories of modern microphysics can provide explanations that comply with the strictly deductivist view of valid inferences that is adopted throughout this book. In chapter 7 I ask whether deductivism is reconcilable with the claim that statements can be rationally accepted into the empirical basis for science, and answer that it is. Such rationally accepted statements need not be restricted to ones about pointer readings, ink bottles, etc. but may include ones about the position of a planet or the electrical current in a wire.

In chapter 8 I ask: given an empirical basis of rationally accepted statements, and given a set of competing theories, is there a method by which we can, at least in a good many cases, identify one of these as the one that best fulfils the above aim? I answer that it is the one, if there is just one, that is better corroborated, very much in Popper's sense, than the others. And this is the essence of my answer to Hume: that all the competing theories in the set are equally uncertain, with no positive probability of being true, does *not* mean that we have no good reason to accept one and reject the others, for we have the best possible reason to accept the one that best fulfils the optimum aim for science. However, one more hurdle remains. There are various more or less well known logical considerations within confirmation theory that suggest either that a candidate for the prize of being the best corroborated one is always accompanied by a host of equally well corroborated alternatives, or else that *no* candidate will ever emerge; so that our method will never enable us to identify just *one* theory as best fulfilling the optimum aim for science. This hurdle is overcome in § 8.4. In the Epilogue I attempt a solution, compatible with Humean scepticism, of the pragmatic problem of induction.

I wanted to write this book clearly enough for it to be accessible to anyone concerned about the rationality of scientific thought, which is perhaps under more severe challenge today than at any time since Hume first pointed out a fatal defect in (what he took to be) the structure of our empirical knowledge, and then advised carelessness, and inattention to it, as the only remedy. In particular, I hoped that some people working in the natural sciences might be tempted to look into it; for it addresses itself to the two questions that, in the passage I have used as a motto, Einstein said must burningly interest him: what is the aim or goal of the science to which he dedicated himself, and what is the cognitive status of its results? However, there are places where I was obliged to introduce a certain amount of technicality. In order that the book should remain accessible to lay readers I adopted the following policy: I warn readers that a technical argument is coming, say what its conclusion will be and where this will be reached. Thus those who wish to do so can skip the technicalities without losing the thread of the main argument. I should add, though, that I do not think that those who do this will get full value for their money. Most of the technicalities occur in chapter 5 where I try to give precision to what I claim to be the optimum aim for science. Here I needed a sharp characterisation of certain key ideas that either have been deemed unanalysable or for which the extant criteria have broken down. The Popperian philosophy of science suffered a severe blow when it turned out that its extant criteria for one theory being *more testable* than another had broken down, at least in cases where one theory revises the other. I try to repair the damage, in § 5.1, with the help of the technical concept of two statements being *incongruent counterparts*. Two interrelated ideas that have been widely assumed to be unanalysable are those of one scientific theory being *deeper* and *more unified* than another. My investigation of these ideas drew me into the problem of distinguishing between "natural" and "unnatural" axiomatisations of a scientific theory. All this was needed if the mists surrounding the idea of the optimum aim for science were to be dispelled and the idea emerge into clear daylight.

It is not for me to say whether these more technical parts of the book succeed; but if perchance they do, then a reader who skips them will miss some of the more worthwhile things in the book. In view of this, I have done my best to make the technicalities as unforbidding as I can, in the, perhaps optimistic, hope that a reader might actually enjoy strolls down these thorny paths.

J.W.

The London School of Economics, April 1983.

ACKNOWLEDGMENTS

My debt to Karl Popper is enormous. The stimulus to write this book came from Adolf Grünbaum who persuaded me, at Kronberg in 1975, that an answer was still wanting to the question: why is the best *corroborated* theory the *best* theory? Some of the arguments of the book were first tried out in public in lectures at Otago University in 1977. I am grateful to Alan Musgrave for his kindness then and for the beneficial criticism he has given me since. The book has benefitted enormously from regular exposure to critical fire at our LSE seminar and to audiences in various parts of the world. Among people who have helped me to improve it are Hartry Field, Jaakko Hintikka, Colin Howson, Irena Lachman, Hugh Mellor, David Miller, Peter Milne, Alan Stuart, and Peter Urbach. Among people who helped me to get the bugs out of the argument in § 2.4 are Sylvain Bromberger, Fred d'Agostino, and Ilkka Niiniluoto. The book would have been far more inelegant than it is but for a mass of helpful criticism from Ernest Gellner. The men who bear the heaviest responsibility for the mistakes that do not remain are Clark Glymour, Graham Oddie, John Worrall, and Elie Zahar.

Katie Platt helped to prepare the Bibliography and Andrzej Lodynski the Index. Janet Chadwick preserved a meticulous typescript through what must have seemed a never-ending process of revision. I am grateful to all these people; and above all to my wife, Micky, for sustaining me during my long preoccupation with this book.

References to authors' works are given by a date, and usually a page number, within parentheses. (There are no footnotes.) The date is normally that of the original publication of the work, but the page number cited may be that of a later version, such as an English translation or the author's collected works. In the case of a book, the date is italicised. Details are given in the Bibliography. (For some time I resisted turning 'Hume's *Treatise*' into 'Hume (*1739–40*)', but in the end consistency prevailed. I have at least avoided making it 'Hume (*1888*)', the date of the first printing of Selby-Bigge's edition, which is the one I use.) Italicised words in quotations were italicised in the original unless the contrary is indicated. In a run of references to the same work, its date is usually dropped after the first reference, only page numbers being given thereafter.

I use the following notation:

$\exists$	the existential quantifier ('there exists . . .')
$\forall$	the universal quantifier ('for all . . .')
$\sim$	negation ('not')
$\vee$	disjunction ('or')
$\wedge$	conjunction ('and')
$\rightarrow$	material implication ('if . . . then . . .')
$\leftrightarrow$	material equivalence ('if and only if').

In the interests of economy I: (1) minimise the use of brackets, for example writing $\exists x\, Fx$ ('there exists an x that is F') rather than $(\exists x)$ (Fx); (2) omit quotation marks around a formula when it is being mentioned rather than used; (3) use '.' for conjunction and '-' for negation in probability formulas: thus $p(a,\, b \cdot \bar{c})$ means 'the probability of a, given b and not–c'. I use '>' and '$\geq$' in the usual way for 'greater than' and 'not less than'. I say that p strictly entails q if q is a logical consequence of p but not vice versa. All other notation is explained when first introduced. The punctuation mark ':–' indicates that what follows it, down to the end of the paragraph, expresses a view to which I may not subscribe myself.

Part One

1

....

Scepticism
and Irrationalism

1.1 Introduction

The scepticism to which I will try to provide an answer in this book is neither Academic scepticism, which said that there is but one thing one can know, namely that one can know nothing else, nor Pyrrhonian scepticism, which said that one cannot even know that, nor Cartesian scepticism, which said that even logic and mathematics are dubious, but Humean scepticism. Humean scepticism allows that each of us can have a good deal of egocentric knowledge about our own beliefs, feelings, and perceptual experiences. It also allows that logical truths can be known. But it denies that one can progress by logical reasoning from perceptual experience to any genuine knowledge of an external world, if there is one. Since its target is not *all* knowledge, but only knowledge of the external world, Humean scepticism is not self-undermining; it does not exclude the possibility that one can *know* (perhaps on logical grounds) that there can be no genuine knowledge of an external world.

Humean scepticism seems to follow from the conjunction of the following three propositions, which I will number since I will often be referring back to them in the course of this book:

   (I)   there are no synthetic a priori truths about the external world;
  (II)   any genuine knowledge we have of the external world must ultimately be derived from perceptual experience;
 (III)   only deductive derivations are valid.

I will also refer to these three propositions as, respectively, the *anti-apriorist* thesis, the *experientialist* thesis, and the *deductivist* thesis. They entail that for any factual statement h to constitute *knowledge*, there must exist true premises e that report perceptual experiences and from which h is logically derivable. But if h speaks about the external world and e speaks only of perceptual experiences, h goes beyond e and therefore cannot be logically derived from e.

1.11 Antisceptical Strategies

I will now attempt to list the main strategies that have been adopted in response to this simple and seemingly cogent argument. These strategies fall into two groups: those that deny one or other of the above three propositions, and those that try to cope with scepticism while retaining all three. I will begin with the former.

Apriorist strategy: deny proposition (I), modifying proposition (II) accordingly, and claim that there are certain truths about the structure of the world (for instance, that the Law of Causality reigns throughout it) that are not analytic but can be known a priori, being necessarily true. This of course was the essence of Kant's answer to Hume.

Transcendental argument strategy: modify proposition (II) and claim that there are certain principles (for instance, a Principle of Induction) that are not themselves established in the ordinary inductive way but can be proved by an argument of the form: 'This principle must be true if scientific knowledge is to be possible; but scientific knowledge exists; therefore this principle is true'.

Conjecturalist strategy: deny proposition (II) and claim that scientific knowledge about the external world is not derived from experience but is essentially conjectural and is only negatively controlled by experience. This of course is the essence of Popper's answer to Hume.

Nondeductivist strategy: deny proposition (III) and claim that an inference does not have to be deductive for it to be valid.

I now turn to strategies that do not (or need not) involve the denial of any of those three propositions.

Probabilist strategy: retain proposition (III) but add that probability logic, or the logic of partial entailment, is a legitimate extension or generalisation of classical logic, and that in the case of a hypothesis h that is not entailed by evidence e we may be able to establish, with the help of probability logic, that h is more or less strongly confirmed by e. This of course is the essence of the solution of the problem of induction offered by Keynes, Jeffreys, Reichenbach, Carnap, Hintikka, and others.

Phenomenalist strategy: close the alleged gap between perceptual experience and hypotheses about the external world by claiming that physical objects just *are* clusters or sequences of actual or possible perceptions. Mach took this position; so did Russell in his (*1914*) and Carnap in his (*1928*).

Vindicationist strategy: vindicate the inductive method by showing, not that it *will* work well, but that, *if* any method for predicting the future will work at all, then an inductive method will work at least as well as any other. Reichenbach, Feigl, and Salmon have taken this position.

Pragmatist strategy: concede that there is no purely epistemological answer to scepticism, but add that we may nevertheless accept and reject hypotheses rationally, in the light of our *utilities.*

Naturalist strategy: concede that there is no epistemological answer to scepticism, but add that scepticism is of only academic interest because nature has endowed us with a cognitive machinery that is too robust and vigorous to be affected by it. Thus we can make our peace with scepticism, acknowledging it in philosophical moments and forgetting it when we hang up philosophy and go out to dine or play backgammon. This was, of course, Hume's "answer" to his own problem. Pascal had previously taken a rather similar view.

Of course, not all of these strategies are mutually exclusive. The transcendental argument and apriorist strategies may be combined, as they were by Kant; so may the probabilist and phenomenalist strategies, and so may the nondeductivist and naturalist strategies. However, I would like to insist that two strategies that are often bracketed together, namely the nondeductivist and probabilist, should be regarded as quite distinct. There are versions of probabilism that come rather close to nondeductivism; I am thinking of those subjective or personalist versions that allow us to distribute initial probabilities according to taste. But the kind of probabilism with which we will be concerned tries to provide a nonarbitrary and impersonal measure of both initial and relative probabilities. A system of probability logic that is to achieve this has to be perfectly explicit. The contrast with nondeductivism could hardly be more glaring. The latter does not try to generalise classical logic: it postulates a peculiar, nondeductive relation between statements whereby *e* can provide inductive proof for *h* without entailing it; and far from being perfectly explicit, nondeductivists maintain a sealed-lips policy concerning the nature of this mysterious relation.

There are some other defences against scepticism to which antisceptics have resorted but which hardly deserve to be called strategies and which I will deal with forthwith in a summary way. One is the claim that scepticism is an *incoherent* position or that "the sceptic" sets a standard for knowledge that is, in P.F. Strawson's words (*1959*, p. 34), 'self-contradictorily high'. Well, I can discern nothing incoherent or self-contradictory in the three propositions that I presented above as posing the problem of scepticism. And I trust that readers will discover nothing incoherent or self-contradictory in the case for scepticism that will be presented in the two chapters after this. I could have made my task easier if I had done what many antisceptics do, which is to set up a dummy called "the sceptic" and make him say various things that can be brushed aside as incoherent, self-contradictory, or whatever; but then the real case for scepticism, which is a cogent one, would have been left unanswered.

Another defence against scepticism on which I will dwell only briefly is what has been called the Paradigm Case Argument. I like to think that I knocked out this argument in (1957), though I should have done it more neatly. The argument runs–some of the descriptive expressions that figure in debates between defenders and critics of common sense have a meaning that cannot be explained purely verbally and can be taught only ostensively, by pointing to exemplars of them (Malcolm, 1942, p. 361). Let F be such an expression, and let it be a tenet of common sense that there are things that are F. Let X be either a sceptic who doubts, or a speculative metaphysician who denies, that anything is F. Now X either (1) does, or (2) does not, understand the meaning of F. If (2), then X does not know what he claims to be doubting or denying. So assume (1). But if X understands F then he must at some time have been directly acquainted with something that was F.

Now I believe that there is one class of descriptive expressions to which this argument does indeed apply, namely those that denote what Locke called simple ideas. I presumably could not understand such expressions as 'mauve patch' or 'musty smell' unless I had at some time seen a mauve patch or smelt a musty smell. But if this argument is to assist defenders of common sense against Humean scepticism it must extend beyond simple ideas to at least some items in the external world. For Humean scepticism does not deny that we have a considerable egocentric knowledge of mental existents. But any attempt to extend the Paradigm Case Argument to expressions referring to things that transcend our subjective experiences of them was forestalled long ago by Frege's sharp distinction between the *sense* and the *reference* of a denoting expression. As he said, 'In grasping a sense, one is not assured of a reference' (1892, p. 58). To which we may add that to know what

6

an expression refers to does not ensure that one has grasped its sense. Consider the expression 'material thing'. Norman Malcolm claimed that this is one of those expressions 'the meanings of which must be *shown* and cannot be explained' (1942, p. 361). If that were so, Berkeley would, presumably, stand refuted by the fact that this expression is meaningful, since the only way in which we could have come to understand it would be by acquaintance with material things. But is Malcolm's twofold claim true? Let us postpone the question whether its meaning cannot be explained and begin by asking whether it *can* be shown.

Suppose that Malcolm had wanted to convey the meaning of 'material thing' to a person X who speaks a language unknown to Malcolm and who understands no English. This should not be a serious obstacle, on Malcolm's view, since the meaning of that expression has in any case to be conveyed in an ostensive and nondiscursive way. Imagine that Malcolm holds up a book, nods his head, and pronounces 'Material thing' slowly and clearly. He repeats this performance with various other objects in the room; then he takes X to the window, points to a tree, nodding and pronouncing 'Material thing', and repeats this with other outside objects. In order to disabuse X of the idea that the expression refers just to visible objects he next blindfolds X and places a heavy object in his hand, pronouncing 'Material thing' as he does so. He then removes the bandage, points to the shadow cast by an object, and shaking his head says '*Not* material thing'. In due course X's eyes shine with comprehension. *He* now starts pointing to things to which Malcolm had not pointed, happily exclaiming 'Material thing' each time, and Malcolm nods approvingly. To Malcolm's delight, X then points to a rainbow and, shaking his head, says '*Not* material thing'. X then turns his attention to a mirror. He points to it from behind and says 'Material thing'; then stepping in front of it he first points to his own eyes, ears and nose, then gestures towards the mirror and says '*Not* material thing'. Malcolm interprets all this as X's way of indicating that while the mirror itself is a material thing, the image of X reflected in it is not; and Malcolm again nods approvingly. Let us suppose that on all subsequent occasions X applies this expression only to things to which Malcolm would apply it. Would all this mean that he has grasped its sense?

Quine envisaged a situation where the natives in a certain area appear to a linguist, who is trying to decode their language, to use the word *gavagai* to refer to what we call rabbits; but perhaps it means something like 'rabbity patch' (*1960*, ch. ii). Rather similarly, X may be, for all we know, an idealist of a Berkeleyan kind who understands 'material

7

things' to mean 'sensible objects that are not intangible'; he may have entirely missed the idea of the *materiality* of material things.

As to the second part of Malcolm's claim, namely that the meaning of 'material thing' cannot be explained, this seems plain false. Descartes said that material things are extended (i.e. three-dimensional) and that no two of them can simultaneously occupy the same place, to which Newton added that they resist displacement (possess inertia) and Locke added that they resist compression. What is wrong with a discursive explanation along those lines? It would have been easier to convey the meaning of this expression to X if it had been possible to explain to him that a molecule that is too small to be sensible is nevertheless a material object, and that the Milky Way, though too vast and far-flung to be tangible, is also a kind of material object. I conclude that Malcolm got it the wrong way round: the sense of 'material thing' can be explained and cannot be shown; and that we can grasp its sense is no refutation of the idealist's claim that it has no reference.

1.12 *Science as a Language Game*

There is a defence, if that is the word, of induction that was nicely summarised by Wesley Salmon thus: 'induction needs no defence because it is indefensible' (1968, p. 24). I suspect that this view can be traced back to Wittgenstein's pronouncement: 'What has to be accepted, the given, is—so one could say—*forms of life*' (*1953*, p. 226e). A form of life, according to Wittgenstein, is informed by rules. Some of its rules will have a more or less subordinate status, and may be defended by appeal to rules higher up. But at the top of the hierarchy of rules there will be a supreme rule (or set of rules) that cannot be defended by appeal to anything higher up; nor should it be defended by appeal to anything outside the form of life, for that would subordinate it to something alien. It is indefensible. Now apply all this to that form of life called science and assume, for the sake of the argument, that its supreme rule is a principle of induction. Then this principle is indefensible just because it reigns supreme. As Ayer put it, inductive reasoning

> could be irrational only if there were a standard of rationality which it failed to meet; whereas in fact it goes to set the standard: arguments are judged to be rational or irrational by reference to it. . . .
>
> [The sceptic's] demand for justification is such that it is necessarily true that it cannot be met. . . . When it is understood that there logically could be no court of superior jurisdiction, it hardly seems

troubling that inductive reasoning should be left, as it were, to act as judge in its own cause. (*1956*, p. 75)

To this Bartley riposted: 'The nub of the skeptical . . . objection was not, after all, simply the argument that comprehensive justification is impossible. It was, rather, that *since* comprehensive justification is impossible, the choice between competing ultimate positions is arbitrary' (1964, p. 15). If science constitutes one form of life, then presumably magic constitutes another; and everything that Ayer said about the former's supreme principle would apply *pari passu* to the latter's. Wittgenstein himself touched on this difficulty in a posthumously published work.

608. Is it wrong for me to be guided in my actions by the propositions of physics? Am I to say I have no good ground for doing so? Isn't precisely this what we call a 'good ground'?
609. Supposing we met people who did not regard that as the telling reason. . . . Instead of the physicist, they consult an oracle. . . . Is it wrong for them to consult an oracle and be guided by it?—If we call this "wrong" aren't we using our language-game as a base from which to *combat* theirs? (*1969*, p. 80e)

Wittgenstein provided no answer to this very pertinent question. For what answer could he have given? Had he said that people are *wrong* to allow themselves to be guided by oracles, he would have conceded that there is at least one form of life, or language game, that can be judged and found wanting by criteria external to it. Then why should not our own language game be judged by criteria external to *it*? Yet how could he say that people are *right* to allow themselves to be guided by oracles, when he himself preferred to be guided by 'the propositions of physics'? The only solution open to him would seem to be the relativist thesis that it is right for anyone playing the language game of magic to be guided by oracles rather than by the propositions of physics, and vice versa for anyone playing the language game of Western science.

Ernest Gellner (1968) emphasised the relativism implicit in Wittgenstein's idea that all forms of life have to be accepted as given, however much mutual conflict there may be between them. And as if in anticipation of the reply, 'But relativism is a price worth paying if it enables us to confound the sceptic', Gellner pointed out that this idea becomes incoherent and self-destructive when applied to any form of life (for instance, the Reformation) that does not accept itself as given but is self-critical and self-reforming. It is as if one adopted the maxim 'When in Rome do as Romans do', only to find, when one got to Rome, that many Romans are challenging the Roman way of life.

There is one answer to Hume for which I have the greatest respect but which I am also going to treat very summarily. For my part I regard Kant's great system as a magnificent ruin: there are still many very good things in it, but because of a central structural weakness it collapsed. I accept the received view that there can be no synthetic a priori propositions in Kant's sense. If a proposition is internally consistent and synthetic then its negation is likewise internally consistent and synthetic. A synthetic proposition holds only in some possible worlds (if it held in all, it would be analytic); and its negation holds in just those possible worlds in which it does not hold. As Hume put it, enquiries concerning only matter of fact and existence

> are evidently incapable of demonstration. Whatever *is* may *not be.* No negation of a fact can involve a contradiction. The non-existence of any being, without exception, is as clear and distinct an idea as its existence. The proposition, which affirms it not to be, however false, is no less conceivable and intelligible, than that which affirms it to be. (*1748*, p. 164)

Anyone who holds that the idea of synthetic a priori truths in something like Kant's sense can be rehabilitated should not proceed with this book, which relies on the contrary assumption.

1.13   *The Word 'Knowledge'*

I turn now to a terminological point. It is desirable, in a book on scepticism, to have clear conventions about how the word 'knowledge' is going to be used. Some words are used as *success*-words, to use Gilbert Ryle's apt term (*1949*, pp. 130f). Others are *not* success-words. Consider 'deduce' and 'infer'. According to standard usage a conclusion cannot be invalidly deduced from premises: an invalid derivation is not a deduction. By contrast, an inference may be invalid. But between these unambiguous cases there is an intermediate class of words that sometimes are, and sometimes are not, used as success-words. Consider the word 'law'. If one says, 'Galileo's law of free fall turned out to be incorrect' one is not using it as a success-word; if one had been using it as a success-word one would have had to say instead something like, 'Galileo's "law" of free fall turned out to be no law'. Another example is the word 'explanation'. If one says, 'But your explanation conflicts with the following fact. . . .' one is not using it as a success-word; but some people prefer to build a requirement of truth into their concept of explanation. They would have to say something like: 'But your "explanation" is no explanation: it *conflicts with the following fact. . . .*' As to 'knowledge': many philosophers treat this as a success-word, so that one can no more speak of 'erroneous knowledge' than of 'invalid deduction'. But in everyday speech it is often used just to refer to some organised body of learning. Someone may say that a vast amount of knowledge is contained in the *Encyclopaedia Brittanica* without meaning that it contains no mistakes. Indeed, someone may say that medical knowledge in the eighteenth century was very imperfect and included much that was false. In this book the word 'knowledge' (and the words 'law' and 'explanation') will normally be used in this latter way. If 'knowledge' is being used as a success-word it will be italicised. Thus there will be no inconsistency in the statement, 'There is a lot of astronomical knowledge, and quite a lot of astrological knowledge, and none of this is *knowledge*.'

1.14 *Scepticism and Doubt*

It is widely supposed that to be sceptical of something is to doubt it, where doubt is a mental state midway between belief and disbelief; hence if a professed sceptic cannot help believing something about which he claims to be sceptical, his scepticism, on this view, is not genuine. People who take this view often add, no doubt correctly, that it is psychologically impossible for, say, a Humean sceptic to doubt all those propositions of which he professes to be sceptical, from which they conclude that a genuine scepticism is impossible. Russell took this view: 'Scepticism, while logically impeccable, is psychologically impossible, and there is an element of frivolous insincerity in any philosophy which pretends to accept it' (*1948*, p. 9).

I see Humean scepticism as a strictly epistemological theory: it says that none of our knowledge of the external world is *knowledge*. It is rather analogous to a moral philosophy that says that there can be no objective justification for any moral system. A moral philosopher who holds this view is not thereby obliged to purge himself of his own moral preferences, though he is obliged to admit that he cannot offer any rational defence of them: they constitute, as it were, his personal moral faith. He may dislike moral faiths that conflict with his but he must admit that there can be no rational discrimination between them. Rather similarly, a sceptic is not obliged to try to anaesthetise his personal

11

beliefs: he may believe ever so many propositions of the types that G.E. Moore never tired of claiming to know, with certainty, to be true (*1959*, chs. 2, 7, 9, 10). But he is obliged to declare that he cannot justify them. The idea that he should doubt any proposition of which he is sceptical overlooks a middle possibility between justified belief and unbelief, namely unjustified belief. George Santayana, near the end of his essay on scepticism, wrote of 'the animal faith I live by from day to day' (*1923*, p. 308). A sceptic may find that he cannot help believing various things for which there is no epistemological justification rather as he cannot help fearing heights; they constitute, as it were, his animal faith.

A word now about the plan of Part One of this book. What I am here calling Humean scepticism is actually only a part of Hume's whole argument for scepticism. So in §1.2 I will make a brief historical excursus and take a critical look at what Hume himself considered the most decisive and radical case for scepticism. In §1.3 I will consider the naturalist strategy. My conclusion will be that Hume was wrong to regard philosophical scepticism as a kind of academic joke (Hume wrote: 'When [a Pyrrhonian] awakes from his dream, he will be the first to join in the laugh against himself, and to confess, that all his objections are mere amusement', *1748*, p. 160); for unless we can find a rational answer to it, Humean scepticism is likely to encourage irrationalism. The nondeductivist strategy will be considered in §1.4, with the conclusion that it too is likely to encourage irrationalism. In §1.5 the pragmatist strategy will be examined and found ineffective. Chapter 2 will examine the probabilist strategy. Its conclusion will be that there can be no probabilities of the kind needed by an antisceptical probabilist. Chapter 3 will examine various attempts to rehabilitate the legitimacy of an inductive ascent from perceptual experience to public knowledge of things and events, of empirical regularities, and of underlying laws of nature. The phenomenalist strategy will be examined in §3.2, with the conclusion that, while it eliminates the vertical gap between experiences and things, it leaves the horizontal gap between present and future experiences as wide as before. The transcendental argument strategy will be examined in §3.32, with the conclusion that it boomerangs against an antisceptic who resorts to it since the sceptic is entitled to invert it. In §3.34 the vindicationist strategy will be examined and found ineffective.

If the negative results here anticipated are correct, then by the end of chapter 3 all the antisceptical strategies listed above will have been defeated with the single exception of conjecturalism. In the remainder of this book I will develop a conjecturalist theory of knowledge that is endorsed by the highest cognitive aim that is available to us and which exhibits the essential rationality of the scientific adventure.

1.2 An Historical Excursus

Answering Hume has become a philosophical industry. Its beginning was slow. Book I of the *Treatise* was published in 1739. It then lay around for some years, like a time bomb quietly ticking away, before public attention was drawn to the danger. The first answer to Hume did not appear until 1764: this was Thomas Reid's *Inquiry into the Human Mind on the Principles of Common Sense.* I shall say something about this book shortly. It was followed by books by James Oswald, James Beattie, and Joseph Priestley in 1766, 1770, and 1774. Then came the greatest Hume-answerer of all, Immanuel Kant. In the Preface to the *Prolegomena* Kant wrote:

> But fate, ever unkind to metaphysics, decreed that he [Hume] should be understood by nobody. One cannot observe without feeling a certain pain, how his opponents Reid, Oswald, Beattie and finally Priestley, so entirely missed the point of his problem. (*1783*, p. 7)

He added:

> I freely admit: it was David Hume's remark that first, many years ago, interrupted my dogmatic slumber and gave a completely different direction to my enquiries. . . . (P. 9)

Kant's answer might be summarised, somewhat picturesquely, as follows–Hume rightly saw that perceptual experience cannot support unaided the towering system of our physical knowledge; what Hume failed to see is that this system is not just a collection of hypotheses: it consists of experience infused and structured by a steely framework of synthetic categories. This framework, whose truth can be known a priori, prevents the system from collapsing.

It is a well-known fact that the framework of categories that Kant supplied has been burst asunder by progress in physics since his day. In his day physics was, of course, essentially Newtonian physics, and his metaphysical framework was adjusted to that. Kant depicted physical space as necessarily Euclidean and in his Analogies of Experience he attempted to prove metaphysical versions of three fundamental principles of the Newtonian 'System of the World', namely: the conservation of matter (permanence of substance), physical determinism (law of causality), and the mutual gravitational attraction between all bodies (all substances coexisting in space are in thoroughgoing reciprocity). All this has been either repudiated or at least brought into question by modern physics.

1.21 Hume's First Principle

Now let us turn back to Hume. For him, the main argument for scepticism was based on the famous principle that he laid down in the opening section of the *Treatise*, 'the principle of the priority of impressions to ideas' (*1739–40*, p. 6). This is a venerable principle; it goes back at least to Epicurus. It was sometimes expressed in the maxim 'There is nothing in the intellect which was not first in the senses'. Let us call this principle *sensationalism*. What Hume did was to draw out the devastating implications of sensationalism for the possibility of knowledge of a physical reality beyond our senses. Consider its bearing on the following statement: 'The sun remains in existence during the night; it is an immense mass, composed mainly of hydrogen, and exerts a gravitational pull that keeps the earth in orbit round it.' Sensationalism implies that this sentence, understood in a literal and realistic way, is empty, fails to express an idea. Take first the second part of the sentence. Neither 'hydrogen' nor 'gravitational pull' denotes an entity of which we can have impressions. According to sensationalism, therefore, we can have no idea corresponding to either of these terms. Hume wrote:

> my intention never was to penetrate into the nature of bodies, or explain the secret causes of their operations. . . . I am afraid, that such an enterprize is beyond the reach of human understanding, and that we can never pretend to know body otherwise than by those external properties, which discover themselves to the senses. (*1739–40*, p. 64)

If anyone claims that he does have an idea answering to some abstract theoretical term, Hume asks him to point out the impression from which he derived it, adding: 'But if you cannot point out *any such impression*, you may be certain you are mistaken, when you imagine you have *any such idea*' (p. 65).

So that leaves us with the first part of the sentence. From my impressions I have derived the idea of the sun as a bright yellow disc. Can I form the idea of a physical sun different from my mental picture of the sun? According to sensationalism I cannot:

> Now since nothing is ever present to the mind but perceptions, and since all ideas are deriv'd from something antecedently present to the mind; it follows, that 'tis impossible for us so much as to conceive or form an idea of anything specifically different from ideas and impressions. (P. 67)

Thus I deceive myself when I believe that I have an idea of the sun as a body with a continuous existence independent of my perceptions.

> We may . . . conclude with certainty, that the opinion of a continu'd and of a distinct existence never arises from the senses. (P. 192)

Again:

> as to the notion of external existence, when taken for something specifically different from our perceptions, we have already shewn its abusrdity. (P. 188)

Let us use *irrealism* as a label for the sceptical thesis that it is impossible for us even to form the idea of a body that has a continued existence independent of our minds, let alone any idea of the inner structure and invisible constitution of such a body. Let us assume, for the moment, that Hume was right to hold that sensationalism implies irrealism. But this presents us with a choice: we can accept sensationalism and hence irrealism, as he did; alternatively, we can reject irrealism and hence sensationalism. Which should it be?

Let us take sensationalism first. Kant said that percepts without concepts are blind: for there to be sense-*experience*, as distinct from a flow of inchoate sensations, the sensations must be structured by prior categories. We could restate this in Hume's language as, 'No *impressions* without antecedent ideas'. And we could restate a criticism by Popper (*1963*, pp. 42–46) of Hume's psychological theory as, 'No *repetitions* of impressions without antecedent ideas'. So we should at least put a question mark against sensationalism.

1.22   *Do Ibeims Exist?*

What about irrealism? It seems to me that this sceptical thesis of Hume's is self-defeating. Consider the following three sentences: 'There are no centaurs', 'There are no round squares', 'There are no ibeims'. As we remarked earlier, Hume held that the idea of the nonexistence of anything is as clear and distinct as the idea of its existence: a positive existential statement is meaningful if we have ideas answering to the terms that occur in it; and if we do have such ideas, then the corresponding negative existential statement is equally meaningful. Conversely, if we have no ideas answering to the terms that occur in an existential sentence, then it will be meaningless for us both in its positive and in its negative form. If we are to understand a positive or negative existential statement, we must be able to form an idea of the kind of thing that it declares to exist, or not to exist. Thus 'There are no centaurs' is meaningful: we can attach ideas (namely, the idea of the upper part of a man and the idea of a horse minus its normal head) to the term 'centaur', and we can amalgamate these ideas in imagination. And 'There are no round squares' is meaningful: it says that nothing is both round and square, and we attach ideas to the terms 'round' and 'square' (though in this case we cannot amalgamate these two ideas in imagination). But what about the third sentence? If you attach no idea to 'ibeim', then 'There are no ibeims' will make no sense to you: you will have no idea as to what this sentence declares to be nonexistent. Now replace 'ibeim' by 'idea of a body existing independently of minds'. The third sentence, thus expanded, expresses Hume's irrealist thesis: 'No idea of *a body existing independently of minds* exists'. But what idea could you attach to the italicised phrase in this sentence? The sentence itself says that there is no idea to be attached to it. So we get a sort of paradox: if Hume's thesis were true, no one could understand it. Imagine that all living creatures were blind with the sole exception of one man, call him David, who publishes a book (in braille) that includes the sentence 'No idea of the colour red exists'. Then his sentence is false since *he* has an idea of the colour red. Now suppose that everybody including David is blind, but that he publishes a book that includes that same sentence. Then neither he nor anyone else could know what it means. Now Hume, unlike some philosophers one could name, was not in the habit of writing sentences so profound that no one, himself included, knows what they mean. He obviously found his irrealist thesis perfectly meaningful. So do I. But this can only imply that it is false. And false it assuredly is. Astronomers who have the idea of "black holes" have an idea that could not have existed independently of minds but an idea of things that, if they do indeed exist, undoubtedly exist independently of our minds.

So it is understandable that the first answer to Hume, that of Thomas Reid, concentrated on this prominent component of his philosophical scepticism, and that his answer boiled down to this: since sensationalism implies irrealism, sensationalism must, after all, be false. Writing to Hume after Hume had perused the manuscript of his (Reid's) *Inquiry*, Reid said:

> Your system appears to me not only coherent in all its parts, but likewise justly deduced from principles commonly received among philosophers; principles which I never thought of calling in question, until the conclusions you draw from them in The Treatise of Human Nature made me suspect them. (1763, i, p. 91)

But does Hume's sensationalism lead inevitably to his irrealism? It depends essentially on his key distinction between the original *impressions* and their derivatives, the *ideas*. And Maurice Mandelbaum has argued convincingly that 'in Hume's original distinction between impressions and ideas—a distinction integral to the whole of his theory of knowledge—a necessary realistic assumption is already contained' (*1964*, p. 157). True, Hume sought to base that distinction on the superior 'force and liveliness' of impressions; but Mandelbaum points out that Hume himself admitted that a feverish nightmare, say, may be as vivid as some impressions. (Consider the impressions a coast-guardsman gets staring out to sea during a foggy twilight.) Mandelbaum claims, and he is surely right, that Hume 'in fact constantly thinks of impressions as arising from physical causes operating on our senses' (p. 159); in 'other words, what characterizes all original impressions . . . is the fact that they must be attributed to causes which lie outside of experience'(p. 162).In short, Hume's sensationalism actually presupposes a certain kind of physical realism.

What I shall take as the core of Humean scepticism can be restated very simply thus: the only way in which a factual statement can be established as true, or at least probable, is by past experience; but no statement that goes beyond past experience can be established as true, or even as probable, by past experience. Let us now begin our examination of answers to Humean scepticism, beginning with Hume's own answer.

1.3 The Naturalist Strategy

Let us consider what the effects are likely to be on the beliefs of someone who is persuaded that the case for Humean scepticism is unanswerable, and who concludes that it is irrational for him ever to accept any hypothesis. (I will use the word 'hypothesis' for any statement that has implications that go beyond his present evidence. Thus 'The sun will not explode tomorrow' and 'No man lives to be 1000 years old' are hypotheses in this sense.) The logical possibilities would seem to be these: (1) he *discards* some or all of the hypotheses he had previously accepted, without accepting any new ones in their stead; (2) he *retains* all his previously accepted hypotheses essentially unchanged, despite the admitted irrationality of retaining them; (3) he *switches* from some (or conceivably all) of his previously accepted hypotheses to new ones, despite the admitted irrationality of doing so.

Now Hume seems not to have entertained possibility (3); and he held a psychological theory that ruled out (1); so he concluded in favour of (2): ''Tis evident, that so extravagant a Doubt as that which Scepticism may seem to recommend, by destroying *every Thing*, really affects *nothing*, and was never intended to be understood *seriously*, but was meant as a *mere* Philosophical Amusement' (1745, p. 20). But I hold that the real issue is not between (1) and (2) but between (2) and (3), and that there are reasons for supposing that (3) is the more likely. Let us look into these three possibilities, beginning with (1).

1.31 *Deductivism Is Deadly*

To give a little colour to possibility (1), imagine that a powder has been invented that is tasteless and generally harmless except that, taken over a period, it renders one increasingly unwilling to make logically invalid inferences (we might call it a "logic powder"). Assume, for the present, that Hume was right in supposing that we form all our beliefs about the world around us by a process of invalid inference from perceptual evidence. Then it would seem that a civilised nation could be brought to its knees by agents of a foreign power introducing this powder into the water supply. At first the effects might not be spectacular; here, a bishop preaching agnosticism; there, a doctor refusing to prescribe ('It's worked in similar cases, but that doesn't mean it'll work in yours'). But as time passed, a creeping paralysis would infect all human activity. But for the saving fact that the belief that water quenches thirst would dissolve, the end would be, to borrow Hume's description of what would happen if Pyrrhonian principles were 'universally and steadily to prevail', that 'All discourse, all action would . . . cease; and men remain in a total lethargy, till the necessities of nature, unsatisfied, put an end to their miserable existence' (*1748*, p. 160).

But Hume was perfectly confident that the effect on someone persuaded of the validity of Humean scepticism would not be at all like the effect of this powder. If someone is persuaded by concrete arguments that his acceptance of some specific hypothesis is irrational, he may discard that hypothesis. But if he is persuaded by abstract arguments that it is irrational for him to accept any hypothesis, then we can hardly expect him simply to discard them all. As Hume said about his own sceptical reflections: 'Very refin'd reflections have little or no influence upon us' (*1739–40*, p. 268). Then what should we expect him to do? According to Hume we should expect him simply to retain them all; for sceptical reasoning is up against something immensely robust and resilient, namely the belief-forming machinery that is part of human nature (or rather, of animal nature, since this is something we share with other animals). This cognitive machinery works in a logically invalid but straightforward way: it generates expectations about the future out of past repetitions in experience. It transmutes perceptions into general beliefs:

> Animals as well as men learn many things from experience, and infer, that the same events will always follow from the same causes. By this principle they become acquainted with the more obvious properties of external objects, and gradually, from their birth, treasure up a knowledge of the nature of fire, water, earth, stones, heights, depths, &c. (*1748*, p. 105)

And this part of our animal nature is quite impervious to the sceptic's demonstration that it functions in a nonlogical way:

> Should it here be ask'd me, . . . whether I be really one of those sceptics, who hold that all is uncertain . . . I shou'd reply, that this question is entirely superfluous, and that neither I, nor any other person was ever sincerely and constantly of that opinion. Nature, by an absolute and uncontroulable necessity has determin'd us to judge as well as to breathe and feel. (*1739–40*, p. 183)

Nature has endowed us with an inductive instinct that drives us across logical gaps. (Pascal had said something rather similar: 'Nature comes to the help of impotent reason', *1670*, p. 151).

Now since our minds, according to Hume, all work in essentially the same way, then insofar as the regularities of nature affect us similarly, our respective beliefs about them will tend to converge and, indeed, to coincide: 'For as the faculties of the mind are supposed to be naturally alike in every individual; . . . it were impossible, if men affix the same ideas to their terms, that they could . . . long form different opinions of the same subject' (*1748*, p. 80). And the consensual system of belief towards which we naturally tend is in fact, despite any logical scruples a philosopher may have, essentially reliable; there is, Hume wrote, 'a kind of pre-established harmony between the course of nature and the succession of our ideas' (*1748*, p. 54).

In short, Hume held first, that there is one natural way of forming beliefs about the world and that we do this as instinctively and inevitably as we breathe and feel; and second, that the general beliefs we form about those natural regularities of which we have all had experience tend to coincide. Of course, when a local regularity in one area differs from its counterpart in another area, people exposed to the former will tend to form expectations different from those of people exposed to the latter: an Eskimo living in the Arctic circle will have expectations about night and day different from those of a native of equatorial Africa. But in so far as people's experiences are similar, their beliefs will be pretty similar too.

1.32   *Strawson's Naturalism*

This *monist* thesis, as we may call it, concerning first the *uniform* way in which members of the human species actually form beliefs, and second, the *consensual* nature of the beliefs thus formed, has been reaffirmed in recent times by several philosophers concerned with scepticism and induction. For instance, with respect to the first part of this thesis, J.M. Keynes declared: 'Inductive processes have formed, of course, at all times a vital, habitual part of the mind's machinery. Whenever we learn by experience, we are using them' (*1921*, p. 217). And with respect to the second part, John Kekes has said that a common-sense world view is forced upon us by nature: common sense 'is physiologically basic' (*1976*, p. 44) and constitutes 'the world view that a human being cannot help having' (p. 48).

But the antisceptical philosopher who has, I think, gone furthest in reviving Hume's naturalist strategy towards scepticism, and the monist thesis of cognitive psychology on which that strategy is based, is P.F. Strawson. The following quotation is from a riposte to Wesley Salmon who had argued (1957) that if the basic canons of induction cannot be given a general justification, then our acceptance of them can only be by conventional choice. Strawson replied:

Hume . . . did not think that induction could be given a general justification. He did not, on this account, think that inductive beliefs were *conventional*; he pointed out that they were natural. He did not think that our 'basic canons' were arbitrarily *chosen*; he saw that this was a matter in which, at the fundamental level of belief-formation, we had *no choice at all*. . . . [O]ur acceptance of the 'basic cannons' . . . is forced upon us by Nature.

Suppose I am convinced that there is nothing to choose, as far as Reason goes, between the 'basic canons' of induction, and a consistent counter-inductive policy. Is an 'arbitrary choice' then really open to me? Is it? (Just try to make it.)

If it is said that there is a problem of induction, and that Hume posed it, it must be added that he solved it. (1958, p. 21)

If an inductive mode of reasoning about matters of fact is forced upon us by Nature, then to protest that our process of belief-formation involves logically unsavoury inferences is rather like protesting that our process of reproduction involves morally unsavoury sex.

But is the first part of Hume's monist thesis *true*, as a matter of psychological fact? Is it the case that the processes of belief-formation, at least of sane people, are universally governed by certain 'basic canons' of induction? This is a question for cognitive psychology. Not being a cognitive psychologist, I will rely upon testimony and evidence provided by others. Of course, how we answer it will depend on what we take the 'basic canons' of induction to be. So I will begin with a fairly strong construal of them, and then turn to successively weaker construals.

Do the 'basic canons' of induction constitute something like the basic rules of Bacon's inductive method? If so, we have the testimony of Bacon himself that our *natural* way of thinking is deplorably different from this (e.g. *1620*, I, lxiv). Is it a 'basic canon' of induction that one forms a general belief only after observations of several positive instances? If so, we have the testimony of Hume himself that we frequently form such beliefs 'on one single experiment' (*1739–40*, p. 131). Is it a 'basic canon' of induction that we arrive at a generalisation or theory only after experience of at least one positive instance? If so, the history of Western cosmological speculation tells heavily against it. Consider an early speculation of Anaximander's: namely that the earth does not fall downwards, not because it is supported by something beneath it as Thales had held, but because, being equidistant from all things, there is no cause for it to move in any particular direction. Popper pointed out that this theory has 'no analogy whatever in the whole field of observable facts' (*1963*, p. 138). It seems to have been the free invention of a speculative mind. Boscovic's idea that matter is composed of unextended point-atoms or force-centres, and the Faraday-Maxwell idea of a field of force dissociated from ponderable matter, are later examples of speculative ideas that could hardly have been engendered by observational experience.

Since Hume held that we learn from experience in essentially the same way as do other animals, it is relevant to mention that Konrad Lorenz's "deprivation experiments" strongly suggest that at least some animals are endowed with "beliefs" or information not derived from their experience. In these experiments, an animal is reared from birth under artificial conditions that deprive it of any possibility of learning some specific thing, call it a, by experience. When released it is watched to see if it behaves as though it nevertheless "knows" a; and often it does. For instance:

> A young swift reared in a narrow cave in which it cannot extend its wings . . . [or] attain a sharp retinal image . . . , nevertheless proves to be perfectly able on the very moment it leaves the nest cavity to assess distances. . . . It can also cope, in its rapid flight, with all the intricacies of . . . turbulence, and air pockets and can "recognise" and catch prey, and finally effect a precise landing in a suitable place. (*1965*, pp. 25–26)

Is it a 'basic canon' of induction that, even if we form a hypothesis on no positive evidence, we will at least abandon such a hypothesis if it is crushingly refuted? If so, we have the testimony of Festinger et al. (*1965*) that someone who has invested heavily in such a hypothesis may cling all the more tenaciously to a version of it modified very slightly to avoid the refutation by a hair's breadth.

I daresay that a defender of the claim that we all form our beliefs in accordance with certain 'basic canons' of induction will succeed in immunising it against all this counter-evidence without rendering it vacuous. But in the meanwhile I will assume that the first part of Hume's monist thesis is empirically refuted.

I turn now to its second part, namely that, despite local variations due to peculiarities in each person's experience, there are essential similarities between people's natural belief-systems, because cognitive faculties are 'naturally alike in every individual' and because there are many natural regularities, relating to 'fire, water, earth, stones, heights, depths, etc.', to which we have all been exposed. Strawson (*1959*) has given a touch of precision to this claim. Beneath any local variations in different people's belief systems there is an unchanging common core: 'there is a massive central core of human thinking which has no history—or none recorded in histories of thought; there are categories and concepts which, in their most fundamental character, change not at all' (p. 10). These categories and concepts constitute a common 'conceptual scheme' that makes human reasoning possible: 'the whole process of reasoning only starts because the scheme is as it is; and we cannot change it even if we would' (p. 35). This obviously implies that this scheme is shared by all human beings capable of reasoning.

This scheme, according to Strawson, gives a central position to two kinds of particulars: *material bodies* and *persons*; and for these particulars there must be *criteria of identity*: 'we must have criteria or methods of identifying a particular encountered on one occasion . . . as *the same individual* as a particular encountered on another occasion. . . . ' (p. 31). And this in turn means that we must endow things and persons with a certain 'stability and endurance' (p. 39). Of course, our conceptual scheme allows that things and people change, but not abruptly into something quite different: they do not undergo sudden metamorphoses.

It may well be true that people living within our Western culture share a conceptual scheme of this kind. But is it shared by all people of whatever culture? Again, this is an empirical question, this time for social anthropology. So let us turn to social anthropologists. According to Lévy-Bruhl, the ontology of what he called "the primitive" does not consist of basic *units* (or 'particulars', to use Strawson's term); these are mere transient vehicles for an all-pervasive invisible stuff, sometimes called *mana* or *imunu*, that flows into and out of them, giving them special powers; and things may change abruptly into other things according to the movements of this stuff:

> Like ourselves, the primitive perceives the general differences between a stone and a tree, or a tree and a fish or bird, but he does not heed them, because he does not feel them as we do. The form of objects interests him only so far as it permits him to divine how much *mana* or *imunu* they may possess. Accordingly he sees no difficulty in metamorphoses which are quite incredible to us: to him, all forms of matter may change their dimensions and their shape in the twinkling of an eye. (*1927*, p. 20)

This conceptual scheme repudiates Bishop Butler's dictum, 'Everything is what it is and not another thing'. Suppose that Strawson's faithful old dog unexpectedly rounds on him and bites him. Then it would be

23

perfectly obvious, to people who think in this way, that it was not Strawson's *dog* who bit him: perhaps it was a malicious Popperian who had cunningly turned himself into the dog. According to Lévy-Bruhl, such metamorphoses as a man turning into a crocodile or a lion or a tree or a bird are not regarded, among native people in Africa, Ecuador, Malaya, New Guinea, and other parts of the non-Western world, as rare and special occurrences. It is taken for granted that they happen all the time, at least if Lévy-Bruhl's account is to be trusted. And this in turn means that these people transgress Strawson's way of demarcating particulars into material bodies and persons:

> Among the things we ascribe to ourselves are things [physical characteristics like height, colouring, shape and weight] of a kind that we also ascribe to material bodies to which we should not dream of ascribing others of the things [actions, intentions, sensations, thoughts, feelings, perceptions and memories] we ascribe to ourselves. (P. 89)

Speaking of the natives of the Dutch East Indies, Lévy-Bruhl said that they ascribe to stones and boulders things, namely spiritual powers, that a European would not dream of ascribing to them:

> Hence the trouble and uneasiness of the natives when they see Europeans attacking stones and boulders with hammers or in any other fashion—feelings which prospectors and miners have to reckon with. . . . Endowed with mystic powers, these boulders or stones (for instance, those of a curious shape or strange position or abnormal size) may exert a favourable or an unfavourable influence upon the native and his family. (P. 27)

How reliable is this testimony? Evans-Pritchard, in his introduction to the English translation of Lévy-Bruhl's (*1927*), wrote:

> The evidence cited by Lévy-Bruhl is impressive and much of it is given by authorities who beyond question must be respected as knowledgeable and trustworthy. So, even if Lévy-Bruhl's conclusions about primitive mentality can no longer be accepted quite in the terms in which he set them forth, it is a plain fact that much of the thought of primitive peoples is difficult, sometimes almost impossible, for us to understand, in that we cannot follow their lines of reasoning because the underlying assumptions on which they are based, while taken for granted by them, are totally alien to us. (p. 6)

Strawson did not raise the question whether his account of that common conceptual scheme that allegedly makes human reasoning possible is compatible with the existence of magical thinking. Nor will I. I will only mention that Evans-Pritchard, in his great (*1937*), exhibited the everyday, practical thinking of the Azande as differing radically from its Western counterparts: their magical world view penetrates into almost every detail of their daily lives. When Evans-Pritchard entered into their way of thinking, the world around him seemed very different from a world seen through Western eyes. At one point, having described how Azande behave when they believe themselves to be bewitched, he went on to raise the question of how they feel.

> I was aided in my understanding of the feeling of bewitched Azande by sharing, at least to some extent, like experiences. I tried to adapt myself to their culture by living the life of my hosts, as far as convenient, and by sharing their hopes and joys, apathy and sorrows. . . . In no department of their life was I more successful in 'thinking black', or as it should more correctly be said 'feeling black', than in the sphere of witchcraft. I, too, used to react to misfortunes in the idiom of witchcraft, and it was often an effort to check this lapse into unreason. (P. 99)

Hume does not seem to have been joking when he wrote: 'would you know . . . the Greeks and Romans? Study well . . . the French and English. . . . Mankind are so much the same, in all times and places, that history informs us of nothing new or strange in this particular' (*1748*, p. 83). It is just as well that Evans-Pritchard did not try to get to know the Azande by studying well the inhabitants of Oxford.

1.33 *Scepticism and the Pleasure Principle*

So it seems that it is not true that there is one world view that a human being cannot help having (Kekes), or one unchanging conceptual scheme that all men share (Strawson), just as it is not true that there is one way (namely, an inductive way) in which we all form our beliefs. The relevance of this negative empirical conclusion to our main problem is that it undermines Hume's own answer to scepticism. I mentioned earlier that the formal possibilities open to someone fully persuaded of the validity of Humean scepticism are: (1) to *discard* some or all of his previous beliefs without replacing them; (2) to *retain* them all essentially unchanged; (3) to *switch* to different beliefs. Hume's monist view of human cognition would, if true, effectively rule out (1) and (3).

25

It denies (1) by saying that we can no more stop judging than we can stop breathing, and it denies (3) by saying that our judgments are formed in certain natural and inevitable ways. Now I am prepared to concede that, as a matter of psychological fact, (1) is hardly a serious possibility: perhaps it is impossible for anyone to sustain a general state of unbelief for long. But once we realise that people confronted by similar evidence may proceed in very different ways to very different interpretations of it, then we must recognise that, even if (1) is not a genuine possibility, (3) most certainly is. Consider an imaginary case. X is a man of strong feelings whose thinking has been governed by what Freud called the reality principle: he has endeavoured to be realistic and rational in his beliefs and opinions, facing up to harsh facts and eschewing wishful thinking. And now X becomes convinced that there is no answer at all to Humean scepticism; all his supposedly realistic beliefs about the world around him are only irrationally adopted hypotheses. He had striven to be intellectually honest and to follow Kant's injunction, *Sapere aude!* And now it turns out that the hypotheses he accepted had no more title to rational credence than those he would have accepted if he had engaged in merely wishful thinking. A cruel disappointment. How will he react? Well, he might become a sort of epistemological hermit for a time, believing almost nothing. But I am assuming that it is psychologically impossible to sustain a state of general unbelief for long; so I assume that X's intellectual vacuum will gradually be filled. But there is no reason to suppose that the beliefs that he now begins to form will be the same as the ones he previously relinquished. His previous submission to the reality principle had cost him a good deal (some of his beliefs about harsh realities were painful) and it now turns out to have been profitless. Why struggle to be realistic if the struggle naught availeth? Why not go over to the pleasure principle? Freud wrote: 'The hermit turns his back on the world and will have no truck with it. But one can do more than that; one can try to re-create the world, to build up in its stead another world in which its most unbearable features are eliminated and replaced by others that are in conformity with one's own wishes' (*1930*, p. 81).

1.34 *Scepticism and Fideism*

When I first read Popkin's classic (*1960*), a good many years ago, I was puzzled by one of its main historical theses, namely that in the late sixteenth and early seventeenth centuries, Pyrrhonian scepticism was used *in support* of the Catholicism of the Counter-Reformation.

26

Popkin's ample evidence left me in no doubt of the truth of this thesis; but how could a very sweeping kind of scepticism be used to justify or encourage acceptance of a theological position? Should it not undermine Catholicism along with everything else? On rereading Popkin I now see that his sceptical fideists could have answered this by addressing a Protestant reformer along the following lines (though I do not claim that any of them would have put it quite like this):–'Your forefathers lived their lives beneath the comforting shelter of the Catholic church. Because they accepted its divine authority unquestioningly, it could give them assured moral and spiritual guidance that they could not have got from a mundane counseling agency. You say that the Church has given no proof of its credentials or reasoned justification of its claim to divine inspiration. And you are right, for there could be no such reasoned justification. You say that no one should accept such a momentous claim blindly: each of us should use the natural reason with which God endowed us to decide for ourselves whether the Catholic church is the true church and whether it is interpreting the Holy Scriptures correctly. But human reason is quite incommensurate with the task you here propose for it. Your unaided reason cannot determine any matter, large or small, that is outside your own limited experience. It cannot even establish that it is more probable than not that water will continue to quench your thrist and bread to nourish you. Your belief that they will continue to do so is instinctive, a matter of faith. Nor can your unaided reason determine whether God exists or whether the Bible is the word of God. You can only believe these things on faith, blind unreasoned faith. So when you exhort your fellow humans to stop believing unquestioningly and to accept only what their reason has established as true, or at least as more probable than not, you are really exhorting them to believe in nothing beyond their immediate experiences. The alternative to a dogmatic Catholicism is nothing but despairing unbelief.'

But this Pyrrhonian defence of the Catholic faith relied on a key assumption, namely that there is only one faith to which a Christian believer can (irrationally) adhere, the Catholic faith. Given that assumption, a would-be Reformer could be confronted by the following dilemma: you can either (1) discard your Christian (= Catholic) faith and try to rely instead on your own puny reason, or (2) retain your Christian faith unchanged. Now this assumption had considerable plausibility so long as the Reformation could be viewed as a reforming movement within Catholicism. But it became untenable once Calvinism had become established as an autonomous, Protestant faith quite independent of Rome. Once that had happened it becomes obvious that there is a third possibility: (3) *switch to a new faith.*

1.35 *Polanyi and Feyerabend*

Let us now turn from the French Pyrrhonians, and their view of the relation between religion and scepticism in the seventeenth century, to Michael Polanyi and his view of the relation between science and scepticism in the twentieth century. Lévy-Bruhl had declared the magical thinking of primitive people to be *prelogical.* But someone who accepts Humean scepticism may retort that the scientific thinking of Western people is also, though in a different way, prelogical. Polanyi combined a passionate personal commitment to Western science with a radical scepticism about the possibility of any objective and impersonal justification for preferring science to magic. He believed in science and disbelieved in magic, but that was his personal faith; objectively and philosophically considered, they are on a par:

> Science and magic are both comprehensive systems of beliefs, possessing a considerable degree of stability, and a comparison of the two systems has shown that the convincing powers of both are derived from similar logical properties of their conceptual frameworks. Yet the two achievements . . . are mutually exclusive. If you accept one system you cannot hold the other, and we today overwhelmingly accept science. (1952, p. 230)

Polyani *welcomed* philosophical scepticism, for two reasons. First, it has actually done nothing to undermine the confidence of Western scientists in their science; Hume's scepticism has caused 'no self-doubt among scientists in the modern free societies of the twentieth century' (*1958*, p. 238); second, it obliges scientists to realise that their adoption of the scientific outlook is akin to the adoption of a religious faith. Polanyi's view, here, is reminiscent of that of seventeenth-century sceptical fideists except that what he wanted accepted on faith was *science.* He seems to have had no doubt that a fideist view of science enhances and deepens its value: science does not consist of careful measurements and cold calculations; it is a vocation involving passion and commitment.

The claim that we have to 'accept science', not on rational grounds but on faith, presupposes that we have to 'accept science'. Now if the choice were between either (1) an intellectual vacuum or (2) a scientific outlook, then, on the assumption that (1) is psychologically impossible, one would indeed be driven by Humean scepticism to adopt (2) as one's personal faith; scepticism would generate a kind of scientific fideism. But (1) is not the sole alternative to (2). As Gellner put it, (*1974*, pp. 125–126), a scientific minded society does not have just one frontier, with uninhabited chaos on the other side; it has frontiers and with inhabited worlds, for instance, the world of magical thinking: 'Chaos is not an option, but *magic* is.' And not only magic. An antiscientific culture has developed in the West. (For some appalling evidence of this, see Rose and Rose, eds., *1976a* and *1976b*. For a critical examination of some of the intellectual sources of the current wave of hostility to science, see Passmore *1978*.) So besides (1) and (2) there is (3), the possibility of opting for some antiscientific ideology. And this makes Polanyi's idea that the tenets of the scientific outlook have to be held like the dogmas of a religious faith rather liable to backfire. Imagine a gifted young man who had been attracted to science, believing it to be a marvellous expression of human rationality; but he becomes persuaded by Polanyi's view. This may have a positive effect. He may say: 'I see now that becoming a scientist is not like becoming an accountant who is only required to think in a rational, accurate way, it is more like entering a priesthood.' But the effect may be negative. He may say: 'If science is an irrational faith, and only one among others, why should I not choose one that offers more warmth and solace?'.

That Polanyi's sceptical-cum-fideist view of science may backfire is borne out by the fact that Feyerabend, who with his (*1975*) became a main spokesman for the antiscientific culture, essentially restates what Polanyi had said (though he makes no acknowledgement to Polanyi) and puts it to an opposite use. Polanyi delighted in drawing attention to cases where the scientific community ignored or waved aside or explained away seeming counter-evidence to accepted theories. He seems to have felt that a scientist would abrogate his personal responsibility for his beliefs if he allowed them to be at the beck and call of experimental results: 'Of this responsibility we cannot divest ourselves by setting up objective criteria of verifiability—or falsifiability, or testability, or what you will' (*1958*, p. 64). Indeed, it was this conviction that scientific theorising is *not* controlled in any objective way by empirical facts that led to his claim that science and magic, for all their difference in content, are on a par when philosophically considered. He saw very clearly from his reading of Evans-Pritchard (*1937*) that a magical system of belief, such as the Zande system, has built-in strategies that enable it to digest each piece of seeming counter-evidence and transform it into a seeming confirmation, thereby strengthening its ability to deal with the next piece. And he insisted that Western science is a system that functions in a similar way: 'For the stability of the naturalistic system which we currently accept instead [of Zande superstitions] rests on the same logical structure. . . . Secured by its circularity and defended further by its epicyclical reserves, science may deny, or at least cast aside as of no scientific interest, whole ranges of experience. . . . ' (1958, p. 292).

All these themes reappear in Feyerabend's (1975). The idea of objective criteria of testability is dismissed as a fairytale (p. 303); the apparent empirical success of quantum theory is diagnosed as 'entirely man-made' (p. 44). And a parallel is again drawn between scientific and 'primitive' thinking: basic scientific ideas are protected by 'taboo reactions which are no weaker than are the taboo reactions in so-called primitive societies . . . the similarities between science and myth are indeed astonishing' (p. 298).

But now comes the big difference: if the scientific outlook is just one nonrational ideology among others, then why, Feyerabend very properly asks, should we retain it? He urges us to 're-examine our attitude towards myth, religion, magic, witchcraft' and to decide for ourselves which ideology suits us best (p. 308). Yes, but what remains by which we can make this choice? Feyerabend answers: 'what remains are our subjective wishes' (p. 285). In other words: decide in accordance with the pleasure principle. Instead of Polanyi's scepticism-cum-fideism we have scepticism-cum-hedonism.

There has been a certain lack of realism in what I have so far said about the likely effect of an unanswered case for Humean scepticism; for I have so far considered its effect just on some individual thinker, whom I have imagined responding to such an intellectual crisis in a thoroughly self-conscious way. But no one, presumably, abruptly decides to abandon the reality principle and go over to the pleasure principle; nor, presumably, do some people record in their diaries such entries as: 'Decided this afternoon to switch from scientific to magical outlook; latter more emotionally satisfying.' Moreover, my account so far ignores the social dimension that the effect of an unanswered scepticism is likely to have. A cultural shift, or a change in intellectual fashion, is not just the sum of a number of individual changes. Once seeded, such a change tends to spread by a kind of osmosis in which imitation, conscious or otherwise, plays a main role. In the present case, the seeding is provided by Humean scepticism's implication that austere scientific standards of reasoning involve a pointless intellectual asceticism, since the hard-won products of such reasoning have no more title to rational credence than the products of alternative kinds of thinking that do not comply with those standards. But I really do not expect this implication to smite first one and then another scientific thinker, resulting in a series of baptisms involving total immersion in an irrational faith. I would expect the questioning and undermining of those standards to proceed rather unobtrusively, without individual traumas and conversions, in an epidemiological way. And this sort of quiet erosion is likely to have a more far-reaching effect. Imagine a community of a hundred scholars in which high intellectual standards are observed; and then one of them goes over to voodooism or whatever. There would now be ninety-nine members of the community plus one dropout, and there is no reason to suppose that the standards observed by the ninety-nine would not be as high as before. But now suppose that, although there is no thoroughgoing repudiation of those standards, there is a certain amount of backsliding from them on the part of some members of the community: they start countenancing inferences that they would previously have condemned as invalid and a certain sloppiness, of a not too obvious kind, begins to infect their reasoning. Especially if their number included some of the more prestigious members of this community, this might prove infectious; and if it did, this might encourage the original backsliders to relax their standards a little more; for once one relinquishes the ideal of deductive inferences there is no natural stopping point lower down at which to say 'Thus far and no further'. And so the creeping process might continue, with standards of rationality slowly deteriorating.

My case against the naturalist strategy boils down to this: the most likely of the possible effects of Humean scepticism, if no rational answer to it can be found, is neither (1) that people will believe less than before, nor (2) that they will believe the same as before, but (3) that their beliefs will become less and less rational. I agree with Kekes when he says: 'Scepticism is not just an epistemological bogey, but a widespread and spreading attitude which ought to be countered. . . . It encourages the appeal to prejudice and the use of force, propaganda, and dogmatism. It is an attack on what is finest in the western tradition' (*1976*, p. 256). And I should perhaps avow that I regard Western science as a part of what is finest in the Western tradition. Nothing human is perfect; but it is as if a comparatively small number of people in a comparatively short time had built a soaring cathedral while the rest of mankind in its long history had built only mud huts.

Hume's scepticism is of course itself a product of that Western tradition of intellectual rationality within which modern science has developed. It was a case, to use Hume's phrase, of *reason subverting itself* (*1739–40*, p. 267). So an answer to Hume must not betray that tradition. Bearing this in mind, let us now turn to the nondeductivist strategy.

1.4   The Nondeductivist Strategy

I mentioned earlier that some antisceptical philosophers, instead of tackling Humean scepticism or Cartesian scepticism or some other strong case for scepticism, tackle a dummy figure called "the sceptic". I should add, though, that this figure of their own devising often displays a remarkable resilience. He sometimes make a comeback on the last pages of a book in which he was supposed to get defeated. Consider David Hamlyn's (*1970*). This book is entitled *The Theory of Knowledge* and the theory of knowledge was said to be, among other things, 'a set of defense-works against skepticism' (p. 9). But I could not discover in it any defense works against Hume or other sceptics in the history of philosophy. Rather to the contrary, there seemed to be a tolerance, sometimes verging on endorsement, of standard sceptical positions. For instance, Hamlyn declared that 'there is no possible way of building up a public world from immediately given sensations' (p. 162); and he pointed to the dilemma posed by the regress involved in any chain of justifications: 'the regress must either go on ad infinitum (a very unsatisfactory idea) or it must come to a stop with a belief or judgment that has no basis (an equally unsatisfactory idea)' (p. 185). However, Hamlyn sought to dispel the appearance of human knowledge crumbling away under the force of sceptical arguments in the following way. First, he invented a figure called "the skeptic" ('our skeptic is a hypothetical animal', p. 9); he then laid down that 'the onus is upon the skeptic to prove his case, not upon the defender of knowledge' (p. 50); all the latter has to do is to put the ball 'firmly in the opponent's court' (p. 290). Finally, Hamlyn saw to it that the ball (which a real sceptic would have slammed back) remains in the opponent's court by making his "skeptic" such an anaemic fellow that he produces no counterplay. The "defender of knowledge" has an easy task: he merely has to dawdle in his court, waiting for a riposte that never comes. Then is he awarded victory by default? Strangely enough, he is not. The umpire's decision, on the last page, seems to tilt against the defender of knowledge: 'Skeptical criticism is always possible and is indeed desirable. But the search for foundations for knowledge has as its aim the provision of an immunity from such criticism. The attempt to find such foundations is thus not only *hopeless*, it is also *undesirable*' (p. 291, my italics).

1.41 Ayer's Method of Descriptive Analysis

Another book on the theory of knowledge in which a figure called "the sceptic" goes through the motions required of him by his author is David Armstrong's (*1973*). This "sceptic" is derided for the views assigned to him ('It is hard to take such a problem seriously', 'the sceptic's position is even more arbitrary than this', p. 218). But at the end of the day it seems that Armstrong does not actually have an answer to the deplorable position adopted by this crazy fellow: 'I do not know how to answer such a desperate man. Do we need to?' (p. 219).

But perhaps the book in which "the sceptic" makes the most remarkable comeback on the last page is Ayer's (*1956*). Commenting on this book later, Ayer said that he had perhaps been too much under the control of commonsense when he wrote it and that he should perhaps have ended on a more sceptical note than he did (1971, pp. 63–64). Well, the book is indeed a defence of common-sense beliefs, especially about other minds and about the past: a method of 'answering the sceptic' is developed and applied. But I do not quite see how Ayer could have *ended* on a more sceptical note than he did. For on the very last page it turns out that the sceptic is *right* 'on the subject of other minds just as he is right on the subject of the past'.

Let us look a little into this book in order to discover what gave Ayer's sceptic his staying power. Ayer depicted four 'different methods of answering the sceptic', the fourth being the method of Descriptive Analysis. He declared this one to be the best, since it 'profits by being the heir to all the rest' (p. 83). He summarised it as follows:

> Finally, there is the method of Descriptive Analysis. Here one does not contest the premises of the sceptic's argument, but only its conclusion. No attempt is made either to close or to bridge the gap [between the evidence and the conclusions we infer from it]: we are simply to take it in our stride. It is admitted that the inferences which are put in question [by the sceptic] are not deductive and also that they are not inductive, in the generally accepted sense. But this, it is held, does not condemn them. They are what they are, and none the worse for that. (P. 80)

Now if the Descriptive Analyst does not contest the sceptic's premises but does contest his conclusions, then, presumably, he contests the validity of the sceptic's inferences from those premises to those conclusions. But here we are in for a shock. Not at all! Ayer repeatedly insisted that the sceptic's logic is faultless: 'Not that the sceptic's argument is fallacious; as usual his logic is impeccable' (p. 68); 'Here again the sceptic makes his point. There is no flaw in his logic' (p. 75). In short, one "answers" the sceptic by striding across logical gaps to conclusions inconsistent with premises that one does not contest.

1.42 *The Ineffable Nature of Inductive Proof*

Strawson championed the nondeductivist strategy in his *(1952)*. He there insisted that, in cases where evidence *e* does not entail a conclusion *h*, *e* may nevertheless be *conclusive* evidence for or an inductive *proof* of *h* (pp. 237–38). How can we tell when this important relation holds? And how can we tell when some weaker relation holds, such as *e* being good though not conclusive evidence for *h*? To these questions Strawson devoted two paragraphs (pp. 245–47). But one wonders why he left those paragraphs in; for he immediately added: 'Now I must emphasize, with all possible force, that the preceding two paragraphs . . . are not to be taken as an accurate description of some standard and familiar process of inductive inference' (p. 247). And he went on to insist that no precise answer can be given to our questions: 'When we bear in mind these things [the great complexity of background beliefs, etc.], it will not seem surprising . . . that *no precise rules of general application can be formulated* for the assessment of evidence' (p. 248, my italics). In short, one counters scepticism concerning inductive inferences by insisting that *there exist* valid inductive inferences though *their nature cannot be divulged*.

Wittgenstein touched on the problem of the nature of inductive grounds for a belief about the future in his *(1953)*: 'If it is now asked: But how *can* previous experience be a ground for assuming that such-and-such will occur later on?—the answer is: What general concept have we of grounds for this assumption?' (§480) But this answer is another question. How was this latter question answered? Well, he gave various negative answers: 'here grounds are not propositions which logically imply what is believed.' Nor is the inference 'an approximation to logical inference' (§481). Nor again is a good ground one that makes the occurrence of the event probable (§482). If a good ground for assuming that such-and-such will occur later is none of these things, what is it? Wittgenstein's positive answer, which I quote in its entirety, reads like this: '483. A good ground is one that looks *like this*.' He had declared earlier that his aim was that 'the philosophical problems should *completely* disppear' (§133). But in this case it seems rather to be the solution that has disappeared.

I insisted earlier that it is a great mistake to bracket the nondeductivist and the probabilist antisceptical strategies together; and I think that this insistence will now be seen to be justified. A probabilist tries to specify, within the confines of a generalised system of deductive logic, the condition under which h is more or less highly confirmed by e. A nondeductivist makes the existential statement that there are valid nondeductive inferences, or that there are circumstances in which it is right to transcend deductive logic, and then either fails to specify any rules of nondeductive inference or even insists, as Strawson did, that it is impossible to specify them.

A nondeductivist might say that everyone can *recognise* a valid nondeductive inference when presented with one, even though he cannot say what makes it valid. But that cannot be true, because many of us (myself included) recognise an inference as valid only if it is deductive. It appears that nondeductivism requires a licensing authority to be set up to decide, in disputed cases, whether or not an inference license is to be issued. In the absence of such an authority it seems likely that the nondeductivist's existential claim that some nondeductive inferences are valid will tend to encourage a sort of inferential licentiousness; when a critic objects, 'But that doesn't follow', the answer may be: 'Are you blind? Can't you *see* that it's a case of a *valid* nondeductive inference?'.

One wonders how wide a logical gap has to be for a nondeductivist to consider it *too* wide for us to 'take it in our stride'. Ayer has set his face against *some* kinds of logically invalid derivation; for instance, he endorsed 'a sound and respectable point of logic which was already made by Hume; that normative statements are not derivable from descriptive statements' (*1959*, p. 22). Yes, but it was also a sound point of logic, made by Hume, that statements about instances of which we have had no experience are not derivable from statements only about instances of which we have had experience. If we can cross this gap, why not the *is-ought* gap? (MacIntyre claimed in his 1959 that both gaps are bridgeable.)

One may even wonder whether an "answer" to scepticism that relies essentially on throwing out proposition (III), the deductivist thesis, may not have implications as sceptical as scepticism itself. Let us epitomise the sceptic's ban on ampliative or nondeductive inferences by 'Nothing goes'. Then we could say that the nondeductivist wishes to insist against this that 'Something goes': from a suitable *e* some appropriate *h* can properly be nondeductively inferred. But suppose that, in a particular case, some cranky person asserts that it is not *h* but some incompatible alternative *h'* that is properly inferrable from *e*, while another makes a similar claim for *h''*. It would seem that the nondeductivist could not restrain this unseemly proliferation so long as he keeps the nature of that legitimising inference, on which his whole defence against scepticism relies, a well-guarded secret. Until he divulges that, his position may be epitomised by 'Anything goes'.

A refrain that runs through the writings of nondeductivists is that the sceptic achieves his easy victories because his standards of reasoning are too high. It might seem that the sceptic could retort that if these antisceptics achieve any victories against him, that is because their standards of reasoning are too low. However, if the argument I have just adumbrated turns out to be correct, the sceptic could make a more biting retort: he could say that they have achieved *only* a lowering of standards; for their lowered standards have let scepticism in by the back door.

I suggested in the previous section that a spreading irrationalism is the most likely effect of a failure to find a rational answer to Humean scepticism; and I now suggest that an "answer" to scepticism that involves throwing out proposition (III), the deductivist thesis, is likely to have a similar effect.

1.5 The Pragmatist Strategy

Whether the strategy to be considered now would, if successful, constitute an answer to scepticism is doubtful. Are wigs a cure for baldness? It has a superficial resemblance to the conjecturalist strategy to be elaborated later; both offer a way of making rational choices among hypotheses that are radically uncertain. But there is a crucial difference: in the case of the conjecturalist strategy, the aim governing rational choices is cognitive whereas the pragmatist aim is noncognitive. By a 'cognitive' aim I mean, for example, aiming at truth, or high probability, or a deeper understanding, or a better explanation. One's aim is noncognitive if it is, for instance, to avoid trouble with the authorities, or to make an impression of profundity, or to be an intellectual hero of the Left.

The basic idea behind the pragmatist strategy is that, given an exhaustive set of mutually exclusive hypotheses, you may still be able to make a rational choice among them even if you are entirely ignorant about their relative chances of being true. Suppose that you are faced by n mutually exclusive and exhaustive hypotheses, $h_1, \ldots h_n$, that you need to act on one of these, and that they are all uncertain. You imagine yourself to be 'playing a game against Nature' in which Nature has n choices (to make h_1 true, or h_2 true ...) and you have n choices (to act on h_1, or on h_2 ...). You now calculate, for each intersection of a choice i of yours and a choice j of Nature's, the payoff or utility u_{ij} to you of acting on h_i if h_j is true. Now it may be that the resulting payoff matrix shows a pattern that enables you to make a pragmatically optimal choice. In the very best case, one of your choices would dominate the rest; that is, by acting on one particular hypothesis, your payoff cannot be worse and may be better than it would be if you acted on any other hypothesis, no matter which hypothesis happens to be true. If no such dominant choice is available, you might apply the famous minimax principle and select the one that minimises your maximum loss. Alternatively, you might assign (equal or possibly unequal) probabilities to Nature's choices, and then select that choice of yours that maximises your expected utility.

A famous example of such a 'game against Nature' is Pascal's wager. Pascal insisted that God is infinitely incomprehensible and that Christians profess a religion for which they can give no reason. All we can say is: either God is, or he is not. Pascal continued: 'But which way shall we lean? Reason can settle nothing here.... A game is on, ... and heads or tails will turn up. What will you wager?' He answered: 'Let us weigh gain and loss in calling heads, that God is. Reckon these two chances: if you win, you win all; if you lose, you lose naught. Then do not hesitate, wager that he is' (*1670*, pp. 117–119). Here there are just two hypotheses, h_1 ('God is') and h_2 ('God is not'). And Pascal's claim is, in effect, that the choice of h_1 dominates that of h_2. For if you choose h_1, then if h_1 is true 'you win all' and if h_2 is true 'you lose naught'; whereas if you choose h_2, then if h_2 is true you win naught and if h_1 is true you lose all.

Before making what I consider the decisive objection to the pragmatist strategy as an answer to scepticism, I will mention two lesser objections. One is that, since the utility rankings of different people may differ, "rational" choices among competing hypotheses become person-relative (Braithwaite, *1953*, p. 252). Another objection is that a pragmatist method breaks down in cases where the utility values, instead of being heavily skewed as in Pascal's wager, vary straightforwardly with the truth values of the hypotheses; that is, in cases where to act on h will bring a gain if h is true and a loss if h is false. For example, in 1944 the German High Command accepted the hypothesis that the main Allied invasion would be in the Pas de Calais, and they clung to this hypothesis until long after the Normany landings, with the result that about half their anti-invasion forces were immobilised in the Pas de Calais area. Acting on this hypothesis would have had a high utility for them had it been true; as it was, it had a high disutility. In cases where the utility of acting on a hypothesis is a straightforward function of its truth value, it is futile to try to proceed to a decision about truth values from a consideration of utility values.

But the most decisive objection to the pragmatist strategy as an answer to scepticism is this. If an agent is to make a pragmatic decision in the above way, *he must already know a lot.* For if he is to draw up a payoff matrix of the required kind, namely one that assigns a definite utility value u_{ij} for each choice by him of an h_i and by Nature of an h_j, then he needs to know, for every i and j, what would be the effect on him of acting on h_i were h_j true. But a sceptical pragmatist is not entitled to assume that all this conditional knowledge is available to the agent. If Pascal meant it when he said that reason can settle *nothing* here, he should not have taken it for granted that reason *can* settle what the outcome for me would be if I wager that God exists and God does exist, namely that I would 'win all'. Many hypotheses concerning that outcome are conceivable: perhaps God (remember that according to Pascal He is infinitely incomprehensible) is indifferent as to what people believe about Him; perhaps He even punishes those who calculate that it is advantageous to wager that He exists, and rewards atheists for their temerity. Who can tell? Not the sceptic. And the suggestion that *the agent can choose pragmatically* among all these latter hypotheses concerning that outcome would obviously open up an infinite regress.

I conclude that, while a pragmatist strategy might enable someone endowed with adequate background knowledge to cope with a little local uncertainty, it is powerless in the face of a general scepticism: without background knowledge its payoff matrices cannot be filled in. I turn now to a strategy that does not replace cognitive by noncognitive values and that does not reject any of the three propositions given in § 1.1.

2

....

Probabilism

2.1 Probabilism versus Personalism

There is much plausibility in the claim that Humean scepticism results from interpreting proposition (III), the deductivist thesis, too simplistically, by taking only classical logic into account, and that it can be dissolved by 'employing a logic of a peculiarly subtle or highly complex character', in the words of George Boole (1854, p. 273), namely probability logic. In classical logic the only relations recognised between statements are entailment, independence, and inconsistency. Thus if we have evidence e, consisting of a conjunction of singular existential statements, and a universal hypothesis h, there are only two possibilities: e and h are either logically independent of one another or else inconsistent with one another; for a universal statement can neither entail nor be entailed by singular existential statements. And e and h will be logically independent if h is a singular existential statement not included in e, for instance because it predicts a future event. So the appearance is created that evidence can provide no support for scientific knowledge that transcends it.

But Keynes (*1921*, p. 115), Jeffreys (*1948*, p. 17), Carnap (*1950*, passim), and other probabilists have sought to generalise classical logic into a much wider and finer-grained system of probability logic. Such a system would, for suitable e and h, enable us to determine the value of r in $p(h, e) = r$ (to be read: 'the logical probability that h is true, given only that e is true, is r') where r takes values between 0 and 1 inclusive. It would also enable us to determine the value of q in $p(h) = q$ (to be read: 'The logical probability that h is true, given no evidence at all, is q'). If e is inconsistent with h we have $p(h, e) = 0$ and if e entails h we have $p(h, e) = 1$. These fix the limits of a continuous scale. If e and h are logically independent we have $0 \leq p(h, e) \leq 1$. But a system of probability logic enables us to associate probabilistic independence with a precise point on this continuous scale: h and e are independent if e neither raises nor lowers the probability of h, or

$p(h, e) = p(h)$ and $r/q = 1$. Assume that $0 < q < 1$. Then we could say that e is the more favourable (unfavourable) to h the higher (lower) is the ratio r/q. Does this not dispel the illusion that evidence can provide no support for scientific knowledge that transcends it?

The posterior probability of h on e is largely a function of the initial probabilities of h and e; for by Bayes's famous theorem

$$p(h, e) = \frac{p(h) \cdot p(e, h)}{p(e)}$$

In a case where e is a predictive implication entailed by h, so that $p(e, h) = 1$, this simplifies to

$$p(h, e) = p(h)/p(e)$$

and the posterior probability is purely a function of the initial probabilities.

For those probabilists who aim at an objective system of probability logic, the initial probabilities of all statements should be determined in some impersonal and nonarbitrary way. But there is a school of probability theorists whose distinguishing feature is that they allow people to distribute initial probabilities according to their subjective expectations and wishes, subject only to the requirement that the distribution is "coherent", or satisfies the axioms of the probability calculus. I will label this approach *personalism*, reserving the label *probabilism* for systems that seek to assess the probabilities of statements in an objective way. In this chapter we will be concerned with probabilism rather than personalism. For no one, I think, would claim that personalism provides any sort of answer to Hume. If a hypothesis h is under a sceptical cloud, one hardly dispels this by providing a very tolerant system in which one is free, if one so wishes, to rig the initial probabilities in such a way that h gets a high posterior probability, though one might equally so rig them that on the same evidence h gets a low, or even zero, probability. Personalists are wont to claim that in the long run the posterior probabilities of people who started with very different initial probabilities will converge, so that the subjectivities are rather harmless. But first, this is unlikely to happen if one person assigns a negligible initial probability to a hypothesis to which another assigns a nonnegligible one. And second, if they assign nonnegligible but very different initial probabilities to it, they may be dead before the run is long enough.

I mentioned earlier that, although I consider it a great mistake to bracket probabilism with nondeductivism, there is some affinity between nondeductivism and personalism. The distinguishing feature of nondeductivism is its claim that it is sometimes legitimate to infer a conclusion *h* from premises *e*, or to regard *e* as an inductive proof of *h*, although *e* does not entail *h* and despite the fact that no rules have been provided for the control of such inferences. One could express the fundamental difference between probabilism and nondeductivism by saying that the former seeks to *generalise* the idea of logical entailment to cover partial entailment, whereas the latter merely *relaxes* it. The intuition behind the former idea is this. If *e* entails *h* then *h* is true in every possible world in which *e* is true and the probability of *h* given *e* is 1; and if *e* entails nearly all of the content of *h*, and is consistent with *h*, then *h* is true in nearly all the possible worlds in which *e* is true and the probability of *h* given *e* is nearly 1. Probabilism aims to provide an objective measure of the logical proximity of *h* to *e* by using the probability calculus in conjunction with a nonarbitrary distribution of initial probabilities over some linguistic representation of "possible worlds".

By contrast, personalism uses the probability calculus in conjunction with arbitrary distributions of initial probabilities; and in cases where a nondeductivist might announce that *e* is an inductive "proof" of *h*, a personalist might so fix it that *h* is highly probable on *e*. Personalism is not so tolerant as nondeductivism. For one thing, personalists have to work hard, under the requirement of coherence, to get the desired results. It is as if they had ever so many valves to adjust to achieve a desired level, and their adjustments had collectively to satisfy an overall formula. But it is too tolerant, as I said before, to provide an answer to Hume.

It will, I think, be helpful to consider probabilism informally, and to review a little of its prehistory, before we turn to contemporary systems of probability logic and the rather technical and tricky question of a nonarbitrary distribution of initial probabilities.

2.2   Underlying Ideas

Probabilism accepts that all hypotheses about the external world that have unverified predictive implications are thereby uncertain. But it insists that this pervasive uncertainty is offset by the possibility of grading hypotheses according to their degree of empirical confirmation.

Some people hold that, if it concedes that all hypotheses with un-verified implications are uncertain, then probabilism is itself only a species of scepticism. But this seems a big mistake; we would surely be out of the sceptical wood if we could, generally, select from a number of alternative hypotheses the one that is most probably true, given our present evidence. Hume declared that his sceptical reflections allowed him to 'look upon no opinion *even as more probable* or likely than another' (*1739-40*, pp. 268-269, my italics). And the view that prob-abilism is opposed to scepticism had been reciprocated from the other side by Joseph Glanvill, who wrote: 'If I should say, we are to expect no more from our experiments and inquiries, than great likelihood, and such degrees of probability, as might deserve an hopeful assent; yet thus much of diffidence and uncertainty would not make me a Sceptick; since *they* taught, that no one thing was more *probable* than an other; and so withheld assent from all things' (*1676*, p. 45).

The map of empirical knowledge, according to Pyrrhonian scepticism, is very simple: it shows just an undifferentiated ocean of uncertainty. According to Humean scepticism, it shows an ocean of uncertainty with a little island of certainty in the middle; this island contains, for any person X at time t, X's egocentric knowledge at t about his own per-ceptual experiences, etc. According to probabilism, the ocean in the vicinity of the island is differentiated by contour lines representing degrees of probability, a line near the shore representing a high degree of probability and the outermost line representing a probability of one half. (We do not need contour lines for probability values of less than one-half, since we can represent the assignment of a low probability to a statement by the assignment of a high probability to its negation.) Locke, who was a fallibilist rather than a sceptic, endorsed this picture:

> most of the propositions we think, reason, discourse, nay, act upon, are such as we cannot have undoubted knowledge of their truth; yet some of them border . . . near upon certainty. . . . [There are] degrees herein, from the very neighbourhood of certainty and dem-onstration, quite down to improbability and unlikeliness . . . ; and also degrees of assent from full assurance and confidence, quite down to conjecture, doubt, and distrust. (*1690*, IV, 15, 2)

He went on to speak of *the several degrees of probability and assent.* And Keynes endorsed this characterisation of probabilism at the be-ginning of his (*1921*). (Instead of speaking of an island of certainty he spoke of 'our ultimate premisses' or of 'our direct knowledge'.) He said:

'Part of our knowledge we obtain direct; and part by argument. The Theory of Probability is concerned with that part which we obtain by *argument, and it treats of the different degrees in which the results* obtained are conclusive or inconclusive' (p. 3). He added: 'Given the body of direct knowledge which constitutes our ultimate premisses, this theory tells us what further rational beliefs, certain or probable, can be derived by valid argument from our direct knowledge' (p. 4).

According to some versions of probabilism it is legitimate to proceed in a *step-by-step* way from our 'ultimate premisses' to an eventual conclusion; thus if *a* is known to be true and *b* is not but is strongly supported by *a*, and if *b* strongly supports *c*, then we may proceed from *a* to *b* and thence to *c*. Keynes seems to have endorsed such stepwise inferences: 'Our [probability] logic is concerned with drawing conclusions by a *series of steps* of certain specified kinds' (*1921*, p. 18, my italics). (We will later examine an attempt of his to justify his own 'Inductive Hypothesis' in a step-by-step way.) Such a step-by-step procedure seems to be called for by the view of science that sees scientific progress as an inductive ascent from observations to low-level generalisations, and thence to broader experimental laws, and thence to sweeping and exact theories. Thus Whewell spoke of 'successive steps of induction' in 'the inductive ascent' (*1837*, ii, p. 139). He also said:

> as our knowledge becomes more sure and more extensive, we are constantly transferring to the class of facts, opinions which were at first regarded as theories.
>
> Now we have further to remark, that in the progress of human knowledge respecting any branch of speculation, there may be *several* such steps in succession, each depending upon and including the preceding. . . . As men rise from the particular to the general, so, in the same manner, they rise from what is general to what is more general. (*1840*, i, p. 46)

We might express this picturesquely as the thesis that the waters around the island of certainty are continually receding; at any given time investigators on dry land are able to catch an uncertain hypothesis only if it is floating near the shore, but with the passage of time it may be left high and dry, and as the area of dry land expands, so it becomes possible to catch hypotheses further and further out.

Other versions of probabilism, however, disallow such step-by-step inferences (where the inferences are not deductive) and require the conclusion to be directly related to the 'ultimate premisses'. Carnap generally adhered to this *one-step* view, though as we shall see he did at one point try to rehabilitate a kind of step-by-step reasoning.

2.21 *Possible Worlds and Complete Sentences*

'We need', Leibniz wrote, 'a new logic in order to know degrees of probability' (1679, p. 339). Now in Leibniz's terminology, if *e* entails *h* then *h* holds in every possible world in which *e* holds; and $p(h, e) = 1$. Conversely, if *e* contradicts *h* then *h* holds in no possible world in which *e* holds; and $p(h, e) = 0$. This rather suggests that if *h* holds in, say, half the possible worlds in which *e* holds, then $p(h, e) = \frac{1}{2}$. But although Leibniz made a beginning in the development of probability logic, it seems to have been Bolzano who took up this idea and inaugurated what is now often called the "range theory" of logical probability. In his (*1837*), Bolzano had introduced the idea of the *satisfiability* of a proposition: if a proposition is universally satisfiable (satisfiable in all possible worlds) its degree of satisfiability is 1 and it is analytic; if it is absolutely unsatisfiable (satisfiable in no possible world) its degree of satisfiability is 0 and it is self-contradictory; and if it is neither universally satisfiable nor absolutely unsatisfiable (satisfiable in some but not all possible worlds) its degree of satisfiability lies between 0 and 1 and it is synthetic (§§ 147–148). From this he proceeded to the idea of the *relative* satisfiability of one proposition with respect to another (§ 161). If *e* entails *h*, then the satisfiability of *h* relative to *e* is 1 (*h* is satisfied in all those possible worlds in which *e* is satisfied); if *e* contradicts *h*, then the satisfiability of *h* relative to *e* is 0 (*h* is satisfied in no possible world in which *e* is satisfied); and if *h* is satisfied in, say, half the possible worlds in which *e* is satisfied, then the satisfiability of *h* relative to *e* is $\frac{1}{2}$. Bolzano said that he would call this relation, of relative satisfiability, *probability*.

The programme inaugurated by Bolzano posed two questions. First, can the vague idea of a possible world be rendered manageable and exact? Can it be captured in language and, more especially, in a language adequate for empirical science? Second, if there are units that can serve as linguistic representations of possible worlds, is there a distribution of initial probabilities over these units, for instance an equal distribution, that is, as it were, natural and nonarbitrary? Only if both these questions can be answered affirmatively is there an objective basis for probability logic as envisaged by Bolzano. This and the next two subsections will be devoted to these two questions. They are not easy questions and the discussion of them will unavoidably be rather technical. My conclusion will be that there are units (namely, degree-d constituents in Hintikka's sense) that will serve as surrogates, in a reasonably rich language, for "possible worlds", and that there is a nonarbitrary probability weighting of these units, though this does not weight them equally. A reader who prefers to avoid technicalities and who is willing to take these conclusions on trust should go to §2.24.

Let L be some rather simple artificial language with a restricted and fixed vocabulary that includes the usual logical constants '$\sim$', '$\vee$' and '$\wedge$' but *not* '$=$'. (Identity, together with the special problems it raises for probability logic, will be introduced later.) We assume that the rules for the formation of sentences in L are well specified so that it is always decidable whether a particular string of words constitutes a (well-formed) sentence in L. Let us say that when two or more such sentences are logically equivalent, they express one and the same proposition (Carnap *1942*, §17). The number of distinct propositions expressible in L will be finite and of these half will be true and half false. So a sentence could be formed in L (though we may not know which it is and it may be enormously long) that expresses *all* and *only* those propositions that are both true and expressible in L. This sentence will describe the *actual* world as completely as L allows: and it will be what Tarski (1930) called a *complete* sentence. A complete sentence in L entails, for every sentence s_i that can be formed in L, either s_i or $\sim s_i$ but not both. Thus a complete sentence is maximally strong in the sense that any attempt to strengthen it further would render it self-contradictory.

This strongest true sentence will be one of the (perhaps enormous, but finite) total number, say K, of complete sentences, say $C_1, \ldots C_K$, in L. Since the one complete sentence that is true describes the actual world as completely as L allows, it seems reasonable to say that C_1, $\ldots C_K$ between them describe all possible worlds as completely as L allows. No doubt there are more things in heaven and earth, and more possible heavens and earths, than are spoken of in L; but so long as we remain in L the nearest we can get to capturing the idea of all possible worlds is with the idea of all the complete sentences in L.

Let us now consider what forms complete sentences can take. I will proceed in stages. I will assume throughout that the language in question possesses, in addition to '$\sim$', '$\vee$' and '$\wedge$', a fixed stock of primitive monadic predicates, $P_1, \ldots P_n$. And I will consider in turn: (i) a language L_1 that has individual constants $a_1, \ldots a_m$ but no quantifiers; (ii) a language L_2 that has variables and quantifiers but no individual constants; (iii) a language L_3 that has both individual constants and quantifiers; and finally (iv) a language L_4 that has constants, quantifiers, and identity.

In L_1 a complete sentence will be what Carnap called a *state-description* (*1950*, pp. 70f). This provides as complete a description as the language allows of all the individuals named in the language. It does this by assigning to each individual constant a Q-predicate. This is a conjunctive predicate in which each of the primitive predicates figures once, either unnegated or negated. Thus it takes the form:

$$\pm P_1 \wedge \pm P_2 \ldots \wedge \pm P_n$$

with $\pm$ replaced in each case by nothing ($=$ yes) or by $\sim$ ($=$ no). If there are n P-predicates there will be 2^n Q-predicates. If (to keep things simple) L_1 contained just two individual constants and three P-predicates, one of its complete sentences (or state-descriptions) would be:

(1) $Q_1 a_1 \wedge Q_8 a_2$

In L_2 a complete sentence will be what Carnap called a *structure-description* (pp. 116 f). This provides a complete description of all the kinds of individuals it asserts to exist without mentioning any individuals by name. It can achieve this either by saying of each Q-predicate in turn either that it is, or is not, instantiated, or by saying that certain Q-predicates are instantiated and adding that only these are instantiated. Thus a complete sentence (or structure-description) in L_2 that corresponds to (1) above would be:

(2) $\exists x Q_1 x \wedge \exists x Q_8 x \wedge \forall x (Q_1 x \vee Q_8 x).$

In L_3 a complete sentence will be a hybrid of a state-description and a structure-description. It *will* assign to each individual constant a Q-predicate; it *may* say that certain further Q-predicates are instantiated; and it *will* add that only the Q-predicates already mentioned are instantiated. Thus one complete sentence in L_3 would be:

(3) $Q_1 a_1 \wedge Q_8 a_2 \wedge \exists x Q_5 x \wedge \forall x(Q_1 x \vee Q_5 x \vee Q_8 x).$

Given that L_3 has just these two individual constants and no identity, it is clear that (3) cannot be strengthened without being rendered self-contradictory: we cannot consistently say anything more about a_1 or a_2 since they are already completely described; and we cannot add a further singular statement since the individual constants have been used

46

up. If we add any existential statement this will be either already entailed by $Q_1a_1 \wedge Q_8a_2 \wedge \exists x Q_5x$ or else inconsistent with $\forall x(Q_1x \vee Q_5x \vee Q_8x)$; conversely, any universal statement will be already entailed by $\forall x(Q_1x \vee Q_5x \vee Q_8x)$ or else inconsistent with $Q_1a_1 \wedge Q_8a_2 \wedge \exists x Q_5x$.

Before proceeding to L_4 and the large new problems raised by the introduction of identity into the language, let us pause to consider what basis we have at this stage for assigning values to q in $p(h) = $ q and to r in $p(h, e) = $ r. Any consistent sentence in our L_3 is logically equivalent to a disjunction of certain of the complete sentences $C_1, \ldots$ C_K of L_3. Such a disjunction is called the 'distributive normal form' or d.n.f. of the sentence (Hintikka, 1965a). If the sentence is analytic, its d.n.f. will consist of all K C_i's (it holds in all "possible worlds'). If the sentence is itself complete, its d.n.f. will consist of one C_i, namely itself. If the sentence is neither analytic nor complete, its d.n.f. will consist of less than K and more than one C_i.

Now it would seem that this provides a very simple and natural basis for probability logic, on the assumption that we are entitled to regard all the C_i's as having the same initial probability. (We will have to reconsider this assumption later.) Let h be some sentence in L_3 and let k be the number of C_i's in the d.n.f. of h. Then we will want the initial probability q of h to vary with the ratio of the number of "possible worlds" in which h holds to the total number of "possible worlds". Then may we not simply equate q with k/K? If h is analytic, $k = K$ and $q = 1$; if h is self-contradictory, $k = 0$ and $q = 0$; and if h is neither analytic nor self-contradictory, $0 < k < K$ and $0 < q < 1$. We will likewise want the relative probability r of h given e to vary with the ratio of the number of "possible worlds" in which both e and h hold to the number in which e holds. Then may we not simply equate r with m/n where m is the number of C_i's in the d.n.f. of $e \wedge h$ and n is the number of C_i's in the d.n.f. of e? If e entails h, $m = n$ and r $= 1$; if e is inconsistent with h, $m = 0$ and $r = 0$; and if e neither entails nor contradicts h, $0 < m < n$ and $0 < r < 1$.

What prospects this simple idea opens up for a probability logic immensely more subtle and comprehensive than classical logic! In classical logic we can say of a sentence h that it is *either* analytic *or* self-contradictory *or* synthetic (neither analytic nor self-contradictory). But probability logic, as here (somewhat naively) envisaged, would provide us with a scale for degrees of content going from $q = 1$ (analytic) to $q = 0$ (self-contradictory). For a maximal synthetic sentence (namely a C_i), q $= 1/K$; and for a minimal synthetic sentence (namely the negation of a C_i), q $= (K - 1)/K$; and every other synthetic sentence can be located at an exact point on this scale simply by counting the C_i's in its d.n.f. and dividing by K.

Again, in classical logic we can say of two consistent sentences e and h that *either* one entails the other *or* they are logically exclusive *or* they are logically independent. But probability logic as here envisaged would enable us, for every pair of consistent sentences e and h in L_3, to measure exactly the logical proximity of h to e, again simply by counting.

2.22 Constituents

However, the prospects for probability logic look less rosy when we turn from L_3 to the language L_4, which has identity. In a language with identity we can of course make statements about the numbers of individuals instantiating a given predicate. And one consequence of this is that a sentence that was complete in a language without identity will, in general, no longer be complete in a language with identity. Consider sentence (3) above in L_3 which ran:

(3) $Q_1a_1 \wedge Q_8a_2 \wedge \exists x Q_5 x \wedge \forall x (Q_1 x \vee Q_5 x \vee Q_8 x)$.

We saw that in L_3 this sentence could not be strengthened without being rendered inconsistent. But in L_4 it can be so strengthened. For instance we could expand it into:

(4) $Q_1a_1 \wedge Q_8a_2 \wedge \exists x \exists y (x \neq y \wedge Q_5 x \wedge Q_5 y) \wedge \forall x (Q_1 x \vee Q_5 x \vee Q_8 x)$.

In contrast with (3) this says that at least two things are Q_5. We could further add that at least one thing other than a_1 is Q_1; and so on. It is clear that we could go on indefinitely strengthening (4) in such ways without inconsistency.

Does this mean that there is nothing like a complete sentence in languages with identity? If so, that would be a serious blow to the probabilist programme. It would mean that the idea of a possible world can be captured only with respect to very primitive languages that are quite inadequate for science. Fortunately for probabilism, a way round this difficulty has been found by Jaakko Hintikka (*1973*, chs. i, xi; *1974*, ch. 7). Hintikka introduced the idea of a sentence of degree-*d*, where *d* is the number of distinct individuals referred to, whether by name or anonymously, by the sentence. If s is a singular sentence (or conjunction thereof), its degree is simply the number of distinct individual constants occurring essentially in it. Thus our sentence (1) above is of degree-2. If s is a general sentence, its degree is the same as its depth, *depth* being defined by Hintikka as the number of overlapping quantifiers occurring in it. Consider the sentence 'All human beings love someone other than themselves'. This could be formalised as: $\forall x\, \exists y\ (x$ loves y and $y \neq x)$, where x and y range over the human race. It has two overlapping quantifiers and is therefore of depth-2. It may seem odd to say that this refers to *two* distinct individuals. Surely it refers to lots of them? But Hintikka would read it as, roughly, 'Each single member x of the human race loves at least one other member y of the human race'. The minimum number of individuals needed to instantiate it is two. Or consider our sentence (2) above, namely: $\exists x\, Q_1 x \land \exists x\, Q_8 x \land (Q_1 x \lor Q_8 x)$. Now it is impossible for one thing to fall under two different Q-predicates. But (2) was formulated in a language without identity in which we could not make this non-identity explicit. In a language with identity we can reformulate (2) thus:

(2′)　$\exists x\, \exists y\ (x \neq y \land Q_1 x \land Q_8 y) \land \forall x\ (Q_1 x \lor Q_8 x)$.

This is of depth-2: $\exists x$ and $\exists y$ are overlapping quantifiers but $\forall x$ does not overlap with them. The minimum number of individuals needed to instantiate (2′) is two; it does not call for additional individuals to instantiate its universal clause. In the case of what I called a hybrid sentence, with a singular and a general component, its degree is $d + e$, where d is the degree of its singular part and e is the depth of its general part. Thus sentence (3) above is of degree-3, and sentence (4) is of degree-4.

I now turn to Hintikka's idea of a *constituent* of degree-d. This is a sentence that is as complete as it can be without referring to more than d distinct individuals. It is not complete in an absolute sense: it could be strengthened without inconsistency. But if it were, it would turn into a sentence of at least degree-$d + 1$. As Hintikka put it, a constituent of degree-d 'is as full a description of a possible world as one can give without ever speaking of more than d individuals at the same time' (*1973*, p. 19). We might call a constituent of degree-d 'quasi-complete' or 'd-complete'.

Now any (consistent) sentence of degree-d or less is logically equivalent to a disjunction of degree-d constituents. Then may we not define initial and relative probabilities within L_4 in just the same way that we

49

defined them for L_3 except that we replace complete sentences by quasi-complete constituents of the requisite degree? Let h and e be sentences of not more than degree-d; and suppose that there are K degree-d constituents, $C_1^d, \ldots C_K^d$, in L_4 and k degree-d constituents in the d.n.f. of h. Then may we not say that $p(h) = k/K$? Again, suppose that there are m degree-d constituents in the d.n.f. of $e \wedge h$ and n in that of e. Then may we not say that $p(h, e) = m/n$?

2.23 *The Weighting of Constituents*

Well, we could say this *if* we could assume that all constituents of a given degree have the same weighting or are equiprobable; and at one time I used to suppose, in my naivete, that there is nothing wrong with this assumption. But Graham Oddie showed me (personal communication) that this was a big mistake; for it can happen that while one degree-d constituent splits into a disjunction of many degree-$d+1$ constituents, another splits into only a few or perhaps just one. Now the initial probability of a degree-d constituent has to be shared out among the degree-$d+1$ constituents into which it splits. Hence weighting all degree-d constituents equally would mean weighting degree-$d+1$ constituents unequally, and vice versa.

Moreover, alerted by this last consideration, we will have to reconsider our earlier supposition that the complete sentences in a language without identity may be weighted equally or ascribed equal initial probabilities. Consider the language L_2 which has existential and universal quantifiers but no individual constants and no identity. Following Hintikka we may define the width w of a constituent in L_2 as the number of Q-predicates that it declares to be instantiated. At one extreme there are width-1 constituents, each of which says that just *one* Q-predicate is instantiated. (No constituent says that *no* Q-predicate is instantiated.) Let us denote by C_{i_1} a maximally narrow constituent that says $\exists x \, Q_i \wedge \forall x \, Q_i x$ where Q_i is one of the K predicates $Q_1, \ldots Q_K$ in L_2. At the other extreme there is a constituent that says that all K Q-predicates are instantiated. Call this C_K. (In my 1978a I called it m^*, a label I now consider unfortunate.) One could regard a C_{i_1} as saying that the world is as orderly, and C_K as saying that it is as disorderly, as it could be; equally, one could regard C_K as a principle of plenitude that says that the world is so rich that every existential possibility (permitted within the language L_2) is realised, whereas a C_{i_1} says that the world is so poor that only one of the existential possibilities (permitted within L_2) is realised. Can we regard a C_{i_1} as having the same initial probability as C_K? Well, in point of falsifiability there is an enormous disparity between them. A C_{i_1} forbids nearly all observational possibilities and is a maximally falsifiable statement within L_2, whereas C_K forbids none and is unfalsifiable. Hilpinen (*1968*, p. 54) pointed out that the equivalent of C_K in Carnap's system has an initial probability that tends to 1 as the number of individuals in the universe of discourse tends to infinity, while that of *all* the others tends to 0. This may seem strange: one may feel that a constituent neighbouring on C_K, namely a constituent that is almost as wide as it and says that all Q-predicates are instantiated except just a particular one, should be allowed some initial probability. Why should C_K get it all and the rest get none? The answer is that in Carnap's system the probability of every (nonanalytic) universal law-statement is 0 in an infinite universe; and every constituent with the sole exception of C_K entails some law-statements, and hence has zero probability. Given that the probability of such a universal law-statement as $\forall x\,(P_1 x \rightarrow P_2 x)$ is zero, then the probability of its negation, namely the existential statement $\exists x\,(P_1 x \wedge \sim P_2 x)$, must be 1. Now C_K is unique in being equivalent to the conjunction of the negations of every law-statement that can be formulated in the language; for C_K is the maximumly strong existential statement $\exists x Q_1 x \wedge \exists x Q_2 x \ldots \wedge \exists x Q_K x$, which entails all lesser existential statements in L_2.

However, in Hintikka's system, as we will see more clearly in § 2.5 below, law-statements can have a positive probability, which means that a constituent entailing law-statements is not condemned to zero probability. Then how should initial probabilities be distributed over the constituents? Our negative finding so far is that they should *not* be distributed equally; for it is impossible that at each depth all the constituents at that depth are equiprobable (since they split unequally at greater depths); there is no deepest level at which the constituents could be taken as equiprobable (for any depth-d, however large d may be, there is always a depth-$d+1$); and we have just seen that depth-1 constituents vary enormously in their degree of falsifiability.

However, to conclude from this negative finding that there can be no nonarbitrary distribution of initial probabilities would be analogous to concluding that there can be no fair system of wartime food rationing on the ground that a system that gave everyone, whether soldiers, miners, expectant mothers, children, or old-age pensioners, exactly equal rations would be unfair. Since we are concerned with *initial probabilities*, or with the probabilities of constituents prior to the arrival of empirical information, a nonarbitrary distribution will have to discriminate between them on the basis of structural or formal differences only. Discrimination between them on the basis of empirical considerations comes later, when we turn from prior to posterior probabilities. Now there can be only one kind of structural difference between two depth-1 constituents (or between any two constituents of an equal depth): either one of them is *wider* than the other or else they are of equal width. If they are of equal width they satisfy the requirements that I will give in § 5.13 for two statements to be what I call *incongruent counterparts*. If two statements are incongruent counterparts, they can be expressed by two sentences that are exactly the same except that in at least one place where one Q-predicate occurs in one of them, a different Q-predicate occurs in the other. Thus one can be converted into the other by a function that sends such disputed Q-predicates in one to the corresponding disputed Q-predicates in the other. For instance, the width-1 constituent $\exists x Q_1 x \wedge \forall x Q_1 x$ can be converted into the width-1 constituent $\exists x Q_2 x \wedge \forall x Q_2 x$ by a function that sends Q_1 to Q_2. Incongruent counterparts are structurally similar. Every consequence of one has a counterpart (not necessarily incongruent) among the consequences of the other and vice versa.

But two depth-d constituents of different widths are structurally different in important ways. For instance, the wider one splits into a larger number of depth-$d+1$ constituents, is less falsifiable and, as we shall see, has less question-answering power. In view of this, I propose that the structural consideration that should govern the distribution of initial probabilities over constituents is differences of width. (This is in line with a suggestion made by Carnap, 1968a, p. 219.) Two constituents of the same width should have the same initial probability and a wider constituent should have a higher initial probability than a narrower one. We might express this a shade more formally as follows. Let $w(C_i)$ denote the number of Q-predicates declared instantiated by constituent C_i. (Thus $1 \leq w \leq K$.) Then the distribution of initial probabilities over the constituents should satisfy the following two requirements:

(1) $\Sigma\, p(C_i) = 1$

(2) $p(C_i) > p(C_j) \leftrightarrow w(C_i) > w(C_j)$.

An advantage of such a distribution is that while the widest constituent C_K will of course have the highest initial probability, it will not grab all the probability, as it tends to do in Carnap's system. A neighbouring constituent, which says that all Q-predicates except a specified one are instantiated, will be less probable than C_K but more probable than a neighbouring constituent on its other side that says that all Q-predicates except two are instantiated; and so on down until we reach a C_{i_1} that says that only one Q-predicate is instantiated and which has a minimal initial probability.

In this chapter I want to give probabilism a run for its money. I said earlier that the probabilist programme calls for (i) propositional units of a kind that can serve as stand-ins for possible worlds, and (ii) a nonarbitrary distribution of initial probabilities over these units. Apropos (i) we have briefly looked at state-descriptions, structure-descriptions, and constituents. The big disadvantage, from a probabilist point of view, of state-descriptions and structure-descriptions is that their number depends not only on the number of primitive predicates in L but also on the number of individuals in the universe of discourse. If no limit is put on this, it will be impossible to prevent, in any reasonable and nonarbitrary way, their initial probabilities from tending to zero. The great advantage of depth-d constituents is that, d being fixed, their number depends on the number of primitive predicates in the language (at least when the language has quantifiers and identity but not individual constants), and is invariant with respect to the number of individuals in the universe of discourse; so the number of constituents can be finite, and each of them can have a positive probability, even when the number of individuals (say, space-time regions) over which the quantifiers in the language range is infinite. So let us fall in with the idea that constituents may serve as the basic units in a system of probability logic. Apropos (ii), we saw that it would be wrong to distribute initial probabilities equally over depth-d constituents. But to conclude from this that a nonarbitrary distribution is impossible, and hence that there can be no objective basis for probability logic, would not only be uncharitable but also, it seems to me, fallacious; for I can see no good reason why their initial probabilities should not be a function, of the kind indicated, of their relative widths. Perhaps I am wrong about this; perhaps there is after all no nonarbitrary weighting of constituents and no nonarbitrary system of probability logic. In that case probabilism is baseless. But I want to examine how well probabilism fares on the assumption that there are nonarbitrary systems of probability logic. So I will assume that requirements (1) and (2) above suffice to induce nonarbitrary orderings of the initial probabilities of constituents. We will revert to this question in §2.53.

2.24   *Absolute Probabilities*

Probability values, according to probabilism, are supposed to control rational belief. But there are disagreements within probabilism both over just which probability values should do so, and over how they should exercise their control. As to the first, some probabilists have held that rational beliefs should be controlled in a direct and straightforward way by what are sometimes called *absolute* probabilities, the absolute probability of a hypothesis *h* for a person *X* at time *t* being the probability of *h* relative to *all* the evidence that *X* knows for sure at *t*. Again it seems to have been Bolzano (*1837*, §317) who originated this idea. His reasoning here was very simple. On the one hand, *X* should not include dubious material in the evidence *E* on which he appraises *h*: so *E* should consist *only* of what he knows for sure. On the other hand, *X* should not omit from *E* any evidence that he does know for sure, because such omissions might significantly affect the probability. Thus *E* should consist of all, and only, what *X* knows for sure at *t*.

This idea has been repeated, with acknowledgment to Bolzano, by Roderick Chisholm:

> Probability is a relation between propositions; and one and the same proposition may have different probabilities in relation to different propositions. Of these indefinitely many probabilities that a given proposition may have, *which* is the one that should guide us when we make a decision with respect to that proposition? The answer is that we should be guided by the *absolute* probability of the proposition—the probability that the proposition has in relation to what we know. (*1966,* p. 188)

If rational beliefs are to be controlled by absolute probabilities, or probabilities on total evidence, the next question is just how that control should be exercised. Perhaps the simplest answer to this question is that, *E* being my total evidence, and *h* being the hypothesis in question, I should accept *h* if and only if $p(h, E) > p(\bar{h}, E)$. Since $p(h, E) + p(\bar{h}, E) = 1$, this means accepting *h* if its absolute probability is greater than one half. Bolzano endorsed this: 'If *h* becomes true in more than half of the cases in which *E* is true, the truth of *E* entitles us to accept *h* as well, otherwise not' (*1837*, p. 239; I have adjusted his notation). And Richard Robinson said this about the way in which 'the reasonable man' forms his opinions:

About synthetic statements he is content with probability. On the considerations available to him at the time concerning a given statement, either a statement is more probable than its contradictory, or it is less probable, or it is equally probable. In the last case he suspends judgment; otherwise he adopts for the present the more probable of the two contradictories. (*1964*, p. 86)

But surely a critical level of one-half is too low? There are slightly more male than female births in the human population. Thus for a pregnant woman who knows this statistical fact and who has no other evidence as to the sex of her unborn baby, the probability that it will be a boy is slightly higher than one-half. But would she not be foolhardy simply to accept, without further ado, that it will be a boy?

But even if we set the level at some arbitrary point, say 0.8, there remains a decisive objection, namely that this may have the result that X ought to accept h_1, and h_2, ... and h_n when he considers each of these separately, but that he ought not to accept the conjunction $h_1 \wedge h_2 \ldots \wedge h_n$; for the probability of the conjunction of two or more statements is, in general, lower than that of each of its conjuncts. This point was made very effectively by Kyburg (1965, 1968) with the help of his "lottery paradox". If 1000 lottery tickets have been sold, one of which will win, then the probability that any particular ticket will not win is 0.999. So X ought to accept h_1, and h_2, ... and h_{1000}, where h_1 says that ticket no. 1 will not win, and so on. Yet he must not accept $h_1 \wedge h_2 \ldots \wedge h_{1000}$ since its probability is zero.

Still another objection to the idea that a hypothesis should be accepted if its absolute probability is above a certain critical level is posed by the fact that a hypothesis may have what I will call a *pseudo* absolute probability of 1 (and its negation, a pseudo absolute probability of 0). By this I mean that for *any E* we automatically have $p(h, E) = 1$ and $p(\bar{h}, E) = 0$. It is a consequence of Bayes's theorem (see §2.1 above) that if the initial probability of h is 0 then its posterior probability will always be 0 and, consequently, that of its negation will always be 1. In many systems of probability logic, including Carnap's, the initial probability of any universal law-statement is zero when the universal quantifier ranges over an infinite domain. Let h_1 say 'There is at least one yellow raven', which is of course the negation of 'There are no yellow ravens' (or 'All ravens are nonyellow'). If the latter's posterior probability is 0 because of its universality, the former's is 1. Let h_2 say 'All ravens are black'. Then we could have $p(h_1, E) = 1$ and $p(h_2,$

E) $= 0$, even if E records ever so many cases of black ravens and no case of a nonblack raven, let alone a yellow one. It is rather as if there were an army in which some soldiers rise to high rank under their own efforts but any royal duke is automatically a field marshal even if he is very lacking in the military virtues. So it seems that we should abandon the idea that rational acceptance of hypotheses should be controlled simply by absolute probabilities.

In the Bayesian approach, the key variable is the *ratio* of a later to a prior probability of a hypothesis. The prior one might be its initial probability and the later one is its absolute probability at the present time. Thus if we have $p(h) = q$ and $p(h, E) = r$, where E is the total evidence, then what concerns Bayesianism is not the value of r as such but the value of r/q. This has large advantages over the previous approach. One is that hypotheses with a pseudo absolute probability of 1 are no longer an embarrassment since they perform very indifferently under this mode of evaluation. If we have $p(h, E) = 1$ merely as a consequence of $p(h) = 1$, then the value of r/q is 1, which is as low as it can get without E actually telling against h. Another advantage is this. If there are a good many unrefuted hypotheses competing with each other in a certain field, it is rather likely that each of them will have a low absolute probability. But this does not embarrass Bayesianism, since in the case of some of them the value of r/q may be high, although that of r is low, just because that of q is very low. Nor does Bayesianism need to lay down some arbitrary critical level that must be attained for a hypothesis to be rationally acceptable. If $h_1, \ldots h_n$ are the competing hypotheses in a certain field, then the one (if there is just one) that has the highest value for r/q is the one, according to Bayesianism, that it is rational to accept.

The prior probability of a hypothesis that is taken into account by Bayesianism need not be its initial probability (or probability relative to a tautology). The prior probability could be its absolute probability before a certain experiment was carried out. Suppose that h_1 and h_2 have at present the same absolute probabilities, or $p(h_1, E) = p(h_2, E)$ where E is the total evidence to date. However, h_2 assigns a high probability to a risky prediction e whereas h_1 leaves unaltered the low probability that e has on E; and this prediction is about to be tested. For ease of computation let us put $p(e, E) = p(e, E.h_1) = .01$ and $p(e, E.h_2) = .99$. By Bayes's theorem

$$p(h, e.E) = \frac{p(h, E) \cdot p(e, E.h)}{p(e, E)}.$$

Thus if the prediction e is verified, the ratio of the posterior to the prior probability is 1 in the case of h_1 but 99 in the case of h_2.

I should emphasise that if a person X is trying to assess a hypothesis h in a Bayesian way then he still needs to operate with absolute probabilities or probabilities on the total evidence known to X at the present time. For suppose we have $p(h) = q$ and $p(h, E) = r$ and the ratio r/q is high, *but* E contains only some of the evidence known to X. The omissions from E may have seriously distorted r. Perhaps there is evidence known to X that actually refutes h. Even personalist versions of Bayesianism, while allowing you to manipulate prior probabilities, do not allow you to pick and choose among the different bits of evidence that come in; that would completely nullify the hope, about which we expressed misgivings earlier, that the effects of divergences between different people's subjectively determined priors will tend to be swamped as more and more evidence comes in.

A probabilist theory typically yields a definition for *degree of confirmation* in which, for the reasons given, absolute probabilities play a key role. Some probabilists hold that, faced with a set of competing hypotheses that are neither verified nor falsified, a rational man should accept whichever is best confirmed. Other probabilists hold that, rather than accept any of them outright, he should apportion his degree of confidence in each of them according to its degree of confirmation. The latter view will tend to merge into the former in cases where one hypothesis is much better confirmed than all its competitors, and the former into the latter in cases where several hypotheses are tied for first place.

I will now try to summarise the foregoing in a proposition, which will, I trust, be acceptable to all shades of probabilist opinion:

(IV)   A rational person's acceptance of, or degree of confidence in, an uncertain hypothesis should be controlled by its degree of confirmation on the total evidence known to him at the time.

This formulation is intended to be neutral between probablists such as Bolzano who hold that, if r is the absolute probability of a hypothesis h, then rational acceptance of (or confidence in) h should be controlled by r, and Bayesians who hold that it should be controlled by r/q where q is some prior probability of h. I label this proposition (IV) because, as we shall see, it adds a highly significant component to the problem already posed by propositions (I) to (III) presented in §1.1 above.

2.3 Probability-Scepticism and Rationality-Scepticism

I will now formulate what I take Humean scepticism to assert against probabilism. I like to think that Hume would have approved my formulation and also the arguments by which I will subsequently uphold this position against probabilism. However, since he did not say a great deal about probability and did not anticipate those arguments, I will introduce a new label, that does not implicate Hume, for this rather more particularized version of Humean scepticism, namely *probability-scepticism.*

Probability-scepticism, as I interpret it, does not disagree with probabilism on all points. It allows, as does Humean scepticism, that a person may possess evidence (about his perceptual experiences etc.) that is certain. And it allows (at least for the sake of the argument against probabilism) that there is (or may be) an objective a priori system of probability logic. This means that a hypothesis *h* may have a determinate initial probability, and also that it may have a determinate probability relative to other statements, including evidence-statements. Nevertheless, probability-scepticism denies that uncertain hypotheses about the external world can be established as more or less probably true by being related by probability logic to evidence that is certain. The ocean of uncertainty around the island of certainty, though it contains a priori differentiations, is not empirically differentiated. There are pseudo but no genuine absolute probabilities. I mentioned earlier that some versions of probabilism allow one to proceed *step-by-step*, while other versions allow only *one* step, from certain evidence to a probable conclusion. According to probability-scepticism, we can make *no* such step. There is no empirical confirmation or justification in any probabilist sense. This negative thesis will be argued for in §2.4.

I wish now to introduce a stronger and, to my mind, much more dangerous sceptical thesis, which I call *rationality-scepticism*; this says that we never have any good cognitive reason to adopt a hypothesis about the external world. (I distinguished between probability-scepticism and rationality-scepticism in 1968b, pp. 278–279.) Peter Unger has similarly distinguished 'scepticism about knowledge' from 'scepticism about rationality' (*1975*, ch. 5). He claims that the former entails the latter, but it seems clear that it is the other way round. If we never had any good cognitive reason to adopt a hypothesis, then one kind of reason we would never have is that a hypothesis is probably true, or more or less well confirmed. So rationality-scepticism does entail probability-scepticism. On the other hand, there might be good cognitive reasons *having nothing to do with probable truth* for adopting a hypothesis (I shall argue in Part Two that this is indeed so). So probability-scepticism does not entail rationality-scepticism. However, probability-scepticism reinforced by proposition (IV) above *does* entail rationality-scepticism; for proposition (IV) says that we have reason to adopt a hypothesis *only* if there is confirming evidence for it. Thus proposition (IV) has an ominous significance: it escalates a kind of scepticism that is compatible with cognitive rationality into one that is not. If we are to avoid rationality-scepticism, it is essential to find a genuine alternative to proposition (IV). It may perhaps help the reader if I now briefly anticipate the later course of the argument in this book by mentioning that I shall propose replacing (IV) by:

(IV*)　it is rational for a person X, faced by evidence e, to adopt an uncertain hypothesis h if h is possibly true and provides the best explanation for e that is available to X.

But I am anticipating. I now turn to what seems to me a decisive argument against probabilism.

2.4　A New Argument for Probability-Scepticism

According to proposition (IV) above it is not rational to accept, or to have some positive degree of confidence in, a proposition unless it is either certainly true or, if uncertain, empirically justified or confirmed to some degree. Suppose that a is an uncertain empirical hypothesis. What conditions should a statement b satisfy if b is to provide some positive justification for a? Well, b must surely (1) *bear favourably* on a and also (2) be *less uncertain* than a. A statement that satisfied (1) but which was no less uncertain than a itself, would do nothing to erase the question mark hanging over a. The strongest way in which b could bear favourably on a is by entailing a. But requirement (2) has the consequence that b must *not* entail a; for if a is uncertain and b entails a, then b is at least as uncertain as a: the content of b would contain all the uncertain content of a (b would be equivalent to $a \wedge (a \rightarrow b)$). So it seems that if b is to satisfy (1) without entailing a then b must raise the probability of a.

Requirement (2) would obviously be satisfied if b were *known* to be true. But suppose that it is not. Requirement (2) would obviously not be satisfied if *b* were a statement for which no justification has been provided and whose status is wholly indeterminate. So it would seem that if *b* is to satisfy (2) without being known to be true, there must be something beyond *b*, call it *c*, that provides some justification for *b*. (A challenge to this view will be mentioned in a moment.) If this *c* is to provide some justification for *b*, it must likewise (1) bear favourably on *b* and (2) be less uncertain than *b*. And if *c* is to satisfy requirement (2) it must either be *known* to be true or else there must, presumably, be something beyond *c*, call it *d*, that provides a justification for *c* . . . It seems clear that the regress that is opening out here can be halted only if it terminates with 'ultimate premises', to use Keynes's term, that are *known* to be true. If *a* is "supported" by *b*, which in turn is "supported" by *c*, but *c* is wholly unsupported and has a status that is quite indeterminate, then *no* support is provided for *a*. The last link in a chain of justifications must be *secure*, otherwise the chain is dangling in limbo. It was considerations like these that led C.I. Lewis to make the famous pronouncement: 'If anything is to be probable, then something must be certain. The data which support a genuine probability must themselves be certainties' (*1946*, p. 186); for if the data, or ultimate premises, were not certain, then since there is nothing beyond them to endow them with a more or less high probability, their status would be quite indeterminate.

If Lewis's thesis is correct, probabilism would collapse if there were no data, or ultimate premises, that are certain. However, it is no more a part of my case against probabilism that there are no certainties than it is that there is no nonarbitrary system of probability logic. On the contrary, I allow, quite in accord with Humean scepticism, that we each have our own little island of certainty. The question is whether we can exploit those certainties, with the help of probability logic, to endow uncertain hypotheses with a determinate probability of being true. Lewis's thesis has in fact been challenged by Richard Jeffrey, who claimed that an evidential statement may have a kind of probability that is not relative to some further evidential statement lying beyond it but is, so to speak, noninferential and intrinsic; and probabilities of this kind may take values that are determinate though less than 1. Jeffrey's aim here was to uphold probabilism in the face of the thesis that the evidence by which we judge hypotheses is never absolutely certain. Since I am not advancing this thesis, there will be no unfairness to probabilism if, for the time being, we suspend judgment on Jeffrey's claim (it will be examined in §3.11). For we are granting that probabilism can rely on evidential statements with a status superior to that of an evidential statement with a Jeffrey-type probability; in line with Keynes, Lewis, and others we are allowing that a justificatory chain may terminate with ultimate premises that are certainly true.

The question to which I now turn is: *how long* may such a chain be? As we saw, Keynes spoke of 'drawing conclusions by a series of steps': *how many* such steps may we take, if the connections are not entailments? Suppose that c is an ultimate premiss that is certainly true, that b is highly probable on c and that a is highly probable on b: does c provide a justification for a?

David Stove (*1973*, p. 60) attributed to Hume a thesis according to which *no* step in such a chain provides any justification; for he claimed that Hume tacitly assumed that if b does not entail a (and also, Stove should have added, does not entail $\sim a$), then $p(a, b) = p(a)$. In other words, if a and b are logically independent, then they are also probabilistically independent. Stove added that no philosopher has actually asserted this (p. 77); which is just as well, since it is a daft thesis. Let a assert, concerning some randomly chosen integer, that it is odd. Then $p(a) = \frac{1}{2}$. Let b assert of this same integer that it is a prime number. This b does not entail this a because not all prime numbers are odd: 2 is a prime number. Nor of course does a entail b; they are logically independent. Yet we obviously have $p(a) < p(a, b)$ here. Indeed, we even have $\frac{1}{2} = p(a) < p(a, b) = 1$; for Euclid's proof established that there are infinitely many prime numbers, and of these just one is even and the infinitely many others are odd: the ratio of even to odd is $1 : \infty$ and the chance that a randomly chosen integer, that happens to be a prime number, is even is infinitesimal. In this case b, in Carnap's terminology (*1950*, pp. 311f), "almost" entails (or "almost" L-implies) a.

Not only did Hume not assert the thesis imputed to him by Stove, but he asserted something that conflicts with it. We will shortly consider a sceptical argument of his that has the following form: a is highly probable on (though *not* entailed by) b, b is highly probable on c, and so on; *but this regress has no end*; hence the probability of a is not raised at all.

2.41   *The Nontransitivity of Nondeductive Justifications*

Let us now revert to the question of how long a justificatory chain may be. If b entails a, and c entails b, then of course c entails a. But we saw that if requirement (2), which requires b to be less uncertain than a, is to be met, then b must not actually entail a given that a is uncertain. The next best thing that b could do would be to "almost" entail a which would give us $p(a, b) = 1$; after that come cases where b gives to a a probability of almost one, or $p(a, b) = 1 - \epsilon$ where ϵ is very small. Salmon (1965) has pointed out that any relation of support weaker than that of entailment is non-transitive (I will silently adjust his notation to ours): 'Given that $p(a, b)$ and $p(b, c)$ are both high— *they may both equal one* provided they cannot be replaced by strictly universal generalizations—it does not follow that $p(a, c)$ is high, *or even that it is different from zero*' (pp. 166–167, my italics). An example he gave of $p(a, b) = p(b, c) = 1$ but $p(a, c) = 0$ is the following. Let a say, as before, of a randomly chosen integer that it is odd, let b say that it is a prime number, and let c say that it is the number 2. We have $p(a, b) = 1$ for the reasons given previously (b "almost" entails a), and $p(b, c) = 1$ because c entails b, and $p(a, c) = 0$ because c entails $\sim a$.

Clark Glymour has suggested (personal communication) that in this example $p(a, b)$ is indeterminate if the integer is chosen from the set of *all* integers, since $p(a, b) = p(a) \cdot p(b, a) / p(b)$ and both $p(b, a)$ and $p(b)$ become zero in the infinite case. Well, we could let the integer be chosen from the first N integers; and instead of writing $p(a, b) = 1$ we could write $p(a, b) \rightarrow 1$ as $N \rightarrow \infty$.

The nontransitivity of nondeductive justifications is sufficiently important to deserve to be added, as a fifth proposition, to the four that have already been presented as posing the problem of scepticism. We could state it as follows:

(V)   given only that a conclusion a is strongly supported by a statement b and that b is strongly supported by a premiss c, we may not conclude that a is supported by c.

I will sometimes refer to this as *the ban on step-by-step justifications*.

Most sceptics and anti-sceptics seem either to have missed the point contained in (V) or to have underestimated its importance. Hume missed it. I alluded earlier to an argument of Hume's in which the thesis imputed to him by Stove is repudiated. Hume there argued that attempts to assess the probability of a judgment involve an infinite regress. Here is an extract:

But this [new] decision, tho' it should be favourable to our preceding judgment, being founded only on probability, must weaken still further our first evidence, and must itself be weaken'd by a [further] doubt of the same kind, and so on *in infinitum*; till at last there remain nothing of the original probability, however great we may suppose it to have been, and however small the diminution by every new uncertainty. (*1739–40*, p. 182)

Without claiming historical faithfulness to Hume's thinking, here, we might bring this into juxtaposition with proposition (V) by reformulating it along the following lines. Suppose that a would be almost but not quite certain if b were certain, and that b would be almost but not quite certain if c were certain, and so on. To keep it simple suppose that we have $p(a, b) = p(b, c) = p(c, d) \ldots = 1 - \epsilon$, where ϵ represents the 'however small' diminution of probability introduced by each new uncertainty. Then Hume may perhaps have supposed that we have $p(a, b) = 1 - \epsilon, p(a, c) = (1 - \epsilon)^2, p(a, d) = (1 - \epsilon)^3 \ldots$, and that the probability of a after n steps is $(1 - \epsilon)^n$; and $(1 - \epsilon)^n \to 0$ as $n \to \infty$. But proposition (V) says that already on the second step the probability of a on c is *quite indeterminate* given only that $p(a, b) = p(b, c) = 1 - \epsilon$.

I say that $p(a, c)$ has an indeterminate value given *only* that the values of both $p(a, b)$ and $p(b, c)$ are very high. Of course, if we are given other information besides this we may be able to proceed to a conclusion about the value of $p(a, c)$. Mary Hesse (*1974*, p. 148) rightly points out that, given (1) that $p(a, b) > 1 - \epsilon$, and (2) that $p(b, c) > 1 - \epsilon$, and moreover (3) that c is 'not negatively relevant' to a given b, or that $p(a, b.c) \geq p(a, b)$, *then* we may conclude that $p(a, c) > (1 - \epsilon)^2$. But in order to know that premiss (3) is satisfied we must already know a good deal about $p(a, c)$, namely, that $p(a, c) \geq p(a, b) \cdot p(b, c) / p(b, a.c)$. In other words, we can proceed to a *conclusion* about the value of $p(a, c)$ from premises about $p(a, b)$ and $p(b, c)$ if we start with an additional *premiss* about the value of $p(a, c)$ itself.

On the antisceptical side, Carnap first presented something very like our proposition (V), and then tried to find a way round it. He wrote:

To a superficial inspection it might appear as if inferences of the following kind were not only frequently used in every day life but also valid: suppose that on the basis of given evidence e, the hypothesis h_1 is highly probable and that h_1 gives high probability to h_2; then h_2 is highly probable on e. But in this form the chain of inference is not generally valid. (*1950*, pp. 319–320)

So far, so good. He then asked: 'Are then the chains of inductive reasoning customarily made in everyday life, in law courts, and in science invalid?' In an attempt to rehabilitate such chains of inductive reasoning,

Carnap then suggested that they are in order if they have the following cumulative form: h_1 is highly probable on e; h_2 is highly probable on $e \wedge h_1$; h_3 is highly probable on $e \wedge h_1 \wedge h_2$; and so on 'provided the chain is not too long' (p. 320).

But this would license a kind of *cheating*. Suppose that I am a keen partisan of a hypothesis h, but find that on the evidence e available to me, h has a disappointingly low probability. Then I may be able to jack up its probability to a flatteringly high level by inventing mediating hypotheses h', $h'' \ldots$ such that $p(h', e)$, $p(h'', e.h') \ldots$ and $p(h, e.h'.h'' \ldots)$ are all high. (A trivial arithmetical example: let e, h', h'', h''' and h say respectively of some integer n: $1 \leq n \leq 10000$; $1 \leq n \leq 8000$; $1 \leq n \leq 6400$; $1 \leq n \leq 5120$; $1 \leq n \leq 4096$. Then $p(h', e) = p(h'', e.h') = p(h''', e.h'.h'') = p(h, e.h'.h''.h''') = .8$. However, $p(h, e) = .4096$.)

It should be added that this idea of Carnap's seems to have been a momentary aberration, prompted by his desire to rescue everyday modes of inductive reasoning from the sceptical implications of proposition (V); for it violates a famous requirement that he had previously laid down, namely his '*Requirement of total evidence*: in the application of inductive logic to a given knowledge situation, the total evidence available must be taken as basis for determining the degree of confirmation' (*1950*, p. 211). Thus if e is the basis for determining the degree of confirmation of h_3, then e must constitute the total evidence; but if e is the *total* evidence, then we may not subsequently treat h_1 and h_2 as if they too were *evidence* for h_3.

Let us recapitulate. Suppose that c supports (without entailing) b, and that b supports (without entailing) a; however, c is unjustified and does not have a determinate probability status. Then no justification has been provided for a. (We might equally have started out from c' and proceeded via b' to a', where a' is incompatible with a.) Very well; let c be certain. But still no justification has been provided for a by this step-by-step sequence. (If we considered a in direct relation to c, we might find that c is adverse to a.) Very well; let us confine ourselves to just two statements, evidence e and hypothesis h, where e is certain and $p(h, e)$ is high. Are we at last home and dry? May we conclude that h, though not certain, is at least probably true?

2.42 *Requirement of Total Evidence*

No, we may not. For it is entirely consistent with e being certainly true and $p(h, e)$ being high that there exists another statement e' that is consistent with *e* and also certainly true, and such that $p(h, e.e')$ is low or even zero. It was this possibility that, as we saw, led Bolzano to introduce the idea of the *absolute* probability of a hypothesis for a person *X* at time *t*, namely its probability on *all* the propositions that *X* knows at *t*. Keynes rather similarly defined the absolute probability of a proposition as its probability relative to what he called the Universe of Reference, 'which contains *all* the propositions which are known to be true' (*1921*, pp. 129–130). And Carnap, as we just saw, laid down a *requirement of total evidence*. This requirement was incorporated in proposition (IV) in §2.24 above.

I now turn to my main argument for probability-scepticism. It becomes a little technical in places, but at bottom it is very simple. It says that no person *X* could ever know that the absolute probability, for him at the present time, of an uncertain hypothesis *h* is high (provided that *h* does not have a *pseudo* absolute probability of 1 merely in virtue of its logical form, as might 'Yellow ravens exist'); for even if *X* knows both that *E* is true and that $p(h, E)$ is high, he has still to establish that *E* is his *total* evidence; and to establish this would require an *extra premiss* that *E* itself cannot provide and which *X* cannot verify. Hence the status of this extra premiss will remain *indeterminate* (unless it turns out to be false). Thus when *X* seeks to locate *h* on a contour line near his island of certainty he will be dependent on a premiss lying right out in the ocean of uncertainty. Or so I will conclude at the end of this section.

The evidence at *X*'s disposal at time *t* will need to be described in statements if *X* is to use it in assessing the probability of *h*, and I will assume that these are singular observation statements. Let *E* be a large conjunction of singular observation statements and let *m* be a statement, in a suitable metalanguage, that says that *E* contains all and only those singular observation statements that are known to be true by *X* at *t*. In other words, *m* says that *E* describes *X*'s total evidence at *t*. Let *h* be a falsifiable hypothesis whose probability on *E* is high. However, *E* does not entail *h*; so there will be singular observation statements, whether true or false, that are consistent with *E* and adverse to *h*. Imagine some of these to have been conjoined to *E* to form the new statement *E'*, the probability of *h* on *E'* being low. And let *m'* say of *E'* what *m* said of *E*.

Now *if X* could verify *m* then he could affirm that the absolute probability of *h* for him now is high. Let us concede that, presented with any observation statement, *X* can tell unhesitatingly whether or not he knows it to be true. Even so, could he verify m? It would be as if, to vary an example of Descartes's, I wished to verify that this basket contains just those of my apples that are not rotten: I would need to check (1) that each apple in it is not rotten *and* (2) that all apples of mine that are not in the basket are rotten. X would need to check (1) that every observation statement in E is known by him (to ensure that E contains *only* statements he knows to be true) and (2) that every observation statement not in E is not known by him (to ensure that E contains *all* statements he knows to be true). In short, he would need to check through *every* (logically independent) observation statement in his language. He obviously could not do this unless his language were severely restricted (a possibility to which we will revert shortly). Just try drawing up a list of all and only those observation statements you know to be true. The mind boggles. And if you did embark on this daunting task you would, presumably, find that the very process of compiling the list was continually generating new items (for instance, that you are now using a ballpoint pen) that need to be added to the list.

At this point someone may object that it is needless and unfair to require X to assess h on his total evidence, which may indeed include a mass of heterogeneous material, much of it quite irrelevant to h; rather, he should assess h on the total *relevant* evidence in his possession. But this modification is of no help at all. Let m now say that E contains all those observation statements that are both relevant to h and known by X at t. Then in order to verify this m, he must again do two things: (1) check every statement in E in order to verify that E contains only statements that are both relevant to h and known by him to be true; (2) check, in the case of every observation statement *not* in E, that it is not known by him or else is not relevant to h, in order to verify that E contains *all* those observation statements that are relevant to h and known by him. His task will now be even more impossible than before, in rather the same way that it is even more impossible for a cat to swim the Atlantic than to swim the North Sea. For he will again have to check through all the observation statements in the language; and this time he will, in addition, need in each case to assess its relevance to h; and this may well be problematic. As any reader of detective stories will agree, it can easily happen that several bits of evidence seem irrelevant to a hypothesis when considered separately, but turn out to be of decisive importance when assembled together.

A probabilist may try to circumvent these difficulties by so restricting X's object-language that the number of observation statements that can be formulated within this language, call it L_o, is small enough for X to be able to check through all of them. But this will shift the conjectural assumption instead of eliminating it. Previously, X could only conjecture that E describes his total evidence. Now, he may verify that it describes that part of his total evidence that can be described in L_o; but he can now only conjecture that L_o is *adequate* for the appraisal of h, in the sense that h would not have a significantly different "absolute probability" in an extension of L_o that had a richer vocabulary that allowed more of the evidence in X's possession to be reported. Consider the heliocentric hypothesis. One might suppose that a language adequate for its appraisal would not need predicates relating to snails and other such lowly creatures. But if the shells of snails in northern latitudes spiral in one direction and those in southern latitudes in the other, that may be very relevant to this hypothesis. Or consider the neo-Darwinian theory of natural selection: could any of us compile a list of all the predicates needed in a language adequate for the appraisal of this theory? The mind boggles.

We considered earlier an argument for *No probable truth without some certainty* that boiled down to this: X knows nothing about the probable truth of a hypothesis h if he knows only that h is highly probable relative to some 'ultimate premiss' e, the status of which is indeterminate. We now have an argument for *And no probable truth even with some certainty* that boils down to this: even if X knows that h is highly probable on E, and that E is certainly true, his E is only a penultimate premiss; to establish that h is, for him at this time, very probably true, he needs a further premiss to the effect that E describes *all* the evidence in his possession that is relevant to h. And for the reasons given, the status of this ultimate premiss will be indeterminate (unless it turns out to be false). In short, he cannot assign a (genuine as opposed to pseudo) absolute probability to h.

This result is equally damaging to probabilists who hold that rational acceptance of, or degree of confidence in, a hypothesis h should be controlled by the value of r in $p(h, E) = r$ and to Bayesians who hold that it should be controlled by the value of r/q where q is some prior probability of h. If the value of r is high, but E does not contain all the evidence known to x at t, then it is entirely possible that the value of r would be 0 if the omitted evidence were included.

2.5 The Probability of Law-Statements

The negative conclusion just arrived at would of course mean that *it is never rational to accept a universal generalisation g* even though there may be evidence E that is certain and relative to which the probability of g is high. In this section, however, we will assume, for the sake of the argument, that X could know of a given E that it describes his total evidence at the present time; and the question we will now consider is whether, if g is a universal generalisation and E seems entirely favourable to g, E could raise the probability of g above its initial level. Could we have $p(g, E) > p(g)$?

In the course of this investigation some unfinished business left over from §2.2 will be cleared up. But I fear that the investigation will become quite technical in places. This is unavoidable because an answer to the question of whether the probability of law-statements, in an infinite universe of discourse, must be zero depends on what answers are given to two underlying questions that were raised in §2.2: namely, are there basic units (somehow corresponding to "possible worlds") on which a system of probability logic could be based, and if so, is there a non-arbitrary distribution of initial probabilities over these units? For the benefit of a reader who does not care to delve into these rather tricky issues I will now summarise in advance the conclusions of the present section. The views of three thinkers will be considered: Popper, Carnap, and Hintikka. For both Carnap and Popper the answer to our question is that, in infinite domains, the probability of any genuine law-statement on any finite evidence, no matter how favourable, always remains obstinately stuck at zero. For Hintikka, on the other hand, a suitable generalisation can attain a positive, indeed a high, posterior probability on suitable evidence. However, in Hintikka's system there is a free parameter; and in order to secure this result this parameter has to be given a value below a certain critical level; and the assignment of such a relatively low value means *inverting* the probability-ordering of the basic units that was proposed in §2.23. Now I will not insist that the ordering there proposed was right. Perhaps there is no such thing as a right ordering (in which case there is no objective basis for probability logic and the probabilist programme collapses). But I will insist that an inversion of this ordering, which has the result that, of the unfalsified constituents, the most falsifiable one becomes not the least but the most probable, cannot be right. If this, to my mind unacceptable, inversion of the ordering is to be avoided, the free parameter I mentioned must be given a value above, rather than below, that critical level; and this has the result that evidence E cannot raise (though it may lower) the probability of a generalisation g above its initial level. This conclusion will be reached at the end of this chapter.

I now proceed to the investigation. Let g be a universal generalisation of the form $\forall x\, Hx$. (A statement of the form $\forall x\, (Fx \rightarrow Gx)$ can always be recast in the form $\forall x\, Hx$ where Hx is a molecular predicate equivalent to $Gx \vee \sim Fx$; if Hx stood for 'x is black or x is no raven', $\forall x\, Hx$ would say that all ravens are black.) Assume that x ranges over a potentially infinite domain of individuals. Let E^n be an evidence-statement, as favourable as it could be to g, which says of some large finite number n of individuals that each of them is H. And let e^{n+1} be a prediction that says that the next individual that will be observed is H.

Now the very least that a probabilist must establish here is that E^n raises the probability of e^{n+1} above its initial level, or:

(i) $p(e^{n+1}, E^n) > p(e^{n+1})$.

We may call the assertion just of (i) a *weak inductivist* thesis and the denial of (i) a *strong anti-inductivist* thesis. Now a probabilist would no doubt prefer that E^n can also raise the probability of g above its initial level. But this will be impossible if the initial probability of g is zero. So this desired result involves the following two claims:

(ii) $p(g) > 0$,

(iii) $p(g, E^n) > p(g)$.

We may call the denial of (iii) a *moderate anti-inductivist* thesis. To deny (iii) is to say that no evidence can be positively relevant to a law-statement. Finally, a probabilist may hope that as n grows very large, the probability of g on E^n will approach (without ever quite attaining) 1, or:

(iv) $p(g, E^n) \rightarrow 1$ as n $\rightarrow \infty$.

We may call the joint assertion of (i), (ii), (iii) and (iv) a *strong inductivist* thesis. Popper (*1959*, appendix *vii) championed the strong anti-inductivist thesis; Carnap (*1950*, §110) settled for the weak inductivist thesis; Hintikka (1965b, 1968) championed the strong inductivist thesis. I will argue for the moderate anti-inductivist thesis.

2.51 *Popper and Carnap: The Zero Probability of Laws*

Popper argued against (i) along Humean grounds: in the absence of some synthetic principle (and there is no place for a *synthetic* principle in probability *logic*) there is no a priori reason to suppose that unobserved instances will tend to conform themselves to observed instances. From a purely logical point of view, all instances should be regarded 'as mutually *independent* of one another' (*1959*, p. 367). Popper claimed that every assumption other than that of independence 'would amount to postulating *ad hoc* a kind of after-effect' (*ibid*). To which Colin Howson riposted that the assumption of independence 'postulates equally *ad hoc* a *lack* of after-effect' (1973, p. 157): that logic gives us no reason to assume dependence does not mean that logic gives us reason to assume independence.

It turns out that a great deal depends, here, on what kind of unit should be taken as basic for probability logic and how initial probabilities are distributed over these units. There have been three main candidates for the role of basic unit: state-descriptions, structure-descriptions, and constituents (of depth-1). These concepts were introduced in §2.21; but which of them is taken as basic is so fraught with diverging implications for our main question that it will be as well to recapitulate what has already been said about them. A state-description describes every individual in the universe of discourse completely by assigning a Q-predicate to each of them. A structure-description creams off, as it were, all the anonymous information from a set of isomorphic state-descriptions. (Two state-descriptions are isomorphic if one could be turned into the other merely by permuting some or all of the individual constants occurring in it.) It is something like a statistical abstract from a state-description: it says how many individuals instantiate each Q-predicate, but it mentions no names. It would be a complete sentence in a language with quantifiers and identity but without individual constants. A constituent (of depth-1) creams off, as it were, all the qualitative information from a set of structure-descriptions that agree over which Q-predicates are instantiated and disagree only over the numbers instantiating them. It says which Q-predicates are instantiated but does not say by how many individuals. It would be a complete sentence in a language with quantifiers but with neither individual constants nor identity.

Carnap's programme in his (*1950*) could be described, in Leibnizian language, as being to find a measure m such that the degree of confirmation c of h given e is a function of that measure of the ratio of the possible worlds in which $e \wedge h$ holds to those in which e holds; or such that $c(h, e) = m(e \wedge h) / m(e)$. One measure he considered, and which he called $m^\dagger$, takes all *state-descriptions* as equiprobable: $m^\dagger(e)$ is simply the ratio of the number of state-descriptions in which e holds to the total number of state-descriptions (for a given N where N is the number of individuals in the universe of discourse). But he rejected this measure precisely because a $c^\dagger$-function based on $m^\dagger$ would lead to the denial of thesis (i) above, or to $c^\dagger (e^{n+1}, E^n) = c^\dagger (e^{n+1})$: this choice 'would be tantamount to the principle never to let our past experience influence our expectations for the future. This would obviously be in striking contradiction to the basic principle of all inductive reasoning' (p. 565).

So he chose instead the measure m^*, which takes all *structure-descriptions* as equiprobable, the probability of any one structure-description being divided equally among all those (isomorphic) state-descriptions that instantiate it. For a c^*-function based on m^* endorses thesis (i) above: the value of $c^* (e^{n+1}, E^n)$ can approach 1. However, Carnap's m^* leads to the denial of both thesis (ii) and thesis (iii) above: for a universal generalisation g, $m^* (g) \rightarrow 0$ as N $\rightarrow \infty$ and $c^* (g, E^n) \rightarrow 0$ as N $\rightarrow \infty$. Thus Carnap's system endorses what I called the weak inductivist thesis. (It is surprising that Stove, *1973*, p. 60, imputes to Carnap, of all people, the thesis that, for any contingent hypothesis h, $0 < p(h) < 1$.)

2.52 *Hintikka on the Confirmation of Laws*

Hintikka (1965b, 1968) sought to rehabilitate probabilism vis-à-vis law-statements by taking *constituents* as the basic units for probability logic. His aim was to uphold the thesis incorporated in our proposition (IV), namely that rational assent to hypotheses is controlled by their degree of confirmation. More specifically, he sought to provide a positive answer to the following question, which he posed at the outset: 'Can the rationality of our acceptance of the generalisations we in fact accept be explained in terms of the high probability of these generalisations in some interesting sense of probability?' (1965b, p. 274). A universal generalisation g is confirmed via the confirmations that accrue to each of the constituents in the d.n.f. of g, its confirmation being the sum of theirs. So what we have to consider is how, if at all, constituents are confirmed. And this raises the key question of the a priori probabilities of the constituents. Are they weighted equally? Hintikka was at first, in (1965b), inclined to say that they are. But he found that this answer yields an overkill solution: 'the inductive behavior which corresponds to the assignments of equal *a priori* probabilities to the different constituents is wildly overoptimistic in that one jumps to generalisations on the slightest provocation' (1968, p. 197). If the constituents have to be weighted unequally, how is this unequal weighting to be determined? In answering this question, Hintikka introduced a parameter α: he proposed assigning 'to each constituent as its *a priori* probability the probability with which it would be true in a completely random finite universe with α individuals' (*ibid*). The value of α will need to be greater than zero and finite; for if $\alpha = 0$ the constituents become equiprobable and one 'learns too fast' from experience, while if $\alpha = \infty$, one does not learn at all from experience. We shall find later that it can make a crucial difference just what finite value is put on α.

Two things are required of an evidence-statement that may confirm a constituent: first, each individual reported in it must be completely described, or have a Q-predicate assigned to it; second, the exact number of individuals reported in it must be specified. We may let E^n stand for such an evidence-statement, where n is the number of individuals reported in it. Besides α and n, the following parameters play important roles in Hintikka's system: the number K of Q-predicates in the language; the width w of a constituent (that is, the number of Q-predicates that it says are instantiated); and the number c of Q-predicates that are instantiated in a given evidence-statement E^n.

How does the confirmation of a constituent of width w, say C_w, proceed? C_w will have the following form:

$$\exists x Q_{i_1} x \wedge \exists x Q_{i_2} x \ldots \wedge \exists x Q_{i_w} x \wedge \forall x (Q_{i_1} x \vee Q_{i_2} x \ldots \vee Q_{i_w} x).$$

We now suppose that evidence starts coming in as one, two, ... n observed individuals are reported. Each such individual has been assigned a Q-predicate. If ever an individual gets a Q-predicate not included in Q_{i_1} to Q_{i_w} then, of course, C_w is falsified. So let us confine ourselves to cases where that does not happen. Let some value greater than 0 and less than ∞ be put on α. So long as n $\leq$ w, some slight confirmation will be given to C_w by each new individual that instantiates one of its w Q-predicates. But suppose that n has grown much larger than w and that so far only c Q-predicates are instantiated in E^n, where $c < w$. In other words C_w's predicates $Q_{i_{c+1}}$ to Q_{i_w} have not yet been instantiated. Then while C_w will be confirmed by the observation of a new individual that is unlike previous individuals and instantiates one of these hitherto uninstantiated predicates, the observation of a new individual that is just like a previous one will slightly *dis*confirm C_w; for the suspicion is beginning to grow that C_w is wrong in its claim that $Q_{i_{c+1}}$ to Q_{i_w} are instantiated. If this pattern persists, the confirmation of C_w will tend to 0 as n grows larger.

Suppose however that that does not happen and that eventually *all* the Q-predicates in C_w (and none of those not in C_w) are instantiated in E^n, so that $c = w$. In that ideal case the confirmation of C_w will tend to 1 as n grows larger, *provided* that $\alpha < $ n. (We will consider the significance of this proviso later.) C_w entails all those law-statements in whose d.n.f. it occurs; and all such law-statements will inherit a high posterior probability from it. (However, there is one constituent, namely the one with maximum width K, which in §2.23 we called C_K, that entails the negations of all law-statements. Were the confirmation afforded by E^n to accrue to *it*, no law-statement would be confirmed. We will revert to this possibility later.)

One may feel that the requirements that Hintikka puts on evidence-statements are a bit stiff. An ornithological field-worker might resist the admonition to give a complete description, including age and sex, of each of the swallows he observed migrating, and to keep an exact tally of their numbers. A more serious criticism of Hintikka's theory of confirmation is that, given a (total) evidence-statement E^n that meets his exacting requirements, it makes it *too easy* to go at once to the best of all the propositions confirmed by E^n. (This objection is analogous to one raised against Carnap by Popper in *1982a*, p. 333.) Why bother to invent a hypothesis and to investigate its confirmation by summing the posterior probabilities of the constituents that occur in its d.n.f., when you can go straight to the constituent that is best confirmed by E^n? This will be the constituent that says that all things whatever fall into just those Q-cells that observations have already shown to be occupied. Thus if E^n says that c Q-cells, from Q_{i_1} to Q_{i_c}, are occupied, the best confirmed constituent will say

$$\exists x Q_{i_1} x \land \exists x Q_{i_2} x \ldots \land \exists x Q_{i_c} x \land \forall x (Q_{i_1} x \lor Q_{i_2} x \ldots \lor Q_{i_c} x).$$

In other words, the world as a whole is a straightforward projection of the world as so far observed.

There is a certain analogy between this theory of confirmation and Hempel's (1945) theory. Hempel's underlying idea there was that evidence e directly confirms a hypothesis h if e entails what he called the *development* of h for e; that is, if e entails what h would say on the

73

supposition that the universe consists just of those things mentioned in e. One criticism raised by Carnap (*1950*, pp. 480–481) to this idea is that h will fail to be confirmed if e consists of numerous favourable instances plus one or two neutral instances. For instance, if h says of a new fertility pill that it will enable any woman under forty to conceive, and e says that this pill was tested on 1,000 hitherto infertile women of whom 999 became pregnant within a year and one entered a nunnery, then e would not confirm h. The equivalent to this in Hintikka's system would be if one or two of the individuals reported in E^n were not fully specified. But there is an important difference between the two systems: in Hintikka's we can go at once from a suitable E^n to the *strongest* general statement confirmed by it, namely that constituent whose existential clauses exactly mirror the evidence. I will later introduce (in §5.1) as a meta-principle that should regulate philosophising about science what I will call an antitrivialisation principle. This says that there must be something wrong with a methodological theory that has the implication that, given an existing scientific theory T, or a body of evidence E, it is always trivially easy to manufacture either a "better" theory or the "best" theory. It seems that Hintikka's (1968) theory of confirmation offends against this principle.

2.53 *The Weighting of Constituents Again*

However, the main question before us is only whether Hintikka has succeeded, where Carnap and others failed, in establishing that we may have $p(g, E^n) > p(g)$, where g is a universal generalisation. Suppose as before that E^n says that c Q-cells are occupied. Then all constituents of a width less than c are refuted. Of the unrefuted constituents, two are of particular interest, namely the narrowest and the widest. The narrowest will be the one constituent, call it C_{l_c}, of width c that remains unrefuted. The widest will be the constituent C_K, which says that all K Q-cells are occupied. In between there will, normally, be various constituents of width $c + 1, c + 2, \ldots K - 2, K - 1$ that remain unrefuted. If C_{l_c} is in the d.n.f. of g, and if the probability of C_{l_c} is raised by E^n, then we will indeed have $p(g, E^n) > p(g)$. Now whether the probability of C_{l_c} is raised by E^n depends on what value is assigned to α. Imagine the unrefuted constituents laid out on a seesaw, with the narrowest, C_{l_c}, at one end and the widest, C_K, at the other. Then as Hilpinen (1968) and Oddie (1979) have pointed out, if we put $\alpha < n$, C_{l_c} goes up (becomes the most probable) and C_K goes down; if we put $\alpha = n$, the seesaw becomes horizontal (the unrefuted constituents are equiprobable); if we put $\alpha > n$, C_{l_c} goes down and C_K goes up. Putting a relatively low value on α is unfavourable to the wider constituents, and especially to the widest one, by introducing a 'paucity of individuals' assumption. A wider constituent requires more individuals to instantiate all its Q-predicates than does a narrower one. (Notice that we could secure a probability of 1 for C_{l_c} straight off by putting $\alpha = c$. All constituents narrower than C_{l_c} have been refuted and their probability has gone to zero. Any constituent of width w where $w > c$ implies the existence of more than c individuals to instantiate its w Q-predicates, and its probability falls to zero on the assumption that the universe contains only c individuals. Thus C_{l_c} would be the only constituent to survive both the a posteriori impact of E^n and the a priori impact of making $\alpha = c$ by fiat.)

Is it legitimate to put $\alpha < n$? Well, E^n tells us that at least n individuals exist; hence to put $\alpha < n$ is to introduce the counterfactual assumption that the universe contains fewer individuals than are recorded in our evidence. In (1965b) Hintikka introduced the fair-seeming assumption that $N \gg n \gg K$, where N is the number of individuals in the universe, n is the number of observed individuals, and K is the number of Q-predicates in the language. On *this* assumption, there is no risk that even the widest constituent is false merely because the number of individuals in the universe is so limited that it is unlikely that all its K Q-predicates are instantiated.

When we considered the *initial* probabilities of constituents in §2.23, two requirements seemed clear: (i) they must not be treated as equiprobable; (ii) their probability should vary with their width (and inversely with their falsifiability). We are now considering posterior probabilities, after various constituents have been refuted (namely, all those of width less than c, all but one of those of width c, and all those of width greater than c that assert a Q-cell to be empty that is declared occupied by E^n). I will now argue, in connection with posterior probabilities, that requirements (i) and (ii) should continue to apply to the unrefuted constituents. There is no disagreement with Hintikka over requirement (i) (though his reason for rejecting an equal distribution is quite different from mine, namely that it would involve learning "too fast"). However, there is a crucial disagreement over requirement (ii). As we saw, to comply with it we have to put $\alpha > n$, which makes C_{l_c}, which is the most falsifiable of the unfalsified constituents, the least probable, and C_K the most probable. By contrast, to get the results he wanted, Hintikka had to put $\alpha < n$. Let us look into this disagreement.

A constituent consists of an existential component complemented by a universal component. The existential component of C_K entails that of any narrower unrefuted constituent; and the universal component of C_{l_c} entails that of any wider unrefuted constituent. This might suggest that C_K is more informative than C_{l_c} in one respect and less informative in another, so that on balance they come out even. Actually, with respect to informative content, the stronger universal component of the narrower constituent *more than offsets* its correspondingly weaker existential component; the questions that C_K can answer are a *proper subclass* of those that C_{l_c} can answer. This has been shown by Oddie (1979). With respect to depth-1 questions of the form 'Is cell Q_i occupied?', they perform equally, C_K answering 'Yes' in all cases and C_{l_c} answering 'Yes' in some and 'No' in the others. But with respect to depth-d questions of the form 'Is cell Q_i occupied by at least d individuals?' (where $d > 1$), C_K answers 'Don't know' in all cases while C_{l_c} answers 'Don't know' in some and 'No' in the others. Again, if the questions are of the form 'Has an individual of type Q_i ever been observed?', C_K answers 'Don't know' in all cases while C_{l_c} answers 'Don't know' in some and 'No' in the others. Another way in which Oddie makes the same point is that the narrower depth-1 constituent splits into a *smaller disjunction* of depth-d constituents where $d > 1$. Let Q_j be a cell that C_{l_c} declares empty but which, of course, C_K declares occupied. Then the depth-d constituents into which C_K branches will between them say that the number of individuals in Q_j is one *or* two . . . *or* at least d, whereas those into which C_{l_c} branches will all say that the number is zero.

A similar point had been made earlier by Illka Niiniluoto (1977, pp. 127–128), but by making the number of P-predicates, rather than the number of individuals, increase with increasing depth. Let L_1 be a language with n P-predicates and let Q_j be a Q-predicate in L_1. Let L_2, L_3 . . . be extensions of L_1 with respectively n $+ 1$, n $+ 2$ . . . P-predicates. In L_2 the Q_j-cell splits into two subcells: $Q_j \wedge P_{n+1}$ and $Q_j \wedge \sim P_{n+1}$; in L_3 it splits into four, and so on. Now C_{l_c}'s assertion in L_1 that the Q_j-cell is empty will not branch at deeper levels, but C_K's assertion that the Q_j-cell is occupied will branch into all the remaining subcells at deeper levels. As Niiniluoto put it: 'All subcells of an empty cell remain empty, no matter how deep we go in the analysis. But almost anything can happen to the subcells of a non-empty cell—as long as one of them remains non-empty' (p. 128).

All these connected considerations unite to show that C_{l_c} makes a stronger assertion when it says $\sim\exists x Q_j x$ than C_K makes when it says $\exists x Q_j x$. As the narrowest unrefuted constituent, C_{l_c} dominates all wider ones with respect to question-answering power, splits into the fewest depth-d constituents, and is the most falsifiable. Conversely, the widest one, namely C_K, has the least question-answering power, splits into the most depth-d constituents, and is unfalsifiable. There can be no doubt that C_{l_c} has the most, and C_K the least, content. Now it is a paramount principle of probability theory that, other things being equal, probability varies *inversely* with content. This means that the seesaw should (i) not tilt as it would if we put $\alpha < n$, giving the highest probability to C_{l_c}, (ii) not be horizontal, as it would be if we put $\alpha = n$, giving all unrefuted constituents an equal probability, but (iii) tilt as it would if we put $\alpha > n$, giving the highest probability to C_K.

This puts a very different light on the question whether evidence can raise the posterior probability of a universal generalisation g. According both to the proposal, made in §2.23, that the initial probabilities of constituents should vary with their width, and to the above paramount principle that they should vary inversely with their content, C_K as the widest constituent with the least content starts with the highest initial probability. And when evidence starts coming in, this can only *raise* the posterior probability of C_K above its initial level; for the evidence is bound to knock out some constituents whose initial probabilities will be redistributed over the unrefuted constituents; and if this is done in accordance with our paramount principle, the shares of this redistributed probability received by the unrefuted constituents should vary with their width. But C_K is the widest constituent and it is always unrefuted. We remarked earlier that the equivalent of C_K in Carnap's system tends to åmass *all* the probability; and while that does not happen in Hintikka's system, something approximating it will happen if our paramount principle is complied with; for C_K starts as the constituent with the highest initial probability and its posterior probability is always being raised more than that of any other unrefuted constituent. But consider what this means concerning the possibility of evidence confirming any universal generalisation g. Let evidence E^n be as intuitively favourable as one could reasonably wish to g. If this evidence is to confirm g in a probabilistic sense, then, as we saw earlier, we need to have both (i) $p(g) > 0$ and (ii) $p(g, E^n) > p(g)$. Now we do indeed have (i) in Hintikka's system. The initial probability of g is equal to the sum of the initial probabilities of all the constituents that entail g, and this will be positive. But can we have (ii)? We saw that the posterior probability of C_K is bound to be higher than its initial probability whatever the evidence may be; so we are bound to have $p(C_K, E^n) > p(C_K)$. But C_K entails the negation of every universal generalisation. So we are bound to have $p(\bar{g}, E^n) \geq p(\bar{g})$ and hence $p(g, E^n) \leq p(g)$; which amounts to what I called the moderate anti-inductivist thesis: no evidence can *raise* the probability of a universal generalisation so long as our paramount principle is complied with.

3

....

The Inductive Ascent

3.1 The First Step

Hume divided sceptical arguments into those that are antecedent, and those that are consequent, to 'science and enquiry' (*1748*, pp. 149–150). The argument for probability-scepticism presented in §2.4 presumably belongs in the first category. The only assumption it makes concerning the de facto state of our actual knowledge is that much of it is uncertain. And its conclusion is that if any of our knowledge is certain, this cannot be used to grade uncertain hypotheses as more or less probably true. The sceptical arguments to be developed in this chapter will belong in Hume's second category. They will make a more specific assumption about the structure of our commonsense and scientific knowledge, namely that it is *multi-leveled*. More particularly, we can single out, from the enormous variety of statements that figure in our factual knowledge, statements occurring at the following levels:

level-0: perceptual reports of a first-person, here-and-now type (e.g. 'In my visual field there is now a silvery crescent against a dark blue background');

level-1: singular statements about observable things or events (e.g. 'There is a new moon tonight');

level-2: empirical generalisations about regularities displayed by observable things and events (e.g. 'A new moon is followed by spring tides');

level-3: exact experimental laws concerning measurable physical magnitudes (e.g. Snell's law of refraction or the gas law of Charles and Gay-Lussac);

level-4: scientific theories that are not only universal and exact but postulate unobservable entities (e.g. the Faraday-Maxwell theory of fields of force).

Now proposition (II), the experientialist thesis, implies that if a statement at one of the higher levels is to constitute *knowledge* and not be

79

a mere free-floating speculation, it must be supported at lower levels and, eventually, by observations. In short, it must be inductively grounded. But inductively grounded on what sort of basis? Must the basis consist of statements that are certainly true? May it consist of level-1 statements about publicly observable things and events, or must we go down to level-0 reports about perceptual experiences? (A non-inductivist construal of the empirical basis will be offered in chapter 7.)

3.11   *A Basis of Certainty?*

At the beginning of §2.4 we considered an argument that strongly suggested that the ultimate premises, or bottommost statements, in a nondeductive structure must be certainly true if they are to provide any justification or inductive support for statements higher up; for since *ex hypothesi* there are no statements beneath them to endow them with a determinate posterior probability, their status would otherwise be perfectly indeterminate. However, this argument has been challenged by Jeffrey (*1965*, ch. 11; 1968), who introduced the idea of an observational statement having a kind of probability that is noninferential yet empirical.

He calls a statement *e observational* for person X at time t if X's sensory experience at t causes him to raise his degree of confidence in *e*. In typical cases his confidence in *e* will be driven right up to 1. For instance, if he is woken by the sun shining on his face, then his degree of confidence in 'The sun is shining' will be 1; he *cannot help* feeling quite certain of this *e*: the process is quite involuntary. But Jeffrey insisted that there are other cases where such an involuntary process leaves X's degree of confidence in *e* at some level below 1. Suppose that X is examining a piece of cloth by candlelight and that *e* is 'This cloth is green': then according to Jeffrey, X's sensory experience might cause him to give a probability of, say, 0.7 to this *e*.

It seems to me that this example is open to an alternative interpretation, namely that X is *quite sure* that the cloth now has, say, a flickering greenish-greyish-blueish appearance, and on the basis of this perceptual evidence he gives to the prediction that the cloth, examined in broad daylight, would appear green, rather than grey or blue, a probability of 0.7. But I will not rule out Jeffrey's kind of noninferential probability. However, we may wonder whether the empirical basis could still provide inductive support for the superstructure if it were infiltrated by a sizeable number of observational statements with a Jeffrey-type probability of less than 1.

Before answering that question, let us first consider the way in which another inductive philosopher allows uncertainty into the empirical basis. Mary Hesse (*1974*, pp. 125f.) distinguished an observation *statement e*, which describes a thing or event, from an associated observation *report e'*, which reports that the observation statement *e* has been made. And she suggested that it is observation *reports* 'which make up the data upon which the community of science works' (p. 127), her idea being that an observation report is much more reliable than the observation statement that it reports. Observation reports 'need not be regarded as absolutely incorrigible, but only as substantially less subject to correction than the observation statements themselves. . . . [I]t is in fact much less likely that mistakes are made in describing reports than in asserting observation statements' (p. 127).

Assume, for the sake of the argument, the absolute reliability of any observation report *e'* saying that the observation statement *e* has been put forward; and suppose that *e* would confirm a hypothesis *h*. Now we obviously cannot proceed from *e'* to the conclusion that *h* has been confirmed unless we first draw some conclusion from *e'* about the truth of *e*. (If *e'* reported that *e* had been stated by a Cretan, St. Paul would not have considered *h* confirmed by *e'*.) To deal with this, Hesse introduced a postulate that says that 'the observation statement corresponding to a given observation report is true more often than not' (p. 128). Suppose now that we have n observation reports, $e_1', e_2' \ldots e_n'$, which we assume to be certainly true, and corresponding to which are the n observation statements $e_1, e_2, \ldots e_n$, which for the moment we may assume to be logically independent of one another. Denote the conjunction of all these observation statements by E^n. Hesse herself raised the question whether the probability of E^n will not be very low. If each of the n observation statements is probabilistically independent of the others we will have

$$p(E^n) = p(e_1) \cdot p(e_2) \cdot p(e_3) \ldots p(e_n)$$

and the probability of E^n will indeed be very low if n is at all large and the probability of each of the observation statements is significantly less than 1. To deal with this, she introduced a further postulate to the effect that the observation statements may strongly reinforce each other. But where are these postulates coming from? This last one is obviously a very strong synthetic assumption, and one that could not itself be

81

confirmed empirically without circularity. In §3.3 I will argue, without any claim to originality, that any postulate that is introduced to legitimise the inductive process will either not be strong enough to do the job or else cannot itself be legitimised.

In the meanwhile it seems clear that an inductivist who does not invoke extralogical postulates to come to his rescue will very much hope that the inductive ascent can proceed from an absolutely secure basis; for if it contains elements of uncertainty, then either their probability status is indeterminate, or else, if they have a determinate probability of less than 1 and there are quite a few of them, the probability of the conjunction of all the statements composing the empirical basis will sink down towards 0 and it will become incapable of supporting any superstructure.

3.12 *Popper's Glass of Water versus Malcolm's Ink Bottle*

I now turn to the question: given that the empirical basis is required to consist of statements that are certainly true, can these be level-1 statements about publicly observable things and events or should they be level-0 statements about perceptual experiences?

Like everyone else, I *feel* certain about ever so many level-1 statements, such as 'This is a glass of water' or 'That is an ink-bottle'. But such a feeling is not an argument, still less a proof. The question is whether a level-1 statement about whose truth I have a subjective feeling of certainty could be shown to be true in some objective way. And there is a straightforward argument that it could not. This argument, which was used by Popper (*1934*, pp. 94–95), is that to describe something as, say, a glass of water, is to impute a certain *dispositional structure* to it, and this imputation carries with it an indefinite array of conditional predictions (it can be safely drunk, it will not burst into flames if a lighted match is dropped into it. . . .), none of which has yet been verified.

Some philosophers who claim that we can, in suitable circumstances, *know* a level-1 statement to be true have responded to this by denying that an assertion of the form 'I *know* that this is a so-and-so' has predictive implications. Thus J. L. Austin wrote:

> If we have made sure it's a goldfinch, and a real goldfinch, and then in the future it does something outrageous (explodes, quotes Mrs. Woolf, or what not), we don't say we were wrong to say it was a goldfinch. . . . When I have made sure it's a real goldfinch (not

stuffed . . . etc.) then I am *not* 'predicting' in saying that it's a real goldfinch, and in a very good sense I can't be proved wrong whatever happens. *It seems a serious mistake to suppose that* . . . language about real things . . . is 'predictive' in such a way that the future can always prove it wrong. (1946, pp. 56–57)

And Norman Malcolm resisted the thesis that, as he put it, 'any empirical proposition whatever *could* be refuted by future experience' (1952, p. 65) along similar lines. Such a statement as 'I know that there is an ink bottle here', made in appropriate circumstances, lies 'beyond the reach of doubt' and provides a fixed point of certainty (p. 69): 'There is nothing whatever that could happen in the next moment or the next year that would by me be called *evidence* that there is not an ink-bottle here now. No future experience or investigation could prove to me that I am mistaken' (pp. 67–68). He added that he would not have been mistaken even if, when he next reaches for this ink bottle, his 'hand should seem to pass *through* it' (p. 66). We might call this the issue of Popper's glass of water versus Malcolm's ink bottle.

What does an utterance of the form 'I *know* that this is a so-and-so' say when construed in this Austin-Malcolm way as carrying no predictive implications? Suppose that I know you to be under doctor's orders to drink nothing but water. I offer you a glass, saying 'I have made sure it's water'. You drink it down and a moment later complain of burning sensations in your mouth, throat, and lower down. If I then calmly declare, 'I was *not* predicting in saying that it's water, and in a very good sense I can't be proved wrong whatever happens', you might reply that in that case my "statement" was unhelpful and it would have been better if I had kept my mouth shut. What would we actually learn from Austin if he pointed at something and told us that he *knows* that it is a goldfinch? Well, we would learn something about *him*, namely that he is determined to call it a goldfinch whatever happens, but we should not form any expectations about *it*. It would be rather as if he had declared that he will go on calling it Goldie whatever happens. Or suppose that a terrorist organization has devised a booby trap in the form of a fake ink bottle that explodes when someone tries to fill his pen from it. If Malcolm pointed at something and told you that he *knows* it to be an ink bottle, you should remember, before you go to fill your pen from it, that according to Malcolm his utterance is consistent with any future experience whatever.

We have found the following pattern: a level-1 statement, as ordinarily construed, has predictive implications for the future and hence cannot now be *known*, with certainty, to be true; in an attempt to restore certainty to them, some philosophical defenders of common sense have construed them in a way that cuts off their predictive implications; but thus construed, such statements become useless for all practical and theoretical purposes. Indeed, they cease being *statements* and become mere personal avowals. Austin himself went on to assimilate 'I know' to 'I promise', 'I swear', 'I guarantee', etc. (pp. 66f; for a criticism of this assimilation see Jonathan Harrison, 1962); and he insisted that to employ such phrases is to perform a ritualistic act rather than to make a true-or-false statement. An empirical basis for common-sense and scientific knowledge cannot be provided by sentences that are neither true nor false but merely express their authors' resolve to pin certain names to certain objects come what may.

If a genuine level-1 statement cannot be certainly true in any objective sense, however subjectively certain you or I may feel about it, and if the empirical basis for an inductively grounded system of empirical *knowledge* needs to be certainly true, then it looks as though proposition (II), the experientialist thesis, will require the basis to be provided by level-0 perceptual reports; for a perceiving subject surely can *know* the truth of a proposition that merely describes a current perceptual experience of his. But now proposition (II) comes into apparent conflict with proposition (III), the deductivist thesis; how could a level-1 statement about something "out there" possibly be deduced from level-0 premises about goings-on "in here"? At this point an inductivist faces a choice. He may retain common-sense realism, according to which the things described by level-1 statements have a physical reality independent of our perceptions, and set aside proposition (III), claiming that the gulf between level-0 and level-1 statements is and ought to be bridged by some kind of nondeductive inference; or he may, in a praiseworthy endeavour to retain proposition (III), set aside physical realism in favour of phenomenalism, claiming that there is no gulf between level-0 and level-1 statements since the latter are constructible from the former. The first of these two options was investigated in §1.4. We must now investigate the second, which I will call the *phenomenalist* strategy.

3.2 The Phenomenalist Strategy

The kernel of the strategy now to be considered can be indicated by juxtaposing two quotations from Russell. The first poses the problem:

the problem of our knowledge of the external world . . . really is: Can the existence of anything other than our own hard data be inferred from the existence of those data? (*1914*, p. 80)

The second, which comes from a paper entitled 'The Relation of Sense-Data to Physics', also written in 1914, suggests how it should be solved:

The supreme maxim in scientific philosophising is this:
Wherever possible, logical constructions are to be substituted for inferred entities. (*1917*, p. 155)

This "supreme" maxim obviously suggests that we should, if possible, treat material objects not as things existing independently of sense-data and whose existence can only be inferred from the latter, but as logical constructions out of those data. In short, this maxim clearly implies, and Russell in (*1914*) endorsed the implication, that we should abandon physical realism in favour of phenomenalism. For a verificationist, direct verification is superior to dubious inference. May not statements about the physical world be verifiable if we understand the "physical" world to be nothing but actual or possible sensations? This verificationist motive was quite transparent in Russell's (*1914*). In answer to the objection 'that the "matter" of physics is something other than series of sense-data' he said:

We have been considering . . . the question of the *verifiability* of physics. . . . In physics, as ordinarily set forth, there is much that is unverifiable [he here enumerates three kinds of unverifiable hypotheses] . . .
If physics is to consist wholly of propositions known to be true [my italics], or at least capable of being proved or disproved, the three kinds of hypothetical entities we have just enumerated must all be capable of being exhibited as logical functions of sense-data. (P. 116)

Russell's (*1914*) programme of reducing the world of physics to sensory concepts inspired Carnap's (*1928*). Before turning to that, let us first look briefly at Berkeley's account of the Creation.

3.21 *Berkeley on the Creation*

Just what, according to Berkeley, was God supposed to have *created* during the first four days of the Creation, before the creation of 'finite spirits'? Berkeley obviously could not allow that He created a physical heaven, earth, etc. Nor at this stage could He create perceptions for there were as yet no 'finite spirits' to have them. Then did He perhaps create ideas in His own mind on which the perceptions of 'finite spirits' would afterwards be modeled? But that answer would imply that His Creation was a blind act, done without foreknowledge of what He was creating, and Berkeley's spokesman Philonous would have none of that. God had always had a perfect foreknowledge of what He was going to create: 'All objects are eternally known by God, or which is the same thing, have an eternal existence in his mind' (*1713*, p. 252). But, Hylas protested, if neither material objects, nor perceptions, nor ideas were created, 'What shall we make then of the Creation?' (p. 253). Philonous answered that what happened then was that God decreed that things that had hitherto existed as eternal ideas in His mind 'should become perceptible to intelligent creatures'. The Creation was, so to speak, publication day for God's ideas. I think that this brings out very well how idealism not only empties the world of a lot of stuff that physicalism puts into it, but secures a happy fit between ontology and epistemology: at the Creation the world was made immediately accessible to the finite spirits to come.

3.22 *Carnap's* Aufbau

Carnap in his (*1928*) was, according to Quine (1951, p. 39), 'the first empiricist who, not content with asserting the reducibility of science to terms of immediate experience, took serious steps toward carrying out the reduction'. That book, which took Russell's "supreme" maxim as its motto, outlined what Carnap called a *constructional system* that was to provide 'a rational reconstruction of the entire formation of reality' (p. 158). Its basic idea was this. Let a, b and c be objects, or classes of objects, such that they would ordinarily be regarded as of different types, or as existing at different levels, c being at a higher level than b and b at a higher level than a. (A mathematical example: a might be the natural numbers, b the rational numbers, and c the irrational numbers.) Then to construct c out of b is to provide a constructional definition whereby any statement about c can be translated into a statement that is about b only: it is to *reduce c to b*. And if b is in its turn reduced to (constructed out of) a, c will have been reduced to a and any statement about c can be translated into a statement about a only. Constructions and reductions are transitive (p. 6). The objects, here designated by a, to which all other objects in the constructional system are reduced are called the basic objects.

Carnap took as the basic objects of his rational reconstruction of the world what he called "my" elementary experiences. I will not say anything here about the impressive logical machinery of his constructional system, but will simply record the series of results it yielded. From "my" experiences are constructed physical objects. One of these plays an especially important role, namely "my body" (§129). Among the various other physical objects that have been constructed, a subclass is picked out consisting of ones that resemble "my body" in certain characteristic ways: these constitute the class of "other persons" (§137). An "other person" exhibits outward behaviour: facial expressions, gestures, utterances, and so forth. From such outward behaviour are constructed "psychological states of the other" which, 'taken as a class, may be called the *mind of the other*' (p. 216). From the thus constructed "psychological states" can be constructed "the world of the other" in a way analogous to that in which "my" world was constructed out of "my experiences" (§145). Between "my world" and "the other's world," there will be a structural analogy which 'amounts to a very far-reaching, but not to a complete, agreement' (p. 224). By discarding items over which they disagree, these two "worlds" can be brought into intersubjective agreement. Repeating this procedure for all other (normal) "persons" known to "me" results in 'a general one-to-one correspondence between . . . all the worlds of all persons (i.e. normal persons known to me), including myself' (p. 228). Thus we arrive at one *intersubjective world* (p. 229), and this is the world of science (§149).

At first sight it may seem that Carnap has here pulled off an astonishing feat. My experiences have, for me, an assured epistemological status; and the constructional process is analytic in the sense that it involves only logic and definitions and introduces nothing contingent or uncertain. Thus it may seem that the world of science can be built up in a way that complies both with proposition (II), the experientialist thesis, and with proposition (III), the deductivist thesis. But we have to remember that, if c has been constructed out of b, which has been constructed out of a, then any statement about c can be translated *without loss* into a statement about a. Thus any statement about the so-called "intersubjective world of science" can be translated without loss into one about the subjective world of my experience. The former is nothing but the latter after various constructional transformations, which add nothing to it, have been performed on it. Carnap himself made this perfectly clear:

> The constructional system shows that all objects can be constructed from "my elementary experiences" as basic elements. In other

words . . . all (scientific) statements can be transformed into statements about my experiences. . . . Thus, each object which is not itself one of my experiences, is a quasi object; I use its name as a convenient abbreviation to speak about my experiences. In fact, within construction theory, and thus within rational science, its name is *nothing but* an abbreviation. (P. 255)

Thus the stars, the sun, the mountains, the oceans, the cities, the people, and everything else that I may have supposed to exist independently of my own very limited and partial experiences of them turn out to be quasi-objects whose names are merely convenient abbreviations for my experiences. One recalls this declaration by Hume: 'Let us fix our attention out of ourselves as much as possible: Let us chace our imagination to the heavens, or to the utmost limits of the universe; we never really advance a step beyond ourselves, nor can conceive any kind of existence, but those perceptions, which have appear'd in that narrow compass' (*1739–40*, pp. 67–68).

Phenomenalism is a metaphysical thesis introduced to solve an epistemological problem. The problem is: 'How can I acquire knowledge of the external world on the basis of my perceptual experience?' And the solution is: 'Nothing is easier; what you call "the external world" is constituted by your perceptual experiences: by having them you automatically acquire knowledge of it.' A very different metaphysical thesis introduced to solve this epistemological problem is Descartes's, to the effect that an undeceiving God keeps the ideas, or at least the clear and distinct ideas, in our minds in correspondence with external realities. Confronted by any such metaphysical thesis, a sceptic is bound to ask, first whether any good reasons, other than its epistemological convenience, have been given for accepting it, and second, whether it really does solve the epistemological problem. Let us begin with the first question.

Apart from the erroneous psychological doctrine, examined in §1.22 and relied on by both Berkeley and Hume, that it is impossible for us even to form the idea of a body existing independently of mind, I know of only two arguments for phenomenalism that do not trade on its epistemological utility. One is that it solves the mind-body problem. For instance, Mach wrote:

Now if we resolve the whole material world into elements which at the same time are also elements of the psychical world and, as such, are commonly called sensations; if, further, we regard it as the sole task of science to inquire into the connexion and combination of these elements . . . ; we may then reasonably expect to build a unified monistic structure upon this conception, and thus to get rid of the distressing confusions of dualism. (*1886*, p. 312)

And Kant at one time took an essentially similar view. (See his *1781– 87*, A383–96; this long passage was omitted from B.) But it could equally be argued that a materialist identification of thought processes with brain processes also "solves" the mind-body problem by reducing one of the two sides to the other.

3.23   *Parsimony versus Plenitude*

The only other nonepistemological argument for phenomenalism known to me is from some principle of parsimony. Thus Russell in (*1914*) justified his phenomenalist approach by an appeal to Occam's Razor, which, he claimed, 'inspires all scientific philosophizing' (p. 112). One rather *ad hominem* criticism that can be made here is that he shrank back, very understandably, from an unrelenting application of Occam's Razor. An unrelenting Occamist, having eliminated all material bodies, including those of other human beings, would have gone on to eliminate other minds as well. This Russell apologetically declined to do (*1917*, pp. 157–158). Moreover, he wished, in a non-Occamist spirit, to hold that tables etc., though immaterial, remain in existence when not being perceived. Thus he equated a table with a series consisting not only of actual sense-data but also of sensibilia, these being all the *possible* appearances or "aspects" it could have presented to hypothetical observers. Moritz Schlick, in his (*1918*) in which he was still a vigorous realist, claimed with considerable plausibility that Russell's view of material objects was *less* economical than 'the plain world view'. For instance, it makes a table, among indefinitely many other things, all the aspects it would present to a bee: 'What an infinite swarm of aspects is posited here as real—an incalculable series, and one not even specifiable! Is this world view really simpler, more economical, provided with fewer dispensable posits than the plain world view . . . ?' (*1918*, p. 207).

But suppose that Occam's Razor does favour phenomenalism: would that be a reason for supposing phenomenalism to be true? Let us, for a moment, consider this question from a theological point of view. If God had been an Occamist who did not multiply entities needlessly, He would, presumably, not have created anything: He would have enjoyed His own perfect existence without distraction. But He did, it seems, create *some* entities. Are we to suppose that He set about this task in a parsimonious spirit, not creating bodies because He saw that illusions of bodies would suffice?

We encountered the principle of Occam's Razor in a different guise in §2.52 when we considered Hintikka's proposal that we should, once certain conditions have been satisfied, regard the most falsifiable of the unfalsified constituents as the constituent with the highest posterior probability on our evidence. This could be restated as: Regard as best confirmed that constituent that introduces no other kinds of entity than those recorded in our evidence. A vulgar objection to this would be that, from the discovery of fire down to that of the neutrino, the hypothesis that all the kinds of entities that exist have already been discovered has been refuted again and again: the ontological variety of the world has repeatedly turned out to be greater than the evidence had previously suggested. And in §2.53 I argued that a fair distribution of probabilities over constituents would oblige us to regard as most probable that one, namely C_K, which postulates *every* kind of entity, C_K being the secular equivalent of the propostion that God created the richest possible world. I hasten to add that I am not advocating that a principle of plenitude rather than a principle of parsimony should 'inspire all scientific philosophising'; I have brought in this counter-Occamist principle only to suggest that, since there are opposing principles here, 'scientific philosophising' should not be under the a priori and monopolising control of one of them.

I turn now to the second question: does phenomenalism really have the epistemological utility it seems to have? Let us recall the epistemological problem before us. According to proposition (II), factual *knowledge* must be based on experience. Perceptual experiences are reported in level-0 reports. And the first step in the inductive ascent must, presumably, be from these to level-1 statements about things and events. But, the argument went, there is no question of deriving a level-1 conclusion from level-0 premises. A level-0 statement made by me is about me; a level-1 statement made by me is not about me but about things external to me. I might as well try to derive a conclusion about cheese from premises about chalk. Moreover, a level-1 statement is always objectively uncertain, however certain we may feel of its truth, because it carries an indefinite array of unverified predictive implications. Does phenomenalism defeat this argument?

Well, it does remove the first stumbling block: what level-0 statements and level-1 statements respectively talk about are no longer as different as chalk and cheese. They both talk about perceptual experience, one overtly, the other in an elliptical, shorthand way. The vertical gap between level-0 and level-1 is eliminated. However, the horizontal gap remains as wide as ever. As construed by phenomenalism, 'That is an ink bottle' no longer posits a physical object but it does posit a coherent ensemble or series (to use Russell's term) of actual or possible perceptual experiences. On the physicalist construal, this statement carries such implications as that the object in question could be picked up (unless, say, it has been glued to the desk) and that it will not explode if someone tries to fill his pen from it. On the phenomenalist construal, it carries such implications as that you will not experience your hand seeming to pass through it and that visual, tactile, and kinaesthetic experiences associated with pen-filling intentions will not be followed by painful flash and bang experiences. To every predictive implication of a physicalist nature on the realist construal, there will correspond a predictive implication of an experiential nature on the phenomenalist construal. Of course, a phenomenalist could follow the Austin-Malcolm policy of cutting off all such predictive implications; but then, 'That is an ink bottle' will again become a useless avowal that merely expresses, in a highly misleading way, the autobiographical information that its author has momentarily had certain perceptual experiences.

I conclude that the idea of a secure empirical basis, to support an inductive structure of common-sense and scientific *knowledge*, is in no better shape under phenomenalism than under physical realism. The latter allows that each of us can attain certainty at level-0, but adds that level-0 statements by themselves are quite incapable of directly supporting any scientific superstructure. Kepler's laws, say, cannot be supported by such merely autobiographical statements as, 'In my visual field a bright spot is now coinciding with the intersection of a vertical and a horizontal line'. To provide any support for them, we must at least get to level-1 statements about telescopes, angles of elevation, etc., and preferably to more sophisticated ones about, say, the relative positions at certain times of Mars, Jupiter, etc. But since level-1 statements transcend level-0 reports both vertically and horizontally, their objective status is essentially unverifiable and conjectural. And under phenomenalism, which does not eliminate this horizontal transcendence, their status remains unverifiable and conjectural.

3.3   The Illegitimacy of Any Inductive Principle

Our conclusions so far are: (1) that inductivism badly needs an empirical basis that is hard and secure; (2) that while level-0 reports may be certain, for their author when he makes them, they could not by themselves directly support a scientific superstructure; and (3) that a basis consisting of level-1 statements would have no objective certainty, even if interpreted in a phenomenalist way. In short, even the first step in the inductive ascent cannot be made. But suppose that it could be made and that level-1 statements could attain objective certainty; could the next step in the inductive ascent, from level-1 to level-2, be made?

There is of course no possibility of deducing a genuinely universal statement at level-2 from a finite conjunction of singular statements at level-1; and if the negative results arrived at in chapter 2 are correct, such a conjunction could not even raise the probability of a level-2 statement. This means that there can be no legitimate ascent from level-1 to level-2 with the help just of logic, whether classical or probability logic, alone. This point may be reinforced by a slight variation of a famous argument due to Nelson Goodman (1947). Let e and h say respectively that all emeralds so far observed are green and that all emeralds are green. Let h', h'', h''' . . . respectively say that all observed emeralds are green and that all unobserved emeralds are green except one/two/three . . . Let $\bar{h}$ say that all observed emeralds are green and that all unobserved emeralds are nongreen, and let $\bar{h}'$, $\bar{h}''$, $\bar{h}'''$ . . . respectively say that all observed emeralds are green and that all unobserved emeralds are nongreen except one/two/three . . . Then, h, h', h'', h''' . . . $\bar{h}'''$, $\bar{h}''$, $\bar{h}'$, is a mutually exclusive and exhaustive array of possible extensions of e. In the absence of some extralogical principle of discrimination, there is no reason why e should favour one of these extensions more than others. But if it is impartial, it can confirm only their disjunction; but this is equivalent to e itself. This ties in with Hume's point that there is no valid inference from instances of which we have experience to instances of which we have no experience.

So it seems that some sort of inductive principle, or *IP* as I will call it for short, is indispensable if there is to be a legitimate progression from evidence-statements at level-1 to generalizations and hypotheses at level-2. This has been stated very clearly by Russell:

Hume's scepticism rests entirely upon his rejection of the principle of induction. . . . If this principle is not true, every attempt to arrive at general scientific laws from particular observations is fallacious, and Hume's scepticism is inescapable for an empiricist. The principle itself cannot, of course, without circularity, be inferred from observed uniformities, since it is required to justify any such inferences. It must therefore be . . . an independent principle not based on experience. . . . But if this one principle is admitted, everything else can proceed in accordance with the theory that all our knowledge is based on experience. (*1946*, p. 699)

Inductivists have not found it easy to formulate an *IP* in a way that would enable it to play the role required of it. But let us suppose that these difficulties have been overcome and that a happy formulation, just right for inductivist purposes, of an *IP* has been found. What would be the epistemological status of this *IP*? So far as I know the following exhaust the answers that have been given to our question:

1. *IP* is synthetic and true a priori.
2. *IP* is synthetic and provable by a transcendental argument.
3. *IP* is synthetic and empirically justified.
4. *IP* is synthetic and, although it cannot be justified either a priori or a posteriori, it can be vindicated.
5. *IP* is self-justifying and needs no external justification or vindication.
6. *IP* is analytic.
7. *IP* is an unjustified postulate.

In this section we will consider these answers in turn. Some philosophers have combined, or oscillated between, more than one of these answers. Kant combined 1 and 2. As we shall see, Russell oscillated between 2 and 7; Keynes tried 3 but also toyed, a shade uneasily, with 1 and 2; and Carnap went for 6 though he also veered towards 3. Answers 2 and 4 will be of particular interest since they represent antisceptical strategies we have not yet examined.

3.31 *Clifford*

Answer 1 is excluded by proposition (I), the anti-apriorist thesis. *If* we had a hot line to the Author of Nature, and if we had a clearly formulated *IP*, an excellent question to put to him would be: Is our *IP* true? If he answered 'Yes' we could happily set a computer to work to print out all those *h*'s that are singled out by our evidence in conjunction with this authoritatively endorsed *IP*. But I am assuming in this book that no shortcuts to knowledge of the external world are available to us. It would be nice if there were a preestablished harmony between certain inborn ideas in our minds and certain structural features of the world, but there is no a priori reason to suppose that this is so. There is, it is true, an a posteriori argument for the adaptation of animals' (including humans') inborn expectations to those features of their environment that have a bearing on their chances of survival, namely the argument from natural selection. So far as I know, the first thinker to consider this argument in connection with the problem of induction was W. K. Clifford. Clifford understood *IP* in much the same way as Mill had done, namely as an assumption concerning the uniformity of nature, but unlike Mill he did not claim that this assumption is itself inductively inferred from experience; an inference from experience to something beyond experience

> depends on the assumption of the uniformity of nature; and what does this rest on? We cannot infer that which is the ground of all inference; but although I cannot give you a logical reason for believing it, I can give you a physical explanation of the fact that we do all believe it. . . . Nature is selecting for survival those individuals and races who act as if she were uniform; and hence the gradual spread of that belief over the civilised world. (*1886*, p. 209)

Notice that Clifford deliberately refrained from offering natural selection as a justification of *IP* ('I cannot give you a logical reason . . . '); and he was obviously right so to refrain. The theory of natural selection is a large hypothesis that goes far beyond whatever evidence may be adduced in its support. Scepticism concerning the inductive justification of science as a whole cannot be dispelled by appeal to one particular part of science.

Carnap (*1950*, p.x) saw that for any two inductive methods, no matter how inadequate the first might seem in comparison with the second, there will be possible worlds in which the first delivers better results than the second. This knocks out the possibility that *IP* could be both synthetic and true a priori.

3.32 *Transcendental Arguments*

So let us turn to answer 2. A transcendental argument for an *IP* would run something like this: (i) scientific knowledge presupposes *IP*, in the sense that the former would be impossible if the latter were not true; (ii) but scientific knowledge exists; (iii) therefore *IP* is true. Such an argument plays into the sceptic's hands. He can happily accept the first premiss; but he may prefer to express it contrapositively and he will point out that in it the word 'knowledge' is being used as a success-word. Thus in conformity with the convention laid down in §1.13 he might reformulate the first premiss thus: (i) if *IP* is false then scientific *knowledge* is impossible. He will then point out that if, in the second premiss, the word 'knowledge' were used as a success-word, then we have no reason to suppose that this premiss is true. He may very well allow that it is an empirical fact that there are indeed textbooks in physics, chemistry, biology, etc. just as there are manuals of astrology, alchemy, phrenology, etc.; but we are empirically justified in putting, in place of (ii) above, only the following much weaker premiss: (ii') so-called scientific "knowledge" exists. And from premises (i) and (ii'), of course, the conclusion (iii) no longer follows. It would obviously be circular first to use some merely postulated and unjustified *IP* to elevate de facto scientific "knowledge" to the de jure status of scientific *knowledge* and then to proceed by a transcendental argument from this alleged *knowledge* to the conclusion that this *IP* must be true. So the conclusion that a sceptic will draw from a transcendental argument is that, in the absence of any independent argument for it, the status of *IP* is highly dubious and hence, by premiss (i), the possibility of scientific *knowledge* is highly dubious. He may even have arguments for the *falsity* of a proposed *IP*, in which case premiss (i) gives him just the handle he needs for inferring the *im*possibility of scientific *knowledge*. We will encounter an example of such an *inversion* of a transcendental argument in the next section.

How liable a transcendental argument is to backfire, so that instead of establishing an *IP* it tends rather to disestablish scientific knowledge, is vividly illustrated by the conclusion of Russell's (*1948*). Russell insisted that we do most certainly need *some* universal *IP* to justify scientific inductions. And he added that, whatever this *IP* may be, it 'certainly cannot be logically deduced from the facts of experience. Either, therefore, we know something independently of experience, or science is moonshine' (p. 524). Well, which is it? Do we *know* this *IP*, or is science moonshine? Russell's hostility to Kantian apriorism excluded that first alternative, and his profound respect for science excluded the second. So he sought a middle way between the horns of his dilemma, suggesting that we 'need only more or less know our postulates' (ibid.) and that we have a funny kind of "knowledge" of them that does not amount to *knowledge* (p. 526).

3.33 *Keynes*

So let us turn to answer 3. On the face of it, this is a nonstarter; for as Russell, Clifford, and others have insisted, 3 surely involves circularity; if we cannot derive any general conclusions from experience without the help of some *IP*, then we cannot do what Mill notoriously tried to do (*1843*, III, iii, 1), namely, derive this *IP* from experience without its help. However, J.M. Keynes claimed on behalf of his *IP* (which he called 'the postulate of limited independent variety' or more simply 'the Inductive Hypothesis') that it *could* be inductively supported without circularity. I will now investigate his claim. The investigation, which will be somewhat technical, will lead to a negative conclusion near the end of this subsection.

Keynes saw that if the initial probability of a hypothesis *h* were zero, then however much evidence *e* comes in and however favourable *e* may be to *h*, *e* would never raise the posterior probability of *h* above zero. Now it would seem that if the set of mutually exclusive and exhaustive alternatives to which *h* belongs is infinite, then the initial probability of each of them will be zero. (That at any rate was how Keynes saw it. As we will see later, Jeffreys took a different view.) So the role of his Inductive Hypothesis, which he denoted by *H*, was to delimit the alternatives to a given *h* to a finite number so that each of them could start with a small but finite initial probability; then if evidence *e* comes in that consistently favours *h*, the posterior probability of *h* will rise above that of its competitors; indeed, the posterior probability of *h* may tend to one as the number of favourable instances recorded in *e* tends to infinity. In the following quotations, I have silently brought his notation into conformity with ours.

> If our conclusion is *h* and our empirical evidence is *e*, then in order to justify inductive methods, our premises must include, in addition to *e*, a general hypothesis *H* such that $p(h, H)$, the *a priori* probability of our conclusion, has a finite value. The effect of *e* is to increase the probability of *h* above its initial *a priori* value, $p(h, e.H)$ being greater than $p(h, H)$. (*1921*, pp. 259–260)

But where has this *H* come from and what justification is there for adopting it? If, Keynes continued,

> we have another general hypothesis *H'* and other evidence *e'*, such that $p(H, H')$ has a finite value, we can, without being guilty of a circular argument, use evidence *e'* by the same method as before to

strengthen the probability of H. . . . [I]t is not circular to use the inductive method to strengthen the inductive hypothesis itself, relative to some more primitive and less far-reaching assumption. . . . [W]e can support the Inductive Hypothesis by experience. In dealing with any particular question we can take the Inductive Hypothesis, not at its *a priori* value, but at the value to which experience in general has raised it.

But where has this H' come from and what is its justification? Do we have up our sleeve some still more primitive and less far-reaching assumption H'', and further evidence e'', such that we can use e'', reinforced by H'', to "strengthen" H'? It looks as though a regress is opening up.

But suppose that H' can be directly supported by experience; then what we have here is a step-by-step chain of nonentailing justifications of just the kind banned by proposition (V), arrived at in §2.41 above. According to the requirement of total evidence, Keynes should have operated, not with discrete bits of evidence e and e' at different stages in the argument, but with one and the same total evidence E, including both e and e', at all stages. Given this emendation, his position becomes the following—(i) E by itself cannot raise the probability either of H or of h: $p(H, E) = p(H)$ and $p(h, E) = p(h)$; (ii) however, E can somehow raise the probability of H': $p(H', E) > p(H')$; (iii) moreover, E reinforced by H' can raise the probability of H: $p(H, E.H') > p(H)$; (iv) furthermore, E reinforced by H can raise the probability of h: $p(h, E.H) > p(h)$. Thus we overcome the awkward fact, implicit in (i), that E is *indifferent* to h by interposing the intermediaries H' and H to enable E to "strengthen" h by a kind of action-at-a-distance.

Keynes himself seems to have been uneasy about his inductive "justification" of his Inductive Hypothesis; for a few pages later he suggested that it has an affinity with other synthetic principles (such as the Law of Causation) which we 'have no adequate inductive reason whatever for believing' but of which we have 'direct knowledge' (p. 263). I interpret this as a shift towards answer 1. He also inclined towards answer 2. He conceded that he had not 'given any perfectly adequate reason for accepting' the Inductive Hypothesis; but we can hardly lay it aside because 'the inductive method can only be based on it or on something like it' (p. 264).

Max Black (1954, 1958, 1966) presented some sophisticated arguments for the claim that *IP* is justified by evidence of its having worked well in the past. But as Salmon (1957) and Stegmüller (*1977*, ii, p. 75) pointed out, a counter-inductive principle would be justified in a perfectly analogous way on the very same evidence. This latter principle says, roughly, that the more frequently have things that are *A* been found, in the past, to be *B*, the more frequently will things that are *A* be found, in the future, to be ~*B*. Applied to itself, this says that the more frequently a counter-inductive policy has been found unsuccessful in the past, the more frequently it will be found successful in the future.

3.34 *Reichenbach*

So let us turn to answer 4. This answer was given by Hans Reichenbach (*1938*, §§ 38–41) and has been defended since by Herbert Feigl (1950) and Wesley Salmon (e.g. 1961, 1963a, 1963b). Like Russell and Popper, Reichenbach took Hume's criticism of induction very seriously: it 'was the heaviest blow against empiricism' (p. 347); and he found it astonishing that it should have been so underrated by so many empiricists; even Hume himself was 'not ready to realise the tragic consequences of his criticism' (p. 345). What did Hume prove? He proved that the claim that an inductive method *will* deliver better predictions than some noninductive method cannot be defended either a priori or a posteriori. Reichenbach fully accepted this: 'we know today that Kant's attempt at rescue failed' (p. 346). However, and this is the original feature of his answer to Hume, an inductive method can be vindicated by the following, much weaker, conditional claim–*if* any method can deliver good predictions, then no noninductive method can be more successful at doing so than an inductive method. Reichenbach's vindication of induction is rather reminiscent of Pascal's wager, which we considered in §1.5; if the world is fundamentally lawless and unpredictable (if God does not exist), then we lose nothing by applying an inductive method to it, for no noninductive method would succeed any better; but if the world is, at least to some degree and in some domains, lawful and predictable (if God does exist), then an inductive method will succeed at least as well as any other. An inductive strategy, we might say, is a dominant strategy in the game-theoretical sense: it may well serve you better and cannot serve you worse than any non-inductive strategy.

Reichenbach formulated the principle of induction as follows. Let *A* be a property that members of a certain reference class (which may be infinite) may or may not have. Let n be the number of observed members and m the number of these found to be *A*. Then m/n is the observed frequency of A (in this reference class). Let us denote this frequency by $F^n(A)$ and the frequency of A in the whole class by $F(A)$. Then the principle of induction says, roughly, that as n increases so does $F^n(A)$ become an ever closer approximation to $F(A)$. (Of course, $F^n(A)$ and $F(A)$ may both be 1, as they might be if A were the property of being black and the reference class were the class of all ravens.)

Let N be some number, greater than n, of members of the reference class. Now it may be that $F^N(A)$, the frequency of A among these N individuals, does not converge to a limit as N → ∞. This would be so if, for instance, the reference class consisted of an indefinitely long sequence of throws of a coin in which the frequency of heads was $\frac{1}{4}$ in the first 100 throws, $\frac{3}{4}$ in the next 1,000, $\frac{1}{4}$ in the next 10,000, and so on. If no limit exists, then of course the inductive method will not ascertain one. But nor, Reichenbach pointed out, will any alternative method. But suppose that a limit exists. Reichenbach conceded that a noninductive method may, in the short run, predict it better than the inductive method; but we know that if we persist with the latter it 'must lead to the true value, if there is a limit at all' (p. 353).

Reichenbach also conceded that there are other methods that may do as well, in the long run, as the inductive method. Suppose that someone takes as his inductive rule $F(A) \approx F^n(A) + k/n$ where k is some arbitrary constant: as n increases the predictions delivered by this rule will approach asymptotically those delivered by the inductive principle. But the latter will not do *worse* than such a rule. So the conclusion stands: you cannot, in the long run, do better, and you may well do worse, than by adhering to the inductive principle.

This thesis of Reichenbach's conflicts with Carnap's claim, mentioned earlier, that for any two prediction methods, one of which seems highly unreasonable while the other does not, there will be possible worlds in which the former works better. Let us now see whether we can construct counter-examples to Reichenbach's thesis in line with Carnap's claim. I will now introduce two highly unreasonable rules and consider whether there are possible worlds in which they would work better than the inductive principle. Rule (i), which is a very crude version of the principle of insufficient reason, says: 'If you do not know, concerning a particular individual, whether it is A or $\sim A$, conclude that the probability of its being A equals that of its being $\sim A$.' Rule (ii), which expresses a counter-inductive policy, says: 'If in an observed sequence of n individuals $F^n(A) = m/n$, conclude that $F(A) \approx (n-m)/n$.'

One might suppose that neither of these rules could ever perform better than the inductive principle. But let us see what a malicious Demon could manage, beginning with rule (i). Consider a potentially infinite sequence of events, the first of which occurs at t_1. Let it be a random sequence in which $F^N(A) \rightarrow \frac{1}{2}$ as N $\rightarrow \infty$. Now preface this sequence with a finite sequence, beginning at t_o, in which A occurs every time. In this expanded sequence, we will again have $F^N(A) \rightarrow \frac{1}{2}$ as N $\rightarrow \infty$, though the convergence will be slower. Now suppose that our Demon introduces two human observers to this sequence during a period between t_o and t_1. One applies to it the inductive principle and concludes that $F(A) = 1$. The other applies our rule (i) to it and concludes that, since he does not *know*, concerning each of the indefinitely many future events in this sequence, whether it will be A or $\sim A$, its probability of its being A is $\frac{1}{2}$; and from this he concludes that $F(A) = \frac{1}{2}$. And he is right. To make rule (ii) give the correct result, let the potentially infinite sequence commencing at t_1 be such that $F^N(A) \rightarrow \frac{1}{4}$ as N $\rightarrow \infty$, and let the frequency of A in the finite sequence between t_o and t_1 be $\frac{3}{4}$. There are possible worlds in which the inductive principle would perform less well than some noninductive principle.

3.35 Strawson

So let us turn to answer 5, the claim that *IP* needs no external justification because it is self-justifying. Strawson insisted that *IP* can be given no external justification: the claim 'that induction is justified by its success in practice . . . has an obviously circular look. Presumably the suggestion is that we should argue from the past 'successes of induction' to the continuance of those successes in the future. . . . Since an argument of this kind is plainly inductive, it will not serve as a justification of induction' (*1952*, p. 260). Then how can this answer, 5, be kept separate from answer 7, that *IP* is an unjustified postulate? After all, any principle is "self-justifying" in the trivial sense that it entails itself; could we not equally say of a counter-inductive principle that it can be given no external justification but this does not matter since it is self-justifying? Strawson answered by differentiating the factual statement 'Induction will continue to be successful' from the 'fundamentally different' statement, 'Induction is *rational*' (p. 261, my italics). To have good reasons for opinions about what lies outside our observations *means* having inductive grounds for them. The statement 'Induction is rational' is true a priori in virtue of the meaning of 'rational' in this context.

This answer is exposed to an objection brought against the so-called naturalistic fallacy by G.E. Moore (*1903*, ch. 1) and R.M. Hare (*1952*, p. 93). Suppose that a utilitarian announces that 'goodness' *means* 'happiness'; then he deprives himself of the ability to commend the promotion of happiness as the way to make the world a better place, for he will now only be saying that the way to make the world a happier place is to make it a happier place. Likewise, if a Zande philosopher announces that 'rational opinion' *means* 'opinion arrived at by consulting the oracle', he deprives himself of the ability to commend consulting the oracle as the way to arrive at rational opinions. And Strawson likewise deprived himself of the ability to commend inductive procedures as rational. As Bartley put it, faced by any such persuasive definition of 'rational', a critic can ask: 'Well, if what you are doing is "being rational", is it *right* to be rational?' (*1962*, p. 131).

3.36 *Carnap*

So let us turn to answer 6. This seems, on the face of it, a nonstarter: if the total evidence E cannot give an inductive lift to a hypothesis h unaided, then surely it still could not do so aided only by an analytic *IP*? I think that this obvious objection is quite correct. However, Carnap claimed to have developed a system of inductive logic that involves no 'synthetic presuppositions like the much debated principle of the uniformity of the world' (*1950*, p.v) but that can nevertheless tell us, in suitable cases, that, say, a prediction about the future has a high degree of confirmation on past evidence.

According to Carnap, if an inductive logician examines two statements, e and h, and then declares that the degree of confirmation of h on e is, say, 0.8, or $c (h, e) = 0.8$, his statement, if true, is purely analytic. After referring to Russell's conclusion that empiricism cannot be consistently maintained because induction presupposes some principle or postulates which cannot be based upon experience, Carnap went on:

Our conception of the nature of inductive inference and inductive probability leads to a different result. It enables us to regard the inductive method as valid without abandoning empiricism. According to our conception, the theory of induction is inductive *logic*. Any inductive statement (that is, not the hypothesis involved, but the statement of the inductive relation between the hypothesis and the evidence) is purely logical. (*1950*, p. 181)

In Carnap's system the value of r in $c(h, e) = r$ depends not only on what h and e respectively say but on the chosen value of his parameter λ. This can take any value between 0 and ∞. Objectively interpreted, λ is something like a measure of cosmic disorderliness: for an utterly chaotic universe, in which no universal law holds, $\lambda = \infty$; for a perfectly homogeneous universe, in which everything is exactly similar to everything else, $\lambda = 0$. Subjectively interpreted, λ is something like an index of caution: the higher the value you put on λ the more cautious you are about generalising and predicting on the basis of past experience. Putting $\lambda = \infty$ yields (*1952*, p. 38) the confirmation function $c\dagger$ that was referred to in §2.51 above and which precludes all learning from experience. Putting $\lambda = 0$ yields (p. 43) a pure "straight rule": any generalisation that holds for the n individuals mentioned in our (total) evidence E^n holds with probability 1 for *all* individuals in the universe. Putting $\lambda = 2$ yields (p. 35) a version of Laplace's Rule of Succession: if m/n is the frequency with which a characteristic occurs in the observed sample, then the probability that an unobserved individual has this characteristic is $\dfrac{m + 1}{n + 2}$. Putting $\lambda = K$, where K is the number of Q-predicates in the language, yields (p. 45) Carnap's (*1950*) confirmation function c^*.

Suppose that I have a choice between one of two actions whose outcomes will be respectively very good and very bad if a certain hypothesis h is true, and very bad and very good if h is false. Let E be my (total) evidence; and suppose that inductive logician A tells me that he puts $\lambda = 2$ which gives $c(h, E) = 0.75$ while inductive logician B tells me that he puts $\lambda = 3K$ which gives $c(h, E) = 0.25$. This leaves me in a dilemma. I want to ask whether one of these two λ-values yields better estimates than the other. Can I ask this? In his (*1952*), Carnap seemed to suggest that by a method of empirical trial and error one may arrive at ever better values for λ:

> after working with a particular inductive method for a time, [a person] may not be quite satisfied and therefore look around for another method. He will take into consideration the performance of a method, that is, the values it supplies and their relation to later empirical results, e.g., the truth-frequency of predictions and the error of estimates. . . . Here, as anywhere else, life is a process of never ending adjustments. (*1952*, p. 55)

Suppose that Mr A tells me that he chose his λ-value merely for ease of calculation whereas Mr B tells me that he reached his λ-value after years of patient research into the comparative performance of different inductive methods involving a wide variety of λ-values. Then it might seem reasonable for me to prefer Mr B's c (h, E) = 0.25 to Mr A's c (h, E) = 0.75.

But this line of thought is carrying us away from answer 6 back to answer 3: a synthetic inductive principle *IP* is emerging, in which λ is assigned an inductively determined value. The idea that Carnap's system of inductive logic, like Mill's, rests eventually on a synthetic *IP* that is itself inductively determined has been championed by John Graves. He says of λ that

> it is a precise, quantitative, numerical, scalar measure of the degree of uniformity of nature. The value we choose represents our estimate of how uniform nature is, at least in the area under investigation. We may make a poor choice, but the ontologically correct value determines the best rule to use. This value can be discovered only inductively, for it reflects our actual world and would be different in other possible worlds. (1974, p. 316)

But it is obvious that this interpretation is wholly at variance with Carnap's insistence that his confirmation function 'is purely logical'.

Did Carnap shift from answer 6 in (*1950*) to answer 3 in (*1952*)? There is a rather perplexing passage in his (1968b) in which he appears to say (i) that he did in (*1952*) adopt answer 3; (ii) that this answer was not wrong; (iii) that this answer was really answer 6 in disguise; and (iv) that answer 6 is correct. The passage (in which I have inserted the square-bracketed numerals) reads:

> [i] I used myself, . . . in 1952, the factor of past experience. . . . If our experiences seem to indicate that our . . . *C*-function does not work well, we might change it on the basis of those experiences. [ii] Today I would not say that it is wrong to proceed in this way. If you want to, you may use past experience as a guide in choosing and changing your *C*-function. [iii] But in principle it is never necessary to refer to experiences in order to judge the rationality of a *C*-function. Think of the situation in arithmetic. You can show to a child that three times five is fifteen by counting five apples in each of three baskets and then counting the total. But . . . this is not necessary. . . . [iv] We regard arithmetic as a field of *a priori* knowledge. And I believe that the same holds for inductive logic. (pp. 264–265)

But the role assigned to experience by answer 3 is quite unlike the role of experience in teaching arithmetic to children. There is no question of our using past experience as a guide in choosing and changing a system of arithmetic. An arithmetical formula is not something with which, after working with it for a time, one 'may not be quite satisfied and therefore look around for another'. Nor can one arithmetical formula supply better values or be in a better 'relation to later empirical results' than another.

According to answer 6, *both* Mr A's value of 0.75 and Mr B's of 0.25, for the degree of confirmation of h on E, may be *absolutely correct.* What they are really saying is: '$c_{\lambda = 2}(h, E) = 0.75$' and '$c_{\lambda = 3K}(h, E) = 0.25$'. There is no more disagreement between them than there would have been if they had respectively said, 'If $n = 1$ then $1 - 2^{n-3} = 0.75$' and 'If $n = 3$ then $2^{1-n} = 0.25$'. And *neither* Mr A's value nor Mr B's gives me the factual information I am anxiously seeking as to the reliability of h. As critics of Carnap (e.g. Nagel 1963, Salmon 1968) have often pointed out, if h predicts a future event, and e is evidence about past events, and $c(h, e) = r$ is analytic, then $c(h, e) = r$ adds nothing to e and therefore tells us nothing new about h. Our initial response to answer 6 was correct. The objection to answer 6 is complementary to the objection to answer 1. If IP is strong enough to do what is required of it, it will not be a priori true; and if IP is a priori true, or analytic, it will not be strong enough to do what is required of it.

3.37 *Peirce*

This leaves answer 7: IP is an unjustified postulate. Concerning this answer one remembers Russell's famous remark: 'The method of "postulating" what we want has many advantages; they are the same as the advantages of theft over honest toil' (*1919*, p. 71). An anti-sceptical strategy, whether inductivist or otherwise, that relies essentially on some *unjustified* postulate or conjecture is like a ship secured to a mooring buoy that is not attached to the bottom. As Peirce put it: 'Well, if that is the best that can be said for it [namely, that the proposition is a "presumption" or postulate of scientific reasoning], the belief is doomed. Suppose it to be "postulated": that does not make it true, nor so much as afford the slightest rational motive for yielding it any credence' (1892, 6.39).

3.4 Simplicity to the Rescue?

Our negative result so far, entirely in line with Hume's, is that without an inductive principle there can be no legitimate ascent from level-0 to level-1 or from level-1 to level-2 and that any inductive principle strong enough to "legitimise" the ascent could not itself be legitimised. If that is so, then it is obvious that there can be no legitimate ascent to still higher levels. So the case for probability-scepticism does not need a consideration of the problem of induction with respect to exact experimental laws. However, these are of great intrinsic interest and they raise problems not raised by nonquantitative generalisations of the 'All emeralds are green' type. A rational handling of them has seemed to many thinkers to require some principle of simplicity. Now our previous verdict against *any* inductive principle makes consideration of this particular one strictly superfluous. However, the idea of simplicity has been so important historically that it deserves separate consideration.

The special difficulty, from an inductivist point of view, raised by exact experimental laws was pinpointed long ago by Poincaré when he wrote: 'But every [experimental] proposition may be generalised in an infinite number of ways' (*1902*, p. 130). Goodman claimed that his "grue"-type hypotheses pose a new problem of induction. But this problem is really only the old curve-fitting (or curve-projecting) problem of induction, but raised in connection with nonquantitative generalisations involving a single discontinuity, e.g. from green to blue at time *t*. The older problem may be indicated in the following way. Consider a down-to-earth experimentalist, Mr *A*, who is plotting measurements on a graph which seem to him obviously to be falling on a straight line. But he has the misfortune to be paired with a wily and perverse curve-fitter, Mr *B*. Mr *B* finds straight lines monotonous. He prefers to connect Mr *A*'s points with curves. If the gap between two points is relatively wide, Mr *B* is content with a curve with a relatively small amplitude; but as the gaps get smaller the amplitude of his curves becomes larger. Mr *A* indignantly seeks to refute each of Mr *B*'s (in his eyes ridiculous) curve-hypotheses; and he invariably succeeds: each of his new measurements fits in with his straight-line hypothesis. But his refutations merely excite the counter-suggestible Mr *B* to ever more alarming hypotheses. It is an unequal battle. As the gap between Mr *A*'s points tends towards zero, the amplitude of Mr *B*'s curves tends towards infinity. And at any given stage, Mr *B*'s latest curve accords with the existing measurements just as well as Mr *A*'s straight line.

Actually, the problem is worse than this imaginary example suggests; for if, as we were supposing, Mr A was measuring values on a continuous scale (as opposed to counting discrete units), then the best he could do would be to locate these values within the small but finite interval of his "experimental error": he would have supplied Mr B with blobs rather than points, and Mr B would have had still more freedom in his curve constructing.

Another way of stating the problem posed by the precision of an exact law is to say that, whereas an ordinary inductive generalisation may transcend whatever evidence is supposed to confirm it only horizontally, an exact law will also transcend it vertically. As C.S. Peirce put it: 'An opinion that something is *universally* true clearly goes further than experience can warrant . . . I may add that whatever is held to be *precisely* true goes further than experience can possibly warrant' (1893, pp. 239–240, second italics mine). Popper has also stressed the importance of this kind of transcendence: 'Now it is incredible that . . . the absolutely precise statements of [Newtonian theory] could be logically derived from less exact or inexact [observation statements]' (*1963*, p. 186).

Let us take as our example of an exact law, Snell's Law, which says

$$\sin \theta_1 / \sin \theta_2 = \mu$$

where θ_1 is the angle of incidence of a ray of light to the surface of a medium, θ_2 is the angle of refraction, and μ is the index of refraction. Let us call this law l and let e be the evidence that is supposed to confirm it. Since l applies to both observed and unobserved cases of refraction, we could play the same Goodmanesque trick on it that we played on 'All emeralds are green' in §3.3 and form a series of ever more deviant variants of it for the confirming force of e to dissipate itself among. But we could also form an endless series l', l'' . . . of variants by tinkering with the formula in ways that are quantitatively too slight to make an observable difference. For instance, we might successively tack on to the right-hand side of Snell's equation '$+ 1/c$', '$+ 1/c^2$', '$+ 1/c^3$' . . . , where c is the velocity of light in metres per second (2.9979×10^8). The confirming force of e will again be dissipated among this potentially infinite set of variants.

This situation seems to cry out for some principle of discrimination that will enable us to single out an alternative as the one best confirmed by e. And the principle that naturally suggests itself is that we should single out the *simplest* one. Indeed, a principle of simplicity has sometimes been taken as central to induction. For instance, Wittgenstein declared: 'The process of induction is the process of assuming the *simplest* law that can be made to harmonize with our experience' (*1922*, 6.363). And Quine has declared simplicity to be the final arbiter (*1960*, p. 20). If the argument of the previous section was correct, we can say in advance that an inductivist cannot justify the adoption of any such principle of selection. Let us treat simplicity as a test case.

3.41 *Boltzmann*

Boltzmann boldly declared that we may call an idea about nature *false* if there are obviously simpler ideas that represent the facts more clearly (1905, p. 105). This offers a radical solution for the problem of induction: if e describes the facts to be represented, and h, h', h'' . . . are alternative representations of them, and h is simpler than h', h'' . . . , then we may dismiss h', h'' . . . as false and focus attention on the bearing of e on h alone. This idea raises two important questions. First, is there an objective criterion by which to assess the comparative simplicity of competing theories? Second, given such a criterion, is there any reason to suppose that the simplest of a set of competing theories is true, or at least more likely to be true than each of its competitors? To the first question Boltzmann himself inclined to a negative answer; and this answer, as he rightly said, has consequences that are 'quite startling':

> However it may be doubtful and in a sense a matter of taste which representation [we consider to be the simpler one that] satisfies us most. By this circumstance science loses its stamp of uniformity. We used to cling to the notion that there could be only one truth, that error was manifold but truth one. From our present position we must object to this view. (P. 106)

This is to embrace scepticism rather than to answer it; each of us is to distribute truth values over alternative 'representations' according to our personal idea of simplicity; what is "true" for me may be "false" for you; in the end it is a 'matter of taste'.

3.42 *Poincaré*

Clearly, anyone who seriously intends to dispel scepticism concerning induction and scientific inference with the help of the idea of simplicity must offer a working criterion for it, so that which of two alternatives is simpler is no longer a matter of variable personal taste.

Very well, let us now suppose that a formally and materially adequate criterion for simplicity is available. (I think that this is indeed the case, as we shall see below.) Now our second question arises: is there reason to suppose that simplicity and truth are related in some way that would allow the idea of simplicity to come to the rescue of scientific inductions? No one, I think, has discussed this question more penetratingly than Poincaré, in whose philosophy of science simplicity plays a central role. He saw the problem very clearly. He accepted something like proposition (II) above, the experientialist thesis: 'Experiment is the sole source of truth. It alone can teach us something new; it alone can give us certainty' (*1902*, p. 140). But science must be allowed to generalise from experimental facts:

> If this permission were refused to us, science could not exist; or at least would be reduced to a kind of inventory . . . of isolated facts. (Pp. 129–130)

> Experiment only gives us a certain number of isolated points. They must be connected by a continuous line, and this is a true generalisation. (P. 142)

But now comes the great difficulty: I have already quoted Poincaré's statement that every experimental proposition 'may be generalised in an infinite number of ways' (p. 130); any number of curves can be fitted over a finite number of isolated points. Then how are we to choose among these?

> The choice can only be guided by considerations of simplicity. (P. 146)

> Why do we avoid angular points and inflexions that are too sharp? Why do we not make our curve describe the most capricious zigzags? (P. 146)

> Among all possible generalisations we must choose, and we cannot but choose the simplest. (P. 130)

But is there any epistemological justification for preferring the simplest curve or generalisation? We proceed, Poincaré said, 'as if a simple law were, other things being equal, more probable than a complex law' (p. 130). But what justification have we for this assumption? Why should there be any correlation between simplicity and probability?

Let us distinguish a methodological, an epistemological, and an ontological thesis concerning simplicity. The *methodological thesis* says that we should prefer, other things being equal, a simpler to a more complex law-statement; the *epistemological thesis* says that, other things being equal, there is reason to believe that the simpler one is more likely to be true or, as Leibniz put it, 'a hypothesis becomes the more probable as it is simpler' (1678, p. 288); and the *ontological thesis* says that the laws that govern natural processes are essentially simple. If we had reason to suppose that the ontological thesis is true, that would provide encouraging support for the epistemological thesis, which would in turn justify the methodological thesis. But have we any reason to suppose 'Nature to be profoundly simple', in Poincaré's phrase (p. 150)? According to proposition (I), the anti-apriorist thesis, we could not know this a priori; and it is perhaps worth remarking that, among those metaphysicians who supposed that there is a priori knowledge of nature, Spinoza and Leibniz proceeded from the premiss that God is infinitely powerful or infinitely good to the conclusion that the world is infinitely *rich*; and they understood this conclusion in a way that totally repudiates the ontological simplicity thesis, since the thesis of infinite richness endorses something like the theory C_K (discussed in §2.23), which says that every Q-predicate is instantiated and hence that no law-statement, however complex, is true. According to proposition (II) any reasons for supposing this ontological thesis to be true must be empirical. But what sort of empirical reasons could there be for it? The only answer I can think of would be to the effect that the empirical sciences are progressively revealing the underlying simplicity of nature. If Poincaré had offered this answer (which he did not), his position would have been suspiciously circular. He held that science would be impossible unless physicists adhere to something like our methodological simplicity thesis: they are 'guided by an instinct of simplicity' (p. 158) and 'cannot but choose the simplest' generalisation consonant with the experimental facts. If that is so, it would hardly be surprising if their theories and law-statements convey the impression that the laws that nature obeys are essentially simple. By adhering to the methodological thesis they would, almost as a matter of course, generate "confirmation" for the ontological thesis. It would obviously be circular to use the thus "confirmed" ontological thesis to justify the methodological thesis.

Actually, the situation as Poincaré saw it was even worse. He was inclined candidly to admit (though he wobbled a bit on this) that, despite the general demand for simplicity in science, the ontological simplicity thesis has actually been discredited by recent scientific developments. Consider the following passages:

> A century ago it was frankly confessed and proclaimed abroad that Nature loves simplicity; but Nature has proved the contrary since then on more than one occasion. We no longer confess this tendency, and we only keep of it what is indispensable, so that science may not become impossible. (P. 130)

> Fifty years ago physicists considered, other things being equal, a simple law as more probable than a complicated law.... But this belief is now repudiated. (P. 206)

> What we thought to be simple becomes complex, and the march of science seems to be towards diversity and complication. (P. 173)

> Our equations become ... more and more complicated, so as to embrace more closely the complexity of nature. (P. 181)

Remember that these are the judgments of someone who held that the regulative idea of simplicity plays an indispensable role in science. It is as if the Pope had vacillated over the existence of God. A quarter of a century later, Bridgman, in a very interesting review of the idea of the simplicity of nature (*1927*, pp. 198f.), concluded even more strongly that recent developments in science had been adverse to this idea: 'It seems to me that as a matter of experimental fact there is no doubt that the universe at any definite level is on the average becoming increasingly complicated, and that the region of apparent simplicity continually recedes' (p. 205).

The upshot seems to be this:–Simplicity must serve as a principle of selection in science if science is not to become impossible. And progress in science increasingly suggests that such a principle is untenable! When he contemplates this self-destructive circle, the sceptic may conclude that his work has been done for him.

3.43 *Jeffreys*

However, one way remains open to an inductivist who seeks to justify a methodological preference for simpler laws on the ground that they are more likely to be true: he may try to justify the epistemological simplicity thesis without appealing to any ontological thesis (rather as Reichenbach tried to show that in any universe an inductive method will deliver predictions that are at least as good as those of any non-inductive method). Harold Jeffreys took this way in his (*1939*) (second edition *1948*, third edition *1961*; since there are considerable variations between them, I will indicate to which of these editions I am referring). Jeffreys insisted that, if a theory of induction such as his is to be adequate, then its principles 'must not of themselves say anything about the world' (*1948, 1961*, p. 7); so there must be no appeal to ontological simplicity.

Jeffreys's work is important, and demands consideration here, for several reasons. But I should warn the reader that it is not easy and that a critical investigation of it is bound to be rather technical. So I will follow my rule and state my conclusion in advance. I mentioned in the previous section that some philosophers have combined or oscillated between two or more answers to the question of the epistemological status of whatever principle or postulate they invoked in seeking to overcome the problem of induction; and this applies to Jeffreys. With regard to the status of his *simplicity-postulate*, he oscillated between something like answer 2, namely that it is established by a transcendental argument, and something like answer 6, that this postulate is part of an analytic system of probability logic. Like Carnap and Keynes, he was a probabilist who accepted our proposition (III), the deductivist thesis; he put forward his theory of probability as 'an extended logic of which deductive logic will be a part' (*1948, 1961*, p. 17). But he also declared that we *have* to accept the simplicity-postulate 'without hesitation' (*1957*, p. 35) because there would be no confirmation of scientific laws without it. What the simplicity-postulate says, in effect, is that a simpler law has a higher initial probability than a less simple law. And my conclusion, which I will reach near the end of §3.44, and which will be very much in line with Popper's (*1959*, app.* viii), will be that the Jeffreys simplicity-postulate leads to inconsistencies within his probability system. Thus we will have here a striking example of a transcendental argument backfiring: if scientific *knowledge* would be impossible unless the simplicity-postulate were true, then since the simplicity-postulate leads to contradictions, scientific *knowledge* is impossible.

One reason why Jeffreys's work demands consideration here is that he (originally in collaboration with Dorothy Wrinch) introduced an admirably clear-cut and workable concept of simplicity. Another is that the problem of induction, as he saw it, and which he sought to overcome, was precisely the problem to which we are addressing ourselves in this section: how can we say that evidence *e* confirms the experimental law *l* if the confirming force of *e* is dissipated among a host of variants *l'*, *l''* ... of *l*? Jeffreys had anticipated Goodman, at least with regard to exact experimental laws. Let l be such a law, expressed as an equation, say of the form $y = f(x)$. And suppose that n experiments on l have been carried out, one after another, at times $t_1, t_2, \ldots t_n$, and that all their outcomes fitted in nicely with the predictions yielded by l. Jeffreys showed that there is a potential infinity of alternative laws, l', $l'' \ldots$, each of which makes exactly the same predictions for those n experiments, but divergent predictions for experiments carried out after t_n. His method for constructing such a deviant law was to tack onto the right-hand side of the equation a correction factor whose value is zero for any t_i where $t_1 \leq t_i \leq t_n$ but nonzero for all other values of t (*1939*, p. 3). An example of such a correction factor is: $' + k(t - t_1)$ $(t - t_2)(t - t_3)\ldots(t - t_n)'$ where k is some arbitrary positive number.

Now if, for any such l, there are infinitely many potential alternatives to it, will not the initial probability of each of them, including l, have to be zero? Like Keynes, Jeffreys accepted that *if* $p(l) = 0$, then for any e, $p(l,e) = 0$. As we saw, Keynes tried to circumvent this with his postulate of limited independent variety, which limited the alternatives to a finite number, say n, over which he made a uniform probability distribution, so that the initial probability of each of them is $1/n$. Jeffreys took a contrary approach: instead of limiting their number, he put a nonuniform probability distribution on them. Let $l_1, l_2, l_3 \ldots$ be an exhaustive set of mutually exclusive alternatives, and let q_i be the initial probability (Jeffreys called it the prior probability) of l_i ($i = 1,2,3 \ldots$). Jeffreys did of course require these q_i's to sum to 1; but he sought to achieve this by letting $q_1 + q_2 + q_3 \ldots$ converge to 1, say by making them $\frac{1}{2} + \frac{1}{4} + \frac{1}{8} \ldots$ But when should we regard one hypothesis as initially more probable than the next one? Jeffreys claimed that we may, indeed must, regard a *simpler* hypothesis as having a higher initial probability than a less simple alternative. Let us picture all these alternative hypotheses as runners in a race that takes place on a course stretching from 0 to 1. Any hypothesis initially placed right back at 0 is a nonstarter. So Keynes had the race start with a finite number n of runners lined up at $1/n$, whereas Jeffreys had an infinite number of runners with the simplest one(s) starting ahead of the next simplest and so on down. The race gets under way when evidence starts coming in. This may of course go against runners who had a good start. But if it is favourable, a front-runner may progressively increase his initial lead over the rest of the field; and although he will never actually reach 1, he may attain a high posterior probability.

When is one quantitative hypothesis simpler than an alternative to it? Jeffreys and Wrinch had introduced a concept of simplicity in (1921). This was later stripped of all inessentials to become: simplicity varies inversely with the number of adjustable parameters; conversely, the 'complexity of a law is now merely the number of adjustable parameters in it, and this number is recognizable at once' (*1948*, p. 100). An adjustable parameter is a constant whose value is not specified by the hypothesis in which it occurs but is left to be determined experimentally. Thus if x and y are variable magnitudes, and a_1 and a_2 are adjustable parameters, and law l_1 says that $y = a_1x$, while law l_2 says that $y = a_1x + a_2/x^2$, then l_1 is simpler than l_2. This concept of simplicity is neutral with respect to the *shape* of the curve(s) yielded by an equation: it does not say that a straight line is simpler than, for instance, a circle or a hyperbola. It is concerned with the relative *determinacy* of the equation. It seems to me entirely right that a concept of simplicity that is intended to play a methodological role in the appraisal of theories should attend to the determinacy of the equations rather than the shape of the curves; for we surely do have reasons to prefer, other things being equal, a more to a less determinate hypothesis, whereas there seems to be no a priori reason why we should prefer, say, a linear to a simple harmonic hypothesis. We might call this an "input-output" measure of simplicity: the fewer the data that have to be fed in to get a unit of information (say, a particular value of y) out of an equation, the simpler the equation is.

3.44 *Popper versus Jeffreys*

Popper in (*1934*) had proposed, with acknowledgements to Hermann Weyl, a "paucity of parameters" view of the empirical simplicity of theories which, as he later acknowledged (*1959*, p. 139n), had been anticipated by Jeffreys and Wrinch in their (1921). But Popper approached the problem of simplicity from a quite different angle. He accepted the methodological thesis that a simpler hypothesis should be preferred, other things being equal, to a less simple one; but he repudiated the epistemological thesis that it should be preferred *because it is more probable*. On the contrary, he held that if l_1 is simpler (has fewer adjustable parameters) than l_2, then l_1 is *less* (or anyway not more) probable than l_2. Yet l_1, so long as it survives tests, is methodologically preferable to l_2 because it is *easier to test*. The larger the number of adjustable parameters that a law contains the larger is the minimum number of measurements needed to falsify it, and the less easy it is to test. Let us look into this disagreement between Popper and Jeffreys.

Let $x, y \ldots$ be variables ranging over magnitudes (e.g. distance, time, velocity $\ldots$); let $a_1, a_2 \ldots$ be adjustable parameters; and let $u, v \ldots$ be variables ranging over the real numbers. Then we could reformulate the above two laws, somewhat pedantically, as follows:

l_1: $\forall x \; \forall y \; \exists u \; [(y = a_1 x) \wedge a_1 = u]$

l_2: $\forall x \; \forall y \; \exists u \; \exists v \; [(y = a_1 x + a_2/x^2) \wedge a_1 = u \wedge a_2 = v]$.

It is obvious that l_1 is simpler, in Jeffrey's sense, than l_2. But does not l_1 strictly entail l_2? If it does, we are bound to have $p(l_1) \leq p(l_2)$, contrary to the simplicity-postulate. We may rewrite l_1 in the following padded out form:

l_1: $\forall x \; \forall y \; \exists u \; \exists v \; [(y = a_1 x + a_2/x^2) \wedge a_1 = u \wedge a_2 = v = 0]$.

And it is now clear that l_1 would entail l_2 unless zero were excluded as a possible value of a_2 in l_2. For if zero is not excluded we could write l_2 as:

$\forall x \; \forall y \; \exists u \; \exists v \; [(y = a_1 x + a_2/x^2) \wedge a_1 = u \wedge a_2 = v \geq 0]$.

As Jeffreys said, 'making [a_2] zero will reproduce the old law' (*1948*, p. 101). So he stipulated that a_2 in l_2 may *not* take the value zero. (Popper, in a discussion of Jeffrey's simplicity-postulate in *1959*, appendix *viii, to which I am much indebted, overlooked this ban on zero.) Correctly stated we have:

l_2: $\forall x \; \forall y \; \exists u \; \exists v \; [(y = a_1 x + a_2/x^2) \wedge a_1 = u \wedge a_2 = v \neq 0]$.

Thus l_1 and l_2 are mutually inconsistent.

But surely l_1 is stronger than l_2, since l_2 tolerates all values for a_2 other than zero, whereas l_1 tolerates no value for it other than zero. How can this simpler and stronger law be *more* probable than the other? Jeffreys met this by laying down that $p(a_2 = 0) = p(a_2 \neq 0) = \frac{1}{2}$: 'half the prior probability is concentrated at [a_2] $= 0$' (p. 101). This has the desired effect of forcing up the prior probability of l_1 to parity with l_2. But this biasing of the prior probabilities leads to inconsistencies, as we will now see.

We may state the Jeffreys simplicity-postulate, which I will abbreviate to (S-P), as follows:

$$(\text{S-P}) \quad S(l_i) > S(l_j) \rightarrow p(l_i) > p(l_j) \text{ and}$$

$$S(l_i) = S(l_j) \rightarrow p(l_i) = p(l_j)$$

where '$S(l_i) > S(l_j)$' is to be read as: 'l_i is simpler (has fewer adjustable parameters) than l_j'. Jeffreys endorsed what we may call the 'equal simplicity, equal probability' postulate, or $S(l_i) = S(l_j) \rightarrow p(l_i) = p(l_j)$, when he wrote: 'laws involving the same number of adjustable parameters can be taken as having the same prior probability' (*1948*, p. 100).

Since our l_1 has one adjustable parameter while l_2 has two, we have

(1) $S(l_1) > S(l_2)$.

And (1) in conjunction with (S-P) entails

(2) $p(l_1) > p(l_2)$.

Now consider alternatives l_1', l_1'' ... to l_1 that can be got from l_1 by replacing the clause '$a_2 = v = 0$' in its padded out version by '$a_2 = v = k_1$' '$a_2 = v = k_2$' ... where k_1, k_2 ... are real numbers greater than zero. Each member of the l_1', l_1'' ... series is like l_1 in having one adjustable parameter, namely a_1, and in putting a fixed value on a_2 (though unlike l_1 they put a value other than zero on it). So we have

(3) $S(l_1) = S(l_1') = S(l_1'')$ ...

And by the 'equal simplicity, equal probability' postulate we may proceed from (3) to

(4) $p(l_1) = p(l_1') = p(l_1'')$ ...

Now each member of the l_1', l_1'' ... series strictly entails l_2; for l_2 says that a_2 has *some* value greater than zero, and l_1', l_1'' ... each *specify* a value greater than zero for a_2. Thus we are bound to have

(5) $p(l_2) \geq p(l_1') = p(l_1'')$ ...

But (5) in conjunction with (4) entails

(6) $p(l_1) \leq p(l_2)$

which contradicts (2) above.

When we considered the question of the initial probability of law-statements in §2.5, we did not go into the additional problem that is created if the law is exact (in the sense that precise values for its independent variables determine a precise value of its dependent variable). Jeffreys himself pointed out something that seems clearly to imply that if l_1 puts the value 0 on an adjustable parameter a_2 that is located only within an interval, say the interval [0, 1], by l_2, then the initial probability of l_1 must be 0. He wrote: 'Suppose that a measurable quantity . . . must lie between 0 and 1 and that the probability that it lies in any interval is proportional (and therefore equal) to the length of the interval. What is the probability . . . that [it] is precisely $\frac{1}{2}$? Clearly 0' (*1957*, p. 29). And the probability that it is precisely 0 is likewise clearly zero.

Another argument for the thesis that the probability of an exact law-statement is always zero is that, as we have just seen in connection with our l_1, it will be one of a potentially infinite set of laws, say all of the form $y = f(x, a)$, where the functions are equally determinate in all of them, but send (x, a) to divergent values. An example given by Popper (*1959*, p. 384) takes l to be $y = ax$ and lets l', l'' . . . be $y = ax^2$, $y = ax^3$ . . . Given the 'equal simplicity, equal probability' postulate, and given that this series can be extended indefinitely, this means, as he pointed out, that

$$p(l) = p(l') = p(l'') \ldots = 0.$$

Once again the sceptic may feel that his work has been done for him. Jeffreys had said that we must accept something like the simplicity-postulate 'without hesitation' (*1957*, p. 35; *1961*, p. 48) because scientific laws can gain no confirmations without it. The sceptic endorses the second part of this transcendental argument, combines it with the finding that the simplicity-postulate leads to contradictions, and concludes that scientific laws never can gain any confirmations.

Nothing said in this section tells against the methodological simplicity thesis, or against Jeffreys's admirably hard criterion for simplicity or, for that matter, against the attempts made by Kemeny (1953) and others to refine that criterion. However, the methodological thesis has been deprived of epistemological support. How adequately it can be justified from a noninductive point of view is a question we will put on the agenda for Part Two of this book. (This question will be taken up in §5.37.)

3.5 Review and Prospect

The case for probability-scepticism that has been developed in Part One of this book, deploying arguments, nearly all of which were already well known, has a fail-safe character: if an earlier argument should break down, a later one will take over. The first hurdle is the progression from level-0 to level-1. But suppose it to be surmounted. There is then the problem of progressing from level-0 via level-1 to level-2. Proposition (V), the ban on step-by-step justifications of a non-deductive kind, does not allow such a progression. Suppose, then, that level-1 statements could somehow achieve objective certainty, so that the inductive ascent can start from here. Then comes the argument, developed in §2.4, that it is never possible to grade uncertain hypotheses as more or less probably true (because we can never know what our total evidence is). Suppose that that argument is invalid. Even so, we cannot ascend to level-2; for we found in §2.5 that even if we could somehow know that E is our total evidence, and even if a universal generalisation g can have a positive initial probability, E cannot *raise* the probability of g if a certain paramount principle of logical probability is complied with. But suppose that that objection is invalid. There remains the basic difficulty that the confirming force of E will dissipate itself over an indefinite host of variants of the generalisation g we would wish it to confirm. That difficulty would be surmounted if there were a legitimate inductive principle strong enough to exclude, or at least downgrade, these variants and enable the confirming force of E to be concentrated either exclusively or at least primarily on g itself. But we found in §3.3 that the status of an inductive principle strong enough to do the job would be that of an unjustified postulate. (If unjustified postulates were permitted, it would be easy to rebut Hume: we need only postulate that God has designed a world amenable to inductive methods.) And we further found that a principle of simplicity can lead to inconsistencies. But I will not rehearse here the additional hurdles that frustrate any inductive progression to exact experimental laws at level-3. As to full-fledged scientific theories at level-4, they soar far beyond the reach of inductive support.

In §1.1 three propositions were presented that jointly pose the sceptical challenge, and two more were later introduced that intensify that challenge. If that challenge is to be met, some of these propositions must go. Which? I consider it hopeless to reject (I), the anti-apriorist thesis, and to repeat Kant's attempt to construct a synthetic a priori framework for human knowledge; and I consider it disastrous, for the reasons given in §1.4, to reject proposition (III), the deductivist thesis, and to license the crossing of logical gaps by pretending that invalid inferences are valid. And proposition (V), the ban on chains of nonentailing justifications, cannot be rejected since it is provable within any system of probability logic.

That leaves proposition (II), the experientialist thesis, and proposition (IV), the thesis that rational assent is controlled by degree of confirmation. Proposition (II) is central to nearly all empiricist philosophies (see for instance Carnap, *1950*, p. 31). Now it is proposition (II) that imposes the requirement that for scientific knowledge to be *knowledge* it must have an inductive structure; and probability-scepticism says, in effect, that if science has an inductive structure then its structure is logically rotten: the empirical foundation cannot be sufficient to sustain the soaring edifice, which collapses. If 'ought' implies 'can' then proposition (II) is refuted by probability-scepticism. The former says that science ought to be inductively grounded, and the latter that it cannot be. Proposition (II) requires that experience should *positively* control the content of scientific hypotheses at higher levels. Thus we can reject it *without* abandoning the idea that scientific hypotheses should be under empirical control by replacing it by a proposition (II*) that requires that experience should *negatively* control scientific hypotheses. That, essentially, was the decisive innovation made by Popper in (*1934*). Clearly, this negativist conception of the role of experience in science needs to be accompanied by a methodological theory that entitles us to judge, in cases where two or more rival hypotheses have all so far passed the test of experience, which of them is best. But if such a methodological theory is not to have an arbitrary or dogmatic or ad hoc character, it must be governed by some overall aim for science; moreover, this aim should itself be nonarbitrary. In the next chapter, we will seek out such an aim.

Proposition (IV), like proposition (II), is one to which nearly all empiricists adhere. Indeed, many of them take it for granted, and as needing no supporting argument, that it is rational to accept only a more or less well confirmed hypothesis, or that one's degree of rational assent to a hypothesis should be controlled by its degree of confirmation ('confirmation' being understood in some quasi-verificationist or probabilist sense). As I said before, proposition (IV) escalates probability-scepticism into rationality-scepticism. As someone who holds the former kind of scepticism to be unanswerable I would, if there were no alternative to proposition (IV), be obliged to look upon science as another ideological system along with magic and the rest, or as a game that has its own rules, just as magic does, and indulgence in which is entirely optional.

However, an alternative to proposition (IV) suggests itself when we ask what we expect from those hypotheses that, by proposition (II*), are now under only the negative control of experience. We cannot expect anything relating to certainty from them. But we can ask that they should provide (conjectural) *explanations*. And this suggests, as I indicated in §2.3, that we put as an alternative to (IV) the proposition (IV*), which says that it is rational, given evidence e, to adopt the hypothesis h if h is unfalsified and provides the best available explanation for e. The sought-for relation between e and h is now inverted: instead of an upward, quasi-verifying inference from e to h we have a downward, explanatory derivation of e from h. Of course, this proposition (IV*) needs to be filled out by a clear specification of what it is for one explanation of e to be *better* than any available alternative. Such a specification will be forthcoming after we have worked out what is the optimum aim for science. To this task we must now turn.

Part Two

CHAPTER

4

....

The Optimum Aim
for Science

4.1 Adequacy Requirements

Speaking of 'the end and purpose of science' Popper declared: 'The choice of that purpose must, of course, be ultimately a matter of decision, going beyond rational argument' (*1934*, p. 37). That appears to suggest that different groups of scientists may adopt different, perhaps even conflicting, aims for their science; but if that happened, instead of one republic of science would there not be different tribes 'a-whoring after other gods'? Moreover, our hopes of defeating rationality-scepticism would fade. Faced by a choice between competing theories, say T_i and T_j, rationality-scepticism says that we never have any good cognitive reason to prefer one to the other. Now if I freely adopt one aim, and you freely adopt another, it might be that *I* have reason to prefer T_i and *you* have reason to prerer T_j; but there would not be good, impersonal reasons for both of us to prefer one to the other. If we are to defeat rationality-scepticism we need a nonarbitrary aim for science, an aim to which all members of the republic of science could subscribe.

But is not the idea of such an aim a utopian dream? Well, I am going to try to turn it into a reality. My first step will be to spell out some adequacy requirements for any proposed aim for science. I admit that if someone were to reject one or more of these, that would leave him free to adopt an aim different from the one I shall propose. But the adequacy requirements that I shall propose will, I trust, be uncontroversial. However, they will still permit a wide range of alternative aims. Now comes the crucial question: is there one of these that *dominates* all the other permitted alternatives? Is there one that contains all the other permitted alternatives as components of itself, so that it is the strongest or most comprehensive and ambitious aim that does not yet exceed the adequacy requirements; or is every permitted aim *partial* in the sense that adopting it means forgoing some legitimate aspiration contained in at least one of the others? If there is such a dominant aim, it is the optimum aim for science.

I will now list five adequacy requirements, offering some explanatory comments afterwards. I suggest that any proposed aim for science must:

1. be coherent;
2. be feasible;
3. serve as a guide in choices between rival theories or hypotheses;
4. be impartial;
5. involve the idea of truth.

As to 1, an aim is incoherent if it has two components that sometimes or always pull in opposite directions, so that a move that is progressive relative to one component is retrogressive relative to the other. A famous example of an incoherent aim is Bentham's "fundamental axiom" that 'it is the greatest happiness of the greatest number that is the measure of right and wrong' (*1776*, p. 1). Concentrating a benefit on a few people might be right by the greatest happiness component, but wrong by the greatest number component (which would presumably require the benefit to be spread, however thinly, over as many people as possible).

As to 2, I say that an aim is infeasible if we know that it cannot be fulfilled. But we must be careful not to create an air of infeasibility by misportraying the aim in question. One's aim may be: (i) to *attain* a certain goal; or (ii) to *progress towards* a certain goal without necessarily attaining it; or (iii) to *progress in a certain direction* without having an ultimate goal that one is progressing towards. A spurious air of infeasibility may be created if a type (ii) aim is misportrayed as a type (i) aim, or a type (iii) aim as a type (ii) or type (i) aim. Consider golf. A golfer's aim is to go round the course with as few strokes as possible. He would go round with as few strokes as possible if he holed-in-one each time. Is it then his aim to hole-in-one each time (Peter Heath, 1955, pp. 155–156)? No, his aim is not the infeasible type (i) aim of holing-in-one each time, but the feasible type (ii) aim of getting as near as he can to doing that. Condorcet, near the end of his (*1795*), seemed to be proposing for medical science the infeasible aim of rendering us immortal; but he was really proposing the type (iii) aim of steadily extending our expected life span. A type (i) aim is infeasible if the goal is known to be unattainable, but type (ii) and type (iii) aims are infeasible only if progress in the intended direction is infeasible. I may perhaps say in advance that the aim for science that will be proposed in this chapter and elucidated in the next will be of type (iii). It will thereby have the advantage of being an aim that can be pursued in the infancy of science as well as in its maturity. If someone sets himself the type (i) aim of walking from London to Edinburgh, then either he does not fulfil it, or else he fulfils it on a certain date, and that is that. Our aim will be more like the type (iii) aim of walking each day somewhat further than one walked on the previous day. A toddler might start fulfilling this aim: seven steps yesterday, twelve today, perhaps twenty tomorrow . . . And he might go on fulfilling it for many years to come.

As to 3, an aim that is coherent and feasible may nevertheless fail to provide any guidance as to the direction in which we should go if we are to pursue it. Suppose that someone makes it his aim to head for that spot on the earth's surface that is directly above the biggest undiscovered diamond. There presumably is such a diamond and hence such a spot and he *might*, rather incredibly, be heading straight for it. So we cannot dismiss his aim as infeasible: progress in the desired direction is not impossible. But his aim gives him no guidance as to the direction in which he should go. Suppose it were proposed that science should progress with ever ϕ-er theories, where ϕ is a property that theories can have to a greater or lesser degree, towards the (perhaps unattainable) goal of the ϕ-est theory of all. Suppose, furthermore, that a clear meaning had been given to claims of the form 'Theory T_j is ϕ-er than theory T_i', but that there is no possibility of assessing the truth or falsity of such a claim. Then this aim would be feasible: it is possible that a sequence of theories T_1, T_2, T_3 . . . is progressing in the right direction. But while we might *hope* that it is, we might equally *fear* that it is not. Such an aim would satisfy conditions 1 and 2, but fail condition 3.

As to 4, the idea is that a proposed aim for science should be above the scientific battle and not favour one side because of a bias towards theories that endorse a particular metaphysical view of the world. This requirement may seem to conflict with the undoubted fact that many great scientists have incorporated some metaphysical idea into their view of what science should achieve. In Galileo's case it was the idea that nature is essentially mathematical. Helmholtz declared it the vocation of science to refer all natural phenomena back to unchangeable attractive and repulsive forces whose intensity depends only on distance. Einstein had a strong predilection for deterministic theories. Now requirement 4 is not of course intended to preclude a scientist's thinking and research being inspired by a governing idea of this kind, or by a full-fledged research programme in Lakatos's sense. But when we turn

125

from the construction to the appraisal of scientific theories, for instance, competing theories thrown up by rival research programmes, in the light of an overall aim for science, requirement 4 says that this aim, and the appraisals it sponsors, should have an impersonal, judicial character and should not be distorted by a partiality for a particular metaphysical view. Requirement 4 was observed by the judges appointed by the French Academy to award the 1818 prize on the subject of Diffraction. Most of them (Laplace, Poisson, and Biot) were keen supporters of the corpuscular theory of light; but they very properly awarded the prize to Fresnel after the dramatic corroboration of his wave theory by the observation of a white spot at the centre of the shadow of a circular screen (Whittaker, *1951*, i, pp. 107–108). Requirement 4 also precludes a bias in favour of a currently reigning theory; in constructing our aim for science we must not make Kant's mistake of elevating certain striking characteristics of a reigning theory into preconditions for any future theory. A theory, like a successful commercial enterprise, may achieve something like monopoly status under conditions of free competition; but it should not then be protected by being granted a legal monopoly.

As to 5, I have to admit that this requirement is not entirely uncontroversial since it has, unless I have misunderstood him, actually been rejected by at least one contemporary philosopher (Laudan, *1977*, pp. 24 and 125–126). But it seems to me that to say that truth is no part of the aim of science is on a par with saying that curing is no part of the aim of medicine or that profit is no part of the aim of commerce. Whether an aim that satisfies condition 5 is reconcilable with the sceptical conclusions of Part One of this book is a question we will consider in § 4.52.

4.2   The Bacon-Descartes Ideal

The question to which we must now turn is whether, given the above adequacy requirements, there is an aim for science that satisfies them and dominates any other aim that satisfies them. I propose to tackle this question in the following way. To begin with, we will set aside these adequacy requirements, together with the sceptical conclusions of Part One, and seek out a thoroughly utopian aim for science, an aim that is as sweeping and ambitious as anyone could ask. At this stage our motto will be 'Nothing but the best', or rather 'The best, the whole best, and nothing but the best'. And we will arrive at an aim with a number of components, each as imperious as it could be.

We will then examine separately each of these components in the light of those of our adequacy requirements that are appropriate to it, but without, at this stage, bringing in the sceptical conclusions of Part One. Not suprisingly, we will find that each of them has to be moderated to bring it into conformity with them. That done, we will then consider these components collectively; do they, taken as a whole, satisfy our adequacy requirements as a whole, particularly the requirement of coherence? We will find that they do not. There will be one component, to be called demand (A), that pulls in one direction, and four components, to be called demands (B1), (B2), (B3) and (B4), that all pull in the opposite direction.

Now comes the big question: are there alternative ways of cutting back this aim so as to bring it into conformity with the adequacy requirements? Or is there one smallest contraction of it that will suffice? At this stage, the sceptical conclusions of Part One will be reintroduced. They will have no adverse implications for the (B) components of our aim but they will bear adversely on demand (A), which will now turn out to be an infeasible demand. However, something feasible and non-trivial can be retrieved from (A) and combined with the four (B) components, resulting in what I will claim to be *the optimum aim for science.*

Let us now embark on this programme. After Hume it was to be 'never glad confident morning again' for those who had believed in the limitless power of human reason to unlock nature's secrets. So let us imagine ourselves back in the glad confident morning of modern science, in the early seventeenth century, participating with Gilbert, Galileo, Kepler, Harvey, and others in the great scientific adventure. Filled with shining optimism, we ask ourselves: What is the goal of the sciences to which we are dedicating ourselves?

We would surely insist that *truth* is part of what we are aiming at. And we would, I think, add that we not only want to get hold of the truth but to *know* that we have done so, so that we need never fear that error has crept in. In short, *certain truth* is one major component of our aim.

But certain truth about what? Well, there are all sorts of singular facts and empirical regularities of which we will want descriptions that are *known* to be true. But let us go at once to our supreme demand. We believe that we have already made quite good progress in the discovery of underlying factors in terms of which empirical regularities can be explained. But these underlying factors themselves now stand in need of explanation in terms of still deeper factors. And our supreme demand is that our science will penetrate deeper and deeper until eventually it achieves *ultimate explanations* of all phenomena.

This would undoubtedly be accompanied by another major demand. At the present (imagined) time, the sciences to which we are dedicating ourselves are rather compartmentalised. Gilbert's theory of the magnet, Galileo's laws of projectile motion, Kepler's laws of planetary motion, Harvey's theory of the motion of the blood—these are all concerned with motions of one kind or another, yet none of them ties in, in any significant way, with any of the others. But we will expect this compartmentalisation to diminish as we penetrate deeper; and when we arrive at the level of ultimate explanations, the various special sciences will surely merge into *one unified science*. (For try supposing that at the end of the day we had two disjoint sciences, say one for celestial phenomena and the other for terrestrial phenomena, with no possibility of uniting them. Then heaven and earth would constitute two independent coexisting worlds; there would be no laws, spanning both worlds, governing interactions between them.) We want our science eventually to unveil *the* plan of the Creation, not two or more uncoordinated plans for different parts of it.

Let us now turn to the predictive capacity to be expected of the science we are aiming at. The ability to foretell future events has aroused awe at least since Biblical times, and we will not want our science to be deficient in this respect. But a scientific prediction about a singular event requires, of course, two kinds of premises: laws, and initial conditions. Now divine omniscience has, presumably, both a vertical dimension (knowledge of the entire structure of the world down to its bottommost level) and a horizontal dimension (an item-by-item coverage of all singular facts). And while we boldly and optimistically want our science to strive for something like vertical omniscience, we recognize that it would be absurd for we humans even to aspire after horizontal omniscience. And this means that however deep our science penetrates, and even if it eventually reaches nature's bottommost level, there will remain indefinitely many future events that we are in practice unable to predict because we lack sufficiently detailed and thorough coverage of the relevant initial conditions. However, we can hope that our science will eventually give us *potentially* complete predictive power, in the sense that, for any possible future event, we could in principle deduce, from a sufficiently precise and complete description of relevant initial conditions in conjunction with our laws, the prediction that it will, or else that it will not, occur.

Implicit in the foregoing is a further demand that we would surely wish to make explicit and formulate separately. If our laws were not perfectly exact there would be events that we would not be able, even in principle, to predict. A law that is only a little bit fuzzy may make possible reasonably definite predictions in the short term; but the interval of imprecision will grow with the time period and it may well be impossible to make any definite predictions for the long term on the basis of such a law. And in any case, we can no longer be content with an essentially vague, qualitative science à la Aristotle now that we have before us the elegantly precise, mathematically expressed laws of Galileo and Kepler.

Finally, we will of course insist that there is nothing loose or slipshod in the connections between the propositions that will constitute our science; these must be perfectly rigorous.

This utopian aim for science may be summarized thus: all empirical phenomena rendered explainable or predictable by deduction from true descriptions of initial conditions in conjunction with universal principles that are certainly true, ultimate, unified, and exact.

We may regard Bacon and Descartes as philosophical spokesmen for the high expectations that science excited in the early seventeenth century, and I do not think that our ideal would have seemed excessive to either of them. Bacon's method was intended 'to penetrate into the inner and further recesses of nature' (*1620*, I, xviii) and discover the ultimate Forms, the very alphabet of nature. True, he rejected the only formal deductive logic available in his day, namely the syllogism. But so did Descartes and for the same reason, namely that it is not content-increasing. And Baconian induction, like Cartesian deduction, was supposed to lead to conclusions whose truth is guaranteed by the premises from which they are induced; it was supposed to be *truth-preserving*. As Lakatos remarked, 'In the seventeenth and eighteenth centuries there was no clear distinction between "induction" and "deduction"' (*1978*, ii, p. 130).

As to Descartes, with his aim of deducing the mathematical laws of motion governing all physical changes from a rational knowledge of the essence of matter, I think that he would have endorsed this ideal. As Husserl put it, it was for Descartes a truism 'that the all-embracing science must have the form of a deductive system, in which the whole structure rests, *ordine geometrico*, on an axiomatic foundation that grounds the deduction absolutely' (*1931*, p. 7). So I shall refer to it as the Bacon-Descartes ideal.

4.21 A "Progressivised" Version

We must now sober up and take a critical look at this ideal in the light of our adequacy requirements. We have so far presented it as a type (i) aim that science will fulfil only if it actually attains the goals of certainty, ultimacy, etc. The feasibility requirement obviously requires us, as a first step, to recast it as the type (ii) aim of progressing towards these far-off goals without necessarily attaining them. Recast in this way the Bacon-Descartes ideal will look like this:

(A) to progress towards certainty with theories that are ever more probable;

(B1) to progress towards ultimate explanations with theories that are ever deeper;

(B2) to progress towards one unified science with theories that are increasingly unified;

(B3) to progress towards potentially complete predictive power with theories that are ever more predictively powerful;

(B4) to progress towards absolute exactitude at all levels with theories that are increasingly exact.

These five demands are amenable to this kind of progressivisation. But the final demand in the Bacon-Descartes ideal, for deductive rigour, is not amenable to progressivisation; one could hardly say of two "theorems" that, while neither is actually deducible from a certain set of axioms, one is more nearly deducible than the other. (I insisted earlier that probability logic does not provide a nondeductive surrogate for deduction; rather, it enlarges very significantly the range of *analytic* formulas that a deductivist can employ.) This demand is already incorporated in our proposition (III), which will henceforth be taken to be part of the Bacon-Descartes ideal.

The question to which I now turn is whether we ought to take the further step of recasting the Bacon-Descartes ideal as a type (iii) aim that calls upon science to progress in certain directions only, discarding certainty, ultimacy, etc. even as eventual goals. I will be taking a hard look at demand (A) later, so let us begin with (B1), which holds out, as an eventual goal for science, the arrival at ultimate explanations. Leibniz might have objected that this violates our impartiality requirement since it takes sides (and moreover the *wrong* side) in an important cosmological issue. This issue was disputed in the seventeenth century; it was afterwards encapsulated in Kant's Second Antimony; and it is still a live issue today. The issue is whether or not there is, in the constitution of the physical world, a bottom layer consisting of irreducible entities (simples, elements, atoms, or whatever) such that: (i) science could, in principle, explain the nature of all things at all layers above this layer in terms of the intrinsic properties of, and mutual relations between, such ultimate entities; (ii) science could not explain their nature in terms of the properties of further entities because there are no further entities. If there is such a layer, then there is a possibility of ultimate explanations; otherwise not. One physicist for whom this is still a perplexing issue is J. A. Wheeler: 'One therefore suspects it is wrong to think that as one penetrates deeper and deeper into the structure of physics he will find it terminating at some nth level. One fears it is also wrong to think of the structure going on and on, layer after layer, *ad infinitum*' (1977, pp. 4–5).

Leibniz held that there is no last layer in nature. Against the atomist thesis that a corpuscle is 'a body of one entire piece, without subdivision,' he declared: 'The least corpuscle is actually subdivided *in infinitum* and contains a world of other creatures' (1715–16, p. 1124). Again: 'matter is not composed of atoms but is actually subdivided into infinity, so that there is in any particle of matter whatever a world of creatures infinite in number' (1710, p. 974). Place a (seemingly) homogeneous and inert drop of blood under a microscope: it will then appear as a pool teeming with life; from this, select one (seemingly) homogeneous bit and place it under a microscope as powerful again: according to Leibniz, *it* will now appear as a pool teeming with life . . . If we could penetrate ever deeper into the microstructure of things, we would always discover worlds within worlds, each as richly complex as the last.

So if (B1) is to be impartial between this metaphysical thesis and the thesis of classical atomism that there are ultimate bits, we should discard the ideal of ultimate explanations and retain that of *ever deeper explanations*. This amendment corresponds to the shift from what Popper called essentialism to his own modified essentialism. He wrote:

If it is the aim of science to explain, then it will also be its aim to explain what so far has been accepted as an *explicans*; for example, a law of nature. Thus the task of science constantly renews itself. We may go on for ever, proceeding to explanations of a higher and higher level of universality—unless, indeed, we were to arrive at an *ultimate explanation*; that is to say, at an explanation which is neither capable of any further explanation, nor in need of it.

But are there ultimate explanations? (1957, p. 194)

Popper rejected the idea of an 'explanation which is not in need of a further explanation' (p. 195), adding:

although I do not think that we can ever describe, by our universal laws, an *ultimate* essence of the world, I do not doubt that we may seek to probe deeper and deeper into the structure of our world or, as we might say, into properties of the world that are more and more essential, or of greater and greater depth. (P. 196)

Consider next (B2), which holds out, as an eventual goal, the subsumption of *all* phenomena under *one* unified science. Like (B1), this takes one side in an important and controversial metaphysical issue, namely that between reductionism, or the thesis that life and consciousness are (or will be) reducible to physics, and antireductionism. As Bas van Fraassen put it:

There have been arguments both for and against the idea that science . . . aims at unity, that the development of a final single, coherent, consistent account incorporating all the special sciences is a regulative ideal governing the scientific enterprise. To some this seems a truism, to others it is mainly propoganda for the imperialism of physics. (*1980*, p. 83)

Descartes would have protested that his name should not be associated with an aim for science that includes this reductionist doctrine. And Locke would have insisted that this reductionist programme could never be carried out:

If we will suppose nothing first or eternal, matter can never begin to be: if we will suppose bare matter without motion eternal, motion can never begin to be: *if we suppose only matter and motion first or eternal, thought can never begin to be.* For it is impossible to conceive that matter, either with or without motion, could have, originally in and from itself, sense, perception, and knowledge. (*1690*, IV, x, 11, my italics)

Yet Descartes certainly believed in the unification of terrestrial and celestial mechanics; and I imagine that both men would have welcomed the incorporation of, say, the theory of heat into a general theory of matter, or the unification of chemistry and physics. In short, whereas the ideal of one completely unified natural science is highly controversial, the idea that as science progresses, its theories should become progressively more unified is not at all controversial.

Consider now (B3), which holds out, as an eventual goal, potentially complete predictive power. Laplace and other classical determinists would have endorsed this. But the thesis it presupposes, namely a complete physical determinism at all levels, has become highly controversial, at least since 1926 with the development of quantum physics. For instance, it comes under sustained attack in Popper (*1982b*). On the other hand, the idea that as the sciences progress they should become more predictively powerful is uncontroversial.

Finally, consider (B4), which holds out, as an eventual goal, absolute exactitude at all levels. This presupposes that there is nowhere any fuzziness in nature. This may be a less fiercely disputed doctrine than the doctrines of physical reductionism and determinism, but it too has been challenged. Popper characterised it as the thesis that all "clouds", i.e. parts of nature that seem rather fuzzy and indeterminate, are really perfectly exact "clockwork" mechanisms (*1966*), and he countered it with the thesis that seemingly clocklike parts of nature are really more or less "cloudy" at the microlevel. He drew attention to C. S. Peirce's suggestion, which was highly unorthodox at the time, that there is 'an element of indeterminacy, spontaneity, or absolute chance in nature ... a certain swerving of the facts from any definite formula' (1891, 6.12). Here too the impartiality requirement obliges us to drop the eventual goal and to demand only progress in a certain direction.

Recast as a type (iii) aim, the Bacon-Descartes ideal calls upon science to progress with explanatory theories that are ever

- (A)   more probable,
- (B1)   deeper,
- (B2)   more unified,
- (B3)   more predictively powerful,
- (B4)   more exact.

I will sometimes refer to component (A) as the 'security-pole' of this ideal and to the (B)-components as its 'depth-pole'. We must now examine this aim for coherence.

4.22   *The Bipolarity of the Bacon-Descartes Ideal*

I need here to anticipate some results of the next chapter. A criterion will there be given for two statements to be *counterparts* of one another. If T and T' are counterparts, then every consequence of T has a counterpart among the consequences of T' and vice versa. This notion will be used to explicate the idea of a theory T_j having more testable content than a theory T_i, to be denoted by $\mathrm{CT}(T_j) > \mathrm{CT}(T_i)$. Suppose that T_j both goes beyond and revises T_i. Then I will say that the above relation holds if there is a T_i' such that:

(i) T_i and T_i' are counterparts;

(ii) every potential falsifer, in Popper's sense, of T_i' is a potential falsifer of T_j but not vice versa.

If (i) holds, we have $\mathrm{CT}(T_i) \approx \mathrm{CT}(T_i')$ since every testable consequence of T_i has a counterpart among the testable consequences of T_i' and vice versa; and if (ii) holds, we have $\mathrm{CT}(T_j) > \mathrm{CT}(T_i')$. And from (i) and (ii) we can proceed to (iii): $\mathrm{CT}(T_j) > \mathrm{CT}(T_i)$.

It will turn out that for T_j to be genuinely deeper than T_i it is a necessary condition that $\mathrm{CT}(T_j) > \mathrm{CT}(T_i)$; and this is again a necessary condition for T_j to be more unified than T_i; and if T_j is more predictively powerful than T_i we obviously have $\mathrm{CT}(T_j) > \mathrm{CT}(T_i)$. And it will turn out, once more, that $\mathrm{CT}(T_j) > \mathrm{CT}(T_i)$ is a necessary condition for T_j to be more exact than T_i.

Thus all the (B)-components of the Bacon-Descartes ideal are calling, directly or by implication, for increases in testable content as science progresses. We might say that what the depth-pole of the Bacon-Descartes ideal requires, for T_j to be unequivocally superior to T_i, is that T_j should be *deeper-and-wider* than T_i; in virtue of its greater depth and unity, T_j should explain and predict more than T_i at the phenomenal level. This part of the Bacon-Descartes ideal is in tune with what Hempel called the 'desire to gain ever wider knowledge and ever deeper understanding of the world' (*1966*, p. 2) and with what Mario Bunge called the need for 'both surface and depth growth' (1968, p. 120).

But component (A) of the Bacon-Descartes ideal is in conflict with this. A true and important tenet of Popper's philosophy of science since (*1934*) is that the demands for increasing probability and increasing content pull in opposite directions. If T_j strictly entails T_i' we are, of course, bound to have $p(T_j, E) \leq p(T_i', E)$ for any E. Component (A) may endorse a sideways move from T_i to T_i' but it cannot endorse the upward move from T_i' to T_j. On the contrary, it will tend to endorse a downwards slide from T_i' to progressively weaker and more probable and safer consequences of T_j.

So something has to give. The question now is whether component

(A), taken by itself, satisfies our adequacy requirements so that we might, at a pinch, reject all four (B)-components and retain just it. If this choice were in fact open to us, then there would not be, among the possible aims permitted by our adequacy requirements, one that dominates the rest; there would be no such thing as the optimum aim for science. Well, we may assume that component (A) would, by itself, satisfy requirements 1, 3, 4 and 5: that it is coherent, can serve as a guide, is impartial, and does involve the idea of truth. But component (A) cannot satisfy requirement 2, the feasibility requirement, if the case for probability-scepticism, developed in Part One, stands. As was remarked in §3.5, that case has a fail-safe character: it will stand unless *all* the well-known arguments for it are invalid. It would, I think, be in order for me to conclude forthwith, on the basis of the sceptical conclusions of Part One, that (A) is an infeasible aim for science.

However, that would be somewhat cavalier and I shall proceed differently. In the late nineteenth and early twentieth centuries there have been several great students of science, from Ernst Mach to Moritz Schlick, who have perceived the radical antagonism between something like the security-pole and the depth-pole of our Bacon-Descartes ideal and who resolutely adhered to the former. This obliged them to try to reinterpret the laws and theories of science in such a way that something like (A) could be regarded as a feasible aim for science thus understood. In §4.3 we will consider a series of increasingly deflationary reinterpretations to which these men were, I will suggest, driven by their adherence to the security-pole and consequent revulsion from the depth-pole. And in §4.4 it will be asked whether their deflationary programme could eventually be carried far enough for (A) to become a feasible aim for whatever emasculated form of science still remained. The answer will be negative. Attempts to propitiate demand (A) have led to the sacrifice of one valuable feature of science after another; explanatory depth, universality, physical realism. And all these sacrifices were in vain. Or so I will argue.

So what has to give, if the Bacon-Descartes ideal is to be rendered both coherent and feasible, is demand (A). But simply to reject (A) altogether would leave a gap that needs filling. For one thing, requirement 5 would not be satisfied since none of the (B)-components involves the idea of truth; and for another, each of the (B)-components makes only what may be called a prior demand on any new theory T_j that is a candidate for superseding an existing theory T_i; each of them calls for some structural superiority that T_j should have; none of them includes what may be called the posterior demand that T_j should go on to perform well under the test of experience. I will try to fill this gap in §4.5. I will there seek to identify the strongest feasible core, which we may denote by (A*), that can be retrieved from the infeasible (A). After the (B)-components have been clarified in chapter 5, we will find that conjoining (A*) with them gives us an aim for science, which I will call (B*), that satisfies all our adequacy requirements and dominates any other aim that would satisfy them.

4.3 The Revulsion Against Depth

If one first considers what was happening *in* science from say, the time of Planck's discovery of the quantisation of energy around 1900 to the discovery of nuclear fission in the late 1930s, and then turns to what was being said *about* science by philosophers in the same period, one is struck by a remarkable contrast. On the one hand it is as if, in that heroic age of scientific revolution and discovery, such men as Planck, Einstein, Rutherford, Bohr, Heisenberg, and Schroedinger were drawn by the depth-pole of the Bacon-Descartes ideal to try to penetrate ever deeper layers of reality. And it is as if at the same time such men as Mach, Duhem, Russell, Bridgman, and leading members of the Vienna Circle were *repelled* by its depth-pole. An antipathy to, even a horror of, depth was a dominant theme in much of the philosophy of science of that period. A main target of the crusade against metaphysics in that period was scientific realism. The message of positivism was that science has to remain on the surface, at the phenomenal level; it must not try to penetrate beneath the phenomena. I do not think that Mario Bunge exaggerated when he spoke of 'the antidepth war waged by radical empiricists' (1968, p. 136).

Ernst Mach was perhaps the most uncompromising and influential proponent of this antidepth philosophy. Science cannot penetrate beneath the phenomena, according to him, for the good reason that there is nothing beneath them for it to penetrate to; sensations—colours, tones, smells, spaces—are the only realities. 'Nature is composed of sensations as its elements' (*1883*, p. 579). Those things we think of as existing 'behind the appearances exist *only* in our understanding' (*1872*, p. 49). The laws of nature laid down by science do not represent structural features of an external reality: they are only restrictions which we prescribe to our expectations (*1906*, p. 351).

When Duhem's (*1906*) appeared, Mach welcomed it with delight. It has, he said,

> given me great pleasure. I had not hoped to find so soon such far-reaching agreement in any physicist. Duhem rejects any metaphysical interpretation of questions in physics. He sees it as the aim of that science to determine the facts in a conceptually economic way. . . . I value the agreement between us all the more because Duhem arrived at the same results quite independently. (*1906*, p. xxxv)

Actually, Duhem's position was less radical than Mach's. Duhem did not deny the existence of 'realities hidden under the phenomena we are studying'; but he did insist that physical theory 'teaches us absolutely nothing' about them (p. 21). A physical theory consists of experimental laws grouped together under a mathematical superstructure. Whereas Mach took an antirealist or instrumentalist or economical view of all universal statements in science, Duhem's antirealism was concentrated upon those in the mathematical superstructure: it is, he insisted, a fatal mistake to regard these, as some great men of science have, as genuinely explanatory, as subsuming experimental laws under 'assumptions concerning the realities hiding underneath sensible appearances' (p. 38). After examining the view of Kepler and Galileo that the experimental laws of astronomy should be grounded in a deeper physical theory, Duhem concluded: 'Despite Kepler and Galileo, we believe today, with Osiander and Bellarmine, that the hypotheses of physics are mere mathematical contrivances devised for the purpose of saving the phenomena' (*1908*, p. 117).

Duhem's view of the contrasting nature of experimental laws and mathematical theories had been anticipated by Poincaré, who likened the former to the books in a library and the latter to the library's catalogue (*1902*, pp. 144–145). Russell (in *1914*) took a position very close to Mach's sensationalism.

Hostility to depth was very strong in the Vienna Circle. Their (1929) manifesto (signed 'For the Ernst Mach Society' by Hans Hahn, Otto Neurath, and Rudolph Carnap), after paying tribute to Mach, Poincaré, Duhem, Russell and others, declared: 'In science there are no "depths"; there is surface everywhere. . . . Everything is accessible to man. . . . ' (p. 306) It went on to say that the statements of science derive their meaning 'through reduction to the simplest statements about the empirically given'; all other statements are 'empty of meaning' (pp. 306–

307). Our proposition (II), the experientialist thesis, was stated in an uncompromising way:

> There is knowledge only from experience, which rests on what is immediately given. This sets the limits for the content of legitimate science ... the meaning of any concept, whatever branch of science it may belong to, must be statable by step-wise reduction to ... concepts of the lowest level which refer directly to the given. (P. 309)

Bridgman's operationalism analogously required every scientific concept to be reducible to operations that we can carry out.

One explanation for the popularity of Mach's philosophy of science around the beginning of this century is the following. During the previous fifty years, physics had undergone revolutionary changes which wiped out many of the familiar landmarks of the old mechanistic world view. A scientific realist would say that during this period, physics had radically revised its ontology. But many contemporary thinkers seem to have supposed that in discarding its *mechanist* ontology physics had discarded its *ontology*: matter had been dematerialised without a new physical stuff taking its place. Thus the very progress of physics itself seemed to them to call for the renunciation of mechanism and materialism in favour of the de-ontologised view of science presented by Mach. This explanation was put forward by Abel Rey: 'Down to the middle of the nineteenth century, traditional physics had assumed that it was sufficient merely to extend physics in order to arrive at a metaphysics of matter. This physics ascribed to its theories an ontological value. And its theories were all mechanistic' (*1907*, p. 16). However, as a result of the increasingly severe criticism of mechanism during the second half of the nineteenth century 'a philosophical conception of physics was founded which became almost traditional in philosophy at the end of the nineteenth century. Science was nothing but symbolic formulas, a method of notation' (p. 17).

Of course, Machism did not go unchallenged. One vigorous assailant was Lenin (*1908*). According to Lenin, a genuine (i.e. dialectical) materialism is not tied to any specific theory of matter: it regards only one property as essential to matter, namely 'the property of *being an objective reality*, of existing outside our mind' (p. 262). And the recent developments in physics do not tell at all against materialism thus understood:

"Matter is disappearing" means that the limit within which we have hitherto known matter is vanishing and that *our knowledge is penetrating deeper*; properties of matter are disappearing which formerly seemed absolute, immutable, and primary (impenetrability, inertia, mass, etc.) and which are now revealed to be relative and characteristic only of certain states of matter. (Ibid., my italics)

There was also vigorous opposition to Machism within physics. Max Planck rejected Mach's idea that the whole business of science is to achieve economy of thought by encapsulating wide ranges of experience in compact formulas. Planck adhered to something like the depth-pole of the Bacon-Descartes ideal: 'If the physicist wishes to further his science, he must be a Realist, not an Economist; that is, in the flux of appearances he must above all search for and unveil that which persists, . . . and is independent of human senses' (quoted in Holton 1970, p. 194; see also Planck *1948*, p. 34). As to Einstein: Holton (1970) has traced his shift from an earlier allegiance to Machism to a scientific realism close to Planck's. In a letter to Schlick, written in 1930 and quoted by Holton (p. 188), Einstein calls himself a 'metaphysicist' and complains that Schlick is 'too positivistic': physics, which attempts to discover the lawful structure of the real world, must get its empirical equations right, but that is the *only* way in which it is chained to experience.

4.31 *Motives Behind the Antidepth War*

It would be interesting to investigate separately the motives of each of these members of the phenomenalist, or instrumentalist, or positivist, or antidepth movement. No doubt they varied widely. In Mach's case (as we noted near the end of §3.22), an explicit concern was to solve the mind-body problem by replacing a psychophysical dualism with his neutral monism of elements that combine both into bodies and into minds. In Duhem's case, though he mentions the dissolution of the freedom-determinism issue as a happy consequence of his instrumentalist interpretation of the principle of the conservation of energy, a more important motive seems to have been to eliminate all possibility of conflict between science and religion in a way that reserves deep-fishing rights to the latter. In Neurath's case, there was the populist idea that science should be available to the proletariat; this required its more arcane concepts to be tied down to common experiences (for references, see my 1974, pp. 348–349).

However, if we ask in a quite general way why those who put a very high value on science should nevertheless interpret it in a way that seems to deprive it of explanatory power and ontological interest, I think that the simplest and most convincing answer is provided by the antagonism between the two poles of the Bacon-Descartes ideal: they are repelled by the depth-pole just because they are attracted by the security-pole. And I think that this answer applies to the thinkers I have mentioned.

4.32 *Mach*

Had we asked Mach to make the thought experiment we made in the previous section and give us a naively utopian ideal for science, his answer would have been along the following lines:–Ideally, science would provide a complete, item-by-item inventory of all sensible facts and no more than this. Science as we have it contains generalisations, hypotheses, theories; but that is because it is far from perfected: 'the more [science] approaches perfection, the more it goes over into description of fact only' (*1906*, p. 181). At another place he wrote: 'If all the individual facts—all the individual phenomena, knowledge of which we desire—were immediately accessible to us, science would never have arisen' (*1872*, p. 54). Instead of saying 'science would never have arisen' he might equally have said: 'science would already be perfected'. He spoke of theories falling away 'like dry leaves' once their work is done; hypotheses have a self-destroying function, becoming increasingly otiose as, one after another, the singular facts, whether refuting or confirming, to whose discovery they lead are actually observed. Mach's ideal is a purely descriptive, error-free science.

4.33 *Duhem*

Duhem was a merciless critic of the idea that physics is inductive. 'The teaching of physics by the purely inductive method such as Newton defined it is a chimera' (*1906*, p. 203); if the student who is taught it in this way 'is endowed with a mind of high accuracy, he will repel with disgust these perpetual defiances of logic' (p. 205). However, Duhem's scepticism as to the possibility of inductive verification did not extend to experimental laws: in this connection he spoke of *observed laws* and of laws *established* by the experimenter (p. 220). He allowed that such a law may not be formulated quite accurately at first, but subsequent corrections of it will usually be very slight. The experimental laws of physics are *true*, or at least very nearly true. Now if we take not an instrumentalist but a realist view of a sequence of superseding theories, what will we find? Each theory will consist of two parts: a 'representative' part consisting of experimental laws, and an 'explanatory' part which proposes to take hold of the reality underlying the phenomena' (p. 32). And when one theory is superseded by another, there is near continuity at the representative level but a radical discontinuity at the explanatory level: 'the purely representative part enters nearly whole in the new theory ... whereas the explanatory part falls out' (p. 32); there is a 'constant breaking out of explanations which arise only to be quelled' (p. 33). In short:

> Everything good in the theory ... is found in the representative part. ... On the other hand, whatever is false in the theory ... is found above all in the explanatory part; *the physicist has brought error into it, led by his desire to take hold of realities.* (p. 32, my italics)

Although this does not amount exactly to demand (A), it again comes very close to it; if one wants physics to be as error-free as humanly possible, one must renounce the idea that it can penetrate to a reality underlying the phenomena. Duhem's ideas are still very much with us. Thus Van Fraassen holds that the aim of science should be limited to giving us theories that are empirically adequate, that is, 'save the phenomena' (*1980*, p. 12); and he advocates 'a resolute rejection of the demand for an explanation of the regularities in the observable course of nature, by means of truths concerning a reality beyond what is actual and observable, as a demand which plays no role in the scientific enterprise' (p. 203; but compare p. 204).

4.34 *Bridgman*

In the opening chapter of his (*1927*), Bridgman declared that Einstein's theories of relativity had been 'a great shock'. In a passage that has often been quoted, he went on: 'We should now make it our business to understand so thoroughly the character of our permanent mental relations to nature that another change in our attitude, such as that due to Einstein, shall be forever impossible' (p. 2). And he held that just this would be achieved by his operationalist view of scientific concepts:

> It is evident that if we adopt this point of view towards concepts ... we need run no danger of having to revise our attitude toward nature.

For if experience is always described in terms of experience, there must always be correspondence between experience and our description of it. (P. 6)

Again I detect the presence here of something like demand (A).

4.35 *Schlick's U-Turn*

The case of Schlick is particularly instructive. At the time of his (*1918*) he had been a full-blooded scientific realist. There he had boldly declared that the empirical sciences 'supply us with knowledge of the essence or nature of objects. In physics, for instance, Maxwell's equations disclose to us the "essence" of electricity, Einstein's equations the essence of gravitation' (p. 242). Science 'makes judgments about the interior of the sun, about electrons, about magnetic field strengths' (p. 203); Mach was wrong; atoms exist just as much as loaves of bread: 'There is not the slightest difference between the two cases' (p. 218). The aim of science is to discover the universal laws by which things-in-themselves are governed. We might describe this book as a magnificent advocacy of the rightful place of (B) in the aim of science. As we shall see, he also gave an important place to (A).

A few years later, however, his view of science underwent a drastic change. He did not merely retreat to the view that the universal statements of science are never about underlying realities but only about surface phenomena. He declared that they are not really *statements* at all: 'on a strict analysis natural laws by no means have the character of statements that are true or false; they represent "directives", rather, for the forming of such statements' (1931, p. 195). The class of genuine scientific statements was now cut right back, to verifiable singular statements about pointer readings, photographic plates and the like (p. 188).

What led to this drastic change? Schlick was by now under the spell of Wittgenstein, to whom, he said, he owed these ideas and terms (p. 188). And by 1931 the position he now adopted, (which goes back to Mill's doctrine that all inference is from particulars to particulars) was popular in empiricist circles. For instance, Ramsey had written: 'When we assert a causal law we are asserting not a fact, not an infinite conjunction, nor a connection of universals, but a variable hypothetical which is not strictly a proposition at all, but a formula from which we derive propositions' (*1931*, p. 251). And Waisman had said that laws of nature are 'the formal structures we insert between the statements which describe our actual observations' (1930–31, p. 20). Schlick referred approvingly to this paper (p. 208, n. 4).

The official reason given by Schlick for withdrawing the title of *statement* from causal laws was their unverifiability. In their case 'a final verification is impossible. From this we gather that a causal claim by no means has the logical character of an *assertion*, for a genuine assertion must ultimately allow of verification' (p. 187). But surely Schlick's conversion to a verificationist theory of meaning is not by itself sufficient to account for his adoption of a dicticidal policy towards propositions of a kind he had valued so very highly in (*1918*). (By a 'dicticidal' policy I mean a positivist attempt to kill off statements that one finds philosophically inconvenient.) After all, Carnap in (1928) had proposed a criterion for distinguishing between genuine and pseudo statements that did not relegate the universal hypotheses of science to the latter category: he laid down that a genuine statement 'is at least indirectly connected with experience in such a way that it can be indicated which possible experience would confirm or refute it; that is to say, it is itself supported by experiences, or it is testable, or it has at least factual content' (p. 328). Carnap added that 'we are using as liberal a criterion of meaningfulness as the most liberal minded physicist or historian would use within his own science'. Why did Schlick adopt a criterion so much less liberal than Carnap's?

I think that the answer is that he had now become very conscious of something of which he had hardly been aware in (*1918*): namely that, to express it in my terms, there is a profound mutual antagonism between (A) and (B), between the desire for something approaching certainty in science and the desire that science should penetrate Nature's depths. In (*1918*) he had endorsed (A) no less than (B): science sets out to obtain 'knowledge that conforms to its own requirements for rigor and certainty' (p. 19). He spoke of 'the desired goal of absolute certainty and precision in knowing', and of scientific knowledge being 'possible in a form practically free from doubt' (p. 27). There are places in the book where this verificationism is a little mitigated and turns into an optimistic probabilism: 'But in the final analysis [the judgments of the factual sciences] remain only hypotheses; their truth is not absolutely guaranteed. We must be content if [their] probability ... assumes an extremely high value' (p. 73). Elsewhere he had said that 'propositions obtained by induction can only claim to hold with *probability* (though with frequently an extremely high one)' (1921, p. 26, n. 1).

My suggestion is that Schlick shifted to his antirealist position when he became aware of the antagonism between (B) and (A). He came to realise that the difficulties highlighted by Hume mean that there is no hope that a universal theory that tries, in Planck's words, 'to unveil that which persists, ... and is independent of human senses', can be empirically established even as highly probable, let alone as certainly true. But suppose we declare that the statements of science consist only of verifiable singular statements and that so-called "theories" are only devices for deriving singular statements from singular statements; then, surely, Hume's problem disappears:

> The informed reader will note that through considerations such as these the so-called problem of 'induction' is also rendered vacuous. ... For the problem of induction consists, of course, in asking for the logical justification of general propositions about reality, which are always extrapolations from particular observations. We acknowledge, with Hume, that there is no logical justification of them; there cannot be any, because they are simply not genuine propositions. (1931, p. 197)

Science cannot be accused of advancing unjustified propositions if all its real propositions are tied down to individual observations. Demand (A) has triumphed over demand (B).

4.4 Can Scientific Theories Be Cut Down to a Confirmable Size?

We have before us some powerful scientific theory, say Maxwell's, replete with a theoretical ontology (in the case of Maxwell's theory, this includes the idea that absolute space is filled by a stationary ether in which electromagnetic and light waves undulate). From the viewpoint of an empiricist who subscribes to component (A) of the Bacon-Descartes ideal, this theory, call it T, as it now stands, is hopelessly top-heavy relative to the totality of evidence, call this E, favourable to it; for T transcends E in its ontology, its universality, and its precision. The question to which I now turn is: could a policy of deflating T, or of replacing it by a cut-down surrogate version, say T^+, if carried far enough, so reduce this top-heaviness that T^+ is well supported by E? If T is, as it were, a fat man who is too heavy to be supported by the relatively flimsy E, is there within T a thin man T^+ wildly signaling to be let out and light enough to be supported by E? To put it more prosaically: given that for T as it now stands we have $p(T, E) = P(T) = 0$; then what an adherent of component (A) hankers after is a replacement T^+ for T that delivers the empirical results of T and whose posterior probability can be high and even tend to 1 as more and more favourable evidence comes in. But we may content ourselves with a less demanding question: is there a T^+ such that (i) T^+ delivers the same predictions at T, and (ii) the posterior probability of T^+ can at least be *raised*, no matter how slightly, by favourable evidence, so that we can have $p(T^+, E) > p(T^+) > 0$? If the answer to this question is no, then (A) is not a feasible aim for science.

The negative results of chapter 2 strongly suggest that the answer will be no. However, I think that the question deserves an independent examination here. Although we have touched on some of the more extreme deflationary measures resorted to by, for instance, Mach and Schlick, we have not so far considered whether these would fulfil the aim of rendering the thus reinterpreted "theories" positively confirmable. Our conclusion in §2.5 means that there is no possibility of $p(T^+, E) > p(T^+)$ if T^+ contains universal generalisations; but as we saw, Mach, Ramsey, Waismann, Schlick and others reinterpreted universal statements as something like rules of inference, and we have not yet considered whether this device yields the needed result. Again, there is a view, associated with Russell, Carnap, and others, according to which we may take the predictions yielded by a theory not en bloc but one by one, attending at any given time just to what the theory predicts for the next instance; and we have still to consider whether this would yield the needed result. So in this section we will review a series of deflationary measures, beginning with a relatively modest one and proceeding to increasingly severe ones, in order to see whether this sort of empiricist policy can eventually satisfy, at least to a modest degree, the verificationist demand embodied in component (A).

Before plunging in, let us first highlight the salient features of a typically powerful scientific theory T, realistically interpreted, that have to be reckoned with by an empiricist pursuing such a policy.

(i) T will have what I call a theoretical ontology. How this should be identified, and how it should be related to the testable content of T, will be considered in §5.23. Here it is enough to say that having a theoretical ontology means that any set of axioms for T will involve essentially theoretical, or non-observational predicates as well as observational predicates.

145

(ii) T entails various level-3 statements, or universal and exact experimental laws.

(iii) T entails what I call singular predictive implications, or SPIs for short. A SPI is what a universal hypothesis says when relativised to a single individual. (Thus 'If Tom is a man then Tom is mortal' is a SPI of 'All men are mortal'.) The SPIs yielded by experimental laws will often involve a time interval. However, I will usually represent a SPI simply by $Fa \rightarrow Ga$ where F and G are observational predicates; such a SPI is testable and its negation, $\sim(Fa \rightarrow Ga)$ or $Fa \wedge \sim Ga$, is what Popper calls a potential falsifier. Since $a_1, a_2 \ldots$ can denote spatiotemporal regions, it is clear that T entails a potentially infinite number of SPIs.

Let us now review the ways in which empiricists have tried to deal with these features of scientific theories.

4.41 *Ramsey-Sentences and Deontologisation*

As his first step in reducing a fat T to a lean T^+, an empiricist will want to eliminate anything that T says about 'realities hidden under the phenomena', to use Duhem's phrase: given that T has a theoretical core characterising unobservable entities, the object of the exercise will be to cut this out but without thereby reducing the predictive content of T. We might call this *Duhem's step*. But how is it to be carried out? Duhem would say that we are to regard any theoretical predicates that occur in T as mere mathematical symbols. And an axiom all of whose predicates are theoretical should be regarded as neither true nor false. But if such an axiom is needed for the derivation of certain experimental laws within the theory, will not those laws cease to be derivable if that axiom is neither true nor false? To put it in Duhem's terminology: how can the 'representative part' of a theory be deductively organised by its 'explanatory part' if the former consists of propositions with truth values and the latter is merely some sort of mathematical apparatus that has no truth value?

It was only after Duhem that a neat method was found for eliminating the theoretical content of a theory without affecting its predictive content. This method was adumbrated by Frank Ramsey in a difficult paper (1931), published posthumously. It went largely unnoticed until Richard Braithwaite, who had edited Ramsey's papers, drew attention to it in (*1953*, pp. 79f.); after that Ramsey-sentences, as they came to be called, quickly became famous. There is an excellent discussion of them in Scheffler (*1963*, pp. 203f.).

Although it is a shade technical, the basic idea, here, is very simple. It requires a second order predicate calculus which, besides the usual variables x, y ... ranging over individuals, has predicate variables ϕ_1, ϕ_2 ... ranging over properties. Consider the statement 'John is a parent of Hugh but not of Millicent', which we may represent by $R(a, b) \wedge \sim R(a, c)$. Then in first-order calculus we may generalise this to $\exists x\, \exists y\, \exists z\, (R(x, y) \wedge \sim R(x, z))$; but in second-order calculus we may generalise it to $\exists\, \phi\, (\phi(a, b) \wedge \sim \phi\, (a, c))$. This says that there is a relational property that holds between John and Hugh but not between John and Millicent. We must also suppose that each of the predicates occurring in T has been classified either as theoretical or as observational. To form a Ramsey-sentence T_R of T we proceed as follows. We first conjoin all the axioms that constitute T into one composite axiom expressed by a single sentence. Then wherever a certain theoretical predicate, say θ_1, occurs in this sentence we replace it by a predicate variable, say ϕ_1; if another theoretical predicate, say θ_2, occurs in it, we likewise replace it wherever it occurs by another predicate variable, say ϕ_2, and so on until all theoretical predicates have been replaced. Finally, we bind these variables by placing the existential quantifiers $\exists\, \phi_1$, $\exists\, \phi_2$ ... in front of the whole expression. Suppose our "theory" were: 'If a piece of amber is rubbed, it becomes *electrically charged* and *polarised* and if there is a small bit of fluff near it this will be *polarised in the opposite sense* and will move towards it', where the italicised expressions are classified as theoretical predicates. We now introduce the following abbreviations:

$$
\begin{aligned}
F_1 x &= && x \text{ is amber} \\
F_2 x &= && x \text{ is rubbed} \\
F_3 y &= && y \text{ is a small bit of fluff} \\
F_4 (y, x) &= && y \text{ is near } x \\
F_5 (y, x) &= && y \text{ moves towards } x \\
\theta_1 x &= && x \text{ is electrically charged} \\
\theta_2 x &= && x \text{ is polarised} \\
\theta_3 (x, y) &= && x \text{ and } y \text{ are polarised in opposite senses}
\end{aligned}
$$

We now restate our "theory" in a single sentence, say the following:

$$
T: \forall x\, \forall y\, ((F_1 x \wedge F_2 x \wedge F_3 y \wedge F_4\, (y, x)) \rightarrow
$$
$$
(\theta_1 x \wedge \theta_2 x \wedge \theta_3\, (x, y) \wedge F_5\, (y, x))).
$$

And we now form the Ramsey-sentence T_R for T by replacing the theoretical predicates in T by predicate variables and then binding these by existential quantifiers, thus:

$$T_R: \exists \phi_1 \, \exists \phi_2 \, \exists \phi_3 \, [\forall x \forall y \, ((F_1 x \wedge F_2 x \wedge F_3 y \wedge F_4 \, (y, \, x)) \rightarrow$$
$$(\phi_1 x \wedge \phi_2 x \wedge \phi_3 \, (x, \, y) \wedge F_5 \, (y, \, x)))].$$

Translated back into ordinary English this says: 'If a piece of amber is rubbed, and there is a small bit of fluff near it, then it and the fluff will acquire certain properties and the fluff will move towards the amber.'

When it became better understood, Ramsey's idea was eagerly seized on by empiricist philosophers of science. For instance Carnap, who devoted a chapter of his *(1966)* to it, called it 'an important insight', adding: 'we can now avoid all the troublesome metaphysical questions that plague the original formulation of theories' (p. 252). However, Ramseyfication leaves our theory as unverifiable as ever. It lowers it from level-4 to level-3, but that is of no help in securing it a positive posterior probability. It still contains experimental laws that are universal and exact.

4.42 *From Laws to Rules of Inference*

Mill wrote: 'All inference is from particulars to particulars: General propositions are merely registers of such inferences already made, and short formulae for making more: The major premise of a syllogism, consequently, is a formula of this description: and the conclusion is not an inference drawn *from* the formula, but an inference drawn *according to* the formula' (*1843,* II, iii, 4). And as we have seen, in the heyday of verificationism it was widely held that only verifiable singular statements are genuine *statements*, natural laws being rules or directives for proceeding from one set of singular statements to another. For instance, whereas Duhem regarded the Snell-Descartes law as 'the fundamental law of refraction,' which had been obtained by 'experiment, induction, and generalization' (*1906*, p. 34), Mach declared: 'In nature there is no *law* of refraction, only different cases of refraction. The law of refraction is a concise compendious rule. . . .' (*1883*, p. 583). More generally, 'the "laws of nature" are restrictions that under the guidance of our experience we prescribe to our expectations' (*1906*, p. 351). So what we had taken to be a theory T containing exact universal law-statements that entail SPIs we are now to regard as a system T^+ of rules of inference that license those same SPIs. This is, apparently, a drastic reduction of content. But apropos the aim of getting an inductively confirmable T^+, it gets us nowhere. *All* the SPIs derivable within T are retrievable within T^+. But their number is potentially infinite. So the ratio of the verified to the unverified SPIs of T^+ is still infinitesimal. Moreover, many of these SPIs are exact. The law of refraction is now being seen as a rule of inference; but given a precise angle of incidence and a precise index of refraction, it still allows us to proceed to a precise angle of refraction. And as we saw earlier, in §3.44, such precision poses a very awkward problem for probabilism.

4.43 *Nature as the Totality of Sensations*

There is a way, and one that was taken by Mach, of eliminating the problem of exactitude. Kant had said that Nature is an aggregate of appearances behind which are things-in-themselves. In a famous footnote (*1886*, p. 30), Mach recorded two decisive events in his intellectual development: at about fifteen he came across Kant's *Prolegomena* in his father's study; and some 'two or three years later the superfluity of the role played by "the thing in itself" abruptly dawned upon me.' The elements of which Nature is composed are sensations; and 'all bodies are but thought-symbols for complexes of elements (complexes of sensations)' (p. 29). It is clear that this view rules out any statement that postulates something to which no sensation could correspond. Consider the statement 'If the temperature of this metal rod is raised 5°C, its length will increase by 1/10,000 of an inch.' A phenomenalist will, presumably, declare with Berkeley: 'There is no such thing as the ten-thousandths part of an inch' (*1710*, p. 100). For him, presumably, the proper interpretation of this spuriously precise statement is: 'If the temperature of this metal rod is raised 5°C, its length will not alter.' I am not an expert on phenomenalist interpretations of scales for magnitudes such as temperature, length, angle of refraction, etc. but I assume that they take a *minimum sensibile* or "just noticeable difference" as its smallest unit; and predictions of very small changes will be accepted at their face value only if the change involves at least one such unit. So on this phenomenalist view, no SPI is too exact to be verifiable; but the T^+ that is to replace the original T still endorses a potentially infinite number of unverified SPIs.

4.44 *The "Next Instance" Thesis*

One last, and seemingly desperate, step is open to an empiricist who wants to achieve the goal of so cutting down T to T^+ that we get

$p\,(T^+,\,E) > p\,(T^+)$. This is to propose that the content of a scientific theory is only what it predicts for *the next instance* (or for the next time interval, or for the next technological application). At one point Russell veered towards this view as offering a solution to the problem of induction:

> The assumption that *all* laws of nature are permanent has, of course, less probability than the assumption that this or that particular law is permanent; and the assumption that a particular law is permanent for all time has less probability than the assumption that it will be valid up to such and such a date. Science, in any given case, will assume what the case requires, but no more. In constructing the *Nautical Almanac* for 1915 it will assume that the law of gravitation will remain true up to the end of that year; but it will make no assumption as to 1916 until it comes to the next volume of the almanac. (*1917*, pp. 196–197)

He added that 'it is better to argue immediately from the given particular instances *to the new instance*' (my italics) than to a universal conclusion.

Carnap switched to a similar position at the end of his (*1950*). In the body of that book, Carnap developed, in ways touched on earlier, a degree of confirmation function $c^*(h,e)$ that would measure the support provided by evidence e for a hypothesis h, where h may be a prediction or a law or any other statement. But it transpired on p. 571 that if the place of h is filled by a universal law l, and the domain of individuals is infinite, then the value of $c^*(l,e)$ is always what it would be if l were inconsistent with e or self-contradictory, namely 0. This led him to take a jaundiced view of scientific laws. Laws, he declared, are not needed for making predictions: 'Thus we see that X need not take the roundabout way through the law l at all, as is usually believed; he can instead go from his observational knowledge . . . directly to the singular prediction' (pp. 574–575). An engineer who relies on a law l in building a bridge is interested in l only so far as it applies to this 'one instance or a relatively small number of instances' (p. 572): he assumes that l will not let him down over this next bridge, or over the next few bridges, but not that it will never let anyone down anywhere.

To speak of scientific theories or laws making predictions concerning *the next instance* is ambiguous: 'the next' could have the meaning it has in (i) 'The next item on the agenda is the treasurer's report', or in (ii) 'The next number in the sequence 1, 2, 4, 8, 16 . . . is always twice the previous number'. This difference is similar to that between 'tomorrow' in (i) 'We're having dinner with the Joneses tomorrow' and in (ii) 'Never do today what you can put off till tomorrow'. In sense (i) 'the next . . .' and 'tomorrow' denote just one thing; in sense (ii) they turn into variables and spread over one damned thing after another. The "next instance" thesis, as I will call this latest deflationary measure, may owe its attraction to an unnoticed oscillation between these two very different meanings: it may suggest (i) that a scientific theory has very little unverified content because it ventures only *one small step* beyond the evidence, and (ii) that its predictive power is, nevertheless, undiminished since it *keeps on going* one small step beyond the evidence. Now the empiricist's task, as here understood, is to reduce a fat T to a thin T^+ in such a way that we can replace $p(T, E) = p(T) = 0$ by $p(T^+, E) > p(T^+) > 0$, for a suitably favourable E, *without* losing any of the predictive content of the original T. And this seems to mean that the "next instance" thesis cannot be taken in sense (i): in *that* sense it would mean that a theory is like a bee that makes one sting and then dies. But if we understand the assumption that Russell imputed to the compilers of the Nautical Almanac in sense (ii), then it expands into the assumption that if the law of gravity has held up to 1914 it will hold forever; for it *will* hold, on this assumption, during the next year, namely 1915; but if it holds up to 1915 then it *will* hold, on this assumption, during the next year, namely 1916; but if it holds up to 1916 . . . A fat T that has been reduced to a thin T^+ in this manner will still have a predictive content that infinitely transcends whatever supporting evidence for it there may be.

An adherent of this approach might deny that it involves an *oscillation* between sense (i) and sense (ii). He might say:–the idea is that although each of the predictions yielded by the original T will still be extractable from such a T^+, we can no longer extract them en bloc but only one at a time: sufficient unto the day is the prediction therefor. *At any given time* the content of T^+ is just what it predicts for the next time period in sense (i). At this time it is not yet predicting anything for the next but one time period. However, when the next time period materialises, then, and only then, will T^+ issue a prediction for the following time period.

I find it a rather absurd idea that T^+ will have a consequence for the day after tomorrow that we *will* be entitled to draw *tomorrow* but must not draw today, as though its predictive consequences had to pass through a turnstile one at a time. But a neurotic desire for completeness impels me to ask whether, if we grant the empiricist this contention, he can at last secure $p(T^+, E) > p(T^+)$.

I concede that he would if T were some banal generalisation such as 'All ravens are black' and T^+ were 'The next raven to be observed will be black'. We saw in §2.51 that Carnap, by taking structure-descriptions as equiprobable, was able to secure $p(e^{n+1}, E^n) > p(e^{n+1})$, where E^n says that n ravens have been observed, all of them black, and e^{n+1} says that the n + 1th raven to be observed will be black. But our T is supposed to be a powerful scientific theory; and it is a striking feature of a theory of this kind that it yields *novel predictions*. (I will leave this concept undefined at present. It will be given a certain precision in §8.23.) Let T be atomic physics as it stood circa 1944. Then a novel consequence of this T is the following: 'Whenever two lumps of a material called U^{235} are slammed together, the lumps being individually less than, but jointly greater than, a specified size (the "critical" size), there will follow an explosion "brighter than 1000 suns"' (to use the title of Jungk's *1956*). Call this consequence g. As it stands it is a universal statement. A proponent of the "next instance" thesis would proceed from it to g^+ by replacing 'whenever' by 'on the next occasion that'. Let g^+ be asserted some time before 16 July 1945 (the date of the first explosion of an atomic device, at Los Alamos).

This example was used by Hilary Putnam (1963, pp. 287–288) against Carnap's claim that, in going from past evidence to a prediction about the next instance, a person 'need not take the roundabout way' through laws and theories but 'can instead go from his observational knowledge ... directly to the singular prediction'. Let E be all the evidence known prior to 16 July 1945 that working physicists would have regarded as confirming (supporting, corroborating) the body of theory T of which g^+ is a dramatic consequence. Putnam claimed, obviously correctly, that scientists would never have obtained g^+ merely from E and without the help of laws and theories.

In his reply to Putnam, Carnap conceded that the degree of confirmation of this g^+ on this E will not have 'a considerable value' (1963, p. 988). But this does not answer the question that needed answering. The question before us is not whether E can raise the probability of some watered down surrogate of a powerful scientific theory to a considerable value but whether it can *raise it at all*. If the predictive content of T is being taken, at the present time, to be encapsulated in the prediction g^+, do we at least have $p(g^+, E) > p(g^+)$? It seems clear that the answer has to be no. It is as if E recorded numerous sightings of white swans, red robins, black ravens, etc. while g^+ predicts that when certain conditions are, for the first time, satisfied, an exotic bird, of a kind that has never been observed before, will make a first appearance.

I conclude that the "next instance" thesis, even on the extremist interpretation here given it (namely, that a theory at the present time is making no prediction about anything beyond just the very next instance), is defeated by those powerful scientific theories that make novel predictions in which the next instance will also be the *first* observed instance of a certain kind of phenomenon.

Component (A) of the Bacon-Descartes ideal is like a beacon that lures mariners to shipwreck. In their attempts to propitiate this innocent-looking but baleful demand, empiricist philosophers of science have renounced one valuable feature of science after another: its ability to explain phenomena by penetrating to realities behind them was renounced at an early stage. But that was not enough. Its universality, even at the level of surface phenomena, had to go. But that still was not enough. And all these sacrifices were in vain, for (A) can be propitiated only by giving up science as we know it altogether.

4.45 *Husserl*

One philosopher who has seen this very clearly is Edmund Husserl. He insisted that 'the Cartesian idea of a science (ultimately an all-embracing science) grounded on an absolute foundation, and absolutely justified, is none other than the idea that constantly furnishes guidance in all sciences' (*1931*, p. 11). But can the sciences as we know them live up to this guiding idea? Husserl recognised that they are quite incapable of doing so. Then which should be renounced, it or them? Apparently without qualm or hesitation, Husserl concluded that they must be renounced in favour of a quite new kind of science: 'A *science whose peculiar nature is unprecedented* comes into our field of vision: . . . a science that forms *the extremest contrast to sciences in the hitherto accepted sense*, positive . . . sciences' (pp. 29–30). Will this remarkable new science investigate the external world? Will it pay any heed to the depth-pole of the Bacon-Descartes ideal? Husserl's answer to both questions is no. Unlike positive science, which 'is a science lost in the world' (p. 157), the new science will involve nothing but a '*radical* self-investigation' (p. 153) on the part of those who engage in it. It will remain resolutely at level-0, where it will be in no danger of getting lost. It will provide 'a system of phenomenological disciplines ... grounded ... on an *all-embracing self-investigation*' (p. 156). Component (A) of the Bacon-Descartes ideal for natural science has here led explicitly to the renunciation of natural science.

4.5 From Certain to Possible Truth

For all the harsh things I have said about it, component (A) of the Bacon-Descartes ideal surely expresses a deep longing of the human mind (analogous perhaps to a longing not to grow old and die). Moreover, it is the one component of that ideal that speaks of *truth*. So if our proposed aim for science is to comply with requirement 5 (see §4.1 above) and involve the idea of truth, and if it is to be the maximumly ambitious aim that is still feasible, it must be expanded to include whatever is the feasible core of (A). But is there anything that is at once compatible with probability-scepticism *and nontrivial* to be retrieved from (A)? After declaring: 'The search for truth should be the goal of our activities' (*1905*, p. 11), Poincaré asked: 'But if truth be the sole aim worth pursuing, may we hope to attain it? It may well be doubted. ... Does this mean that our most legitimate, most imperative aspiration is at the same time the most vain?' (p. 12). It might seem that a proponent of probability-scepticism can only answer yes. I will suggest a different answer.

We might restate the original and strongest version of (A) thus:

(A)   Science aspires after proven truth. The system of scientific propositions adopted by a person X at any one time should be certainly true for him, in the sense that he knows it to be verified by evidence available to him.

In its "progressivised" form, (A) relegated the goal of certainty to a far-off and perhaps unattainable ideal; we may still aspire after proven truth but must be content to *approach* this ultimate goal via hypotheses that have a progressively higher absolute probability on the total evidence. We found (in §2.4) that this demand is infeasible; and we further found (in §2.5 and again in §4.4) that it remains infeasible when reduced to the weaker demand that accepted scientific hypotheses should have had their probability raised, no matter how slightly, by the available evidence. How should (A) be weakened further, to be rendered feasible? Having retreated from 'certainly true for X' to a weak version of 'probably true for X', it would seem that the next retreat must be to 'possibly true for X. How should this be understood? We might try the following formulation: a system of scientific hypotheses S is possibly true for X at t if X knows that S is: (i) internally consistent and (ii) consistent with the total evidence E available to X at t. But to demand that the scientific hypotheses accepted by X should be possibly true in this sense would again be an infeasible demand, for the following reasons. As to (i): first, it is doubtful whether X could actually provide a complete formulation of all the scientific hypotheses currently accepted by him; second, even if he could, it is well-nigh certain that he could not *know* that his formulation is complete; third, even if he could know this, it is well-nigh certain that the system would be too large and heterogeneous for it to be possible for him to carry out any sort of consistency proof for it. As to (ii): first, we saw in §2.42 that it is impossible that X could ever know of some body of observational propositions E that E constitutes his total evidence; second, even if he could, it is well-nigh certain that he could not establish that there are no inconsistencies between S and E; for S will have a potential infinity of logical consequences of which he could formulate only a finite number; and it is always possible that lurking among the unformulated ones there are some that conflict with E.

Then should we be content with the following formulation: a system of scientific hypotheses S is possibly true for X at t if X is not aware of any inconsistencies either within S or between S and evidence available to him at t? No; for whereas the previous formulation was too demanding, this one is too lax: X might be a lazy fellow who shuns the labour of searching for inconsistencies, or he might be someone who refrains from looking down a telescope if there is a danger that what he would see might conflict with a cherished hypothesis of his. If we are to retrieve the most that is feasible from (A), we need to strengthen the above formulation by adding that X should have done his best to search out any internal inconsistencies within S and any conflicts between S and evidence available to him. (And evidence may be *available* to him, for instance if he is in a position to look down a telescope trained on the moon, which is not yet in his possession. The significance of this point will be brought out in §8.2.) I suggest that the strongest feasible core that can be retrieved from (A) is the following:

(A*)   Science aspires after truth. The system of scientific hypotheses adopted by a person X at any one time should be possibly true for him, in the sense that, despite his best endeavours, he has not found any inconsistencies in it or between it and the evidence available to him.

(How the term 'evidence' should here be understood will be considered in chapter 7.) We might call (A*) a Popperian version of (A).

4.51 *'Accept' and 'Work On'*

When I speak of a scientist X *adopting* (or accepting or preferring) one of a set of competing theories, I do *not* mean a decision to *work on* this theory. I mean that he regards it (i) as *possibly true* in the sense indicated, and (ii) as *the best* of the theories in its field, as they now stand. It is entirely possible that he accepts T_j as the best extant theory, but decides to work on the inferior T_i (which he rejects as it now stands) because he hopes to be able to develop it into a new and better theory T_i' which will be superior to T_j. (By way of illustration, T_j might be a sophisticated version of the Ptolemaic system, and one which was in very good accord with astronomical observations, and T_i an early and simplistic version of the Copernican theory, which put the sun at the dead centre of the solar system and had the planets orbiting with uniform circular motion around it, and which conflicted with astronomical observations.) I personally hold that while methodologists or philosophers of science may have something worthwhile to say about the comparative appraisal of the products of scientific research, it is not their business to advise scientists how to go about their research. In particular, it is not for them to tell scientists what they should or should not work on. But all I want to insist on here is that the question 'Which of these theories now before us is the best?' is entirely separate from the question 'Which of these theories is the most promising one for me to work on?' I will now argue that confounding these two questions can have unhappy consequences.

They are, unfortunately, confounded in Lakatos's methodology of scientific research programmes. Our (A*) includes the requirement that there should be no known inconsistencies in the system of hypotheses accepted by X at a given time. This is in line with the following passage in Popper's (*1934*): 'The requirement of consistency plays a special role among the various requirements which a theoretical system . . . must satisfy. It can be regarded as the first of the requirements to be satisfied by *every* theoretical system' (pp. 91-192). This passage, it seems to me, should be entirely uncontroversial. Yet Lakatos *attacked* it—because he confounded 'accept' and 'work on': 'Moreover, for Popper, an *in-*

consistent system does not forbid any observable state of affairs and *working on it* [my italics] must be invariably regarded as irrational' (1974, p. 147). According to Lakatos, then, Popper held a view that implies, among other things, that once Russell had discovered contradictions in the foundations of mathematics, he should have turned his back on mathematics, since working on it must be irrational. Popper of course held the opposite view. In (1940) he wrote: 'Having thus correctly observed that contradictions . . . are extremely fertile, and indeed the moving forces of any progress of thought, dialecticians [wrongly] conclude . . . that there is no need to avoid these fertile contradictions' (p. 316). And he went on to say that they are fertile 'only so long as we are determined not to put up with contradictions': if our theory turns out to contain contradictions, we must *work on it* to eliminate them.

Since it may be entirely rational to work on a theory that contains contradictions or has sustained refutations, to equate X's accepting (adopting, preferring) a theory as the best in its field as things now stand with X's selecting a theory to work on is all too likely to lead to the conclusion that X may accept (adopt, prefer) a theory that is known to be inconsistent or empirically false. Lakatos declared that consistency should not be construed as 'a precondition of acceptance' (1968, p. 176) and that '*one may accept$_1$ and accept$_2$ theories even if they are known to be false*' (p. 178; to 'accept$_1$ a theory' means to consider it a promising candidate for some stronger kind of acceptance, and to 'accept$_2$ a theory' means to accept it into the body of science).

What has been said about 'accept' and 'work on' applies mutatis mutandis to 'reject' and 'cease working on'. These two quite different things were likewise confounded by Lakatos: 'in the methodology of research programmes, the pragmatic meaning of "rejection" [of a programme] becomes crystal clear: it means *the decision to cease working on it*' (1968, p. 70n).

One obvious objection to this is that someone may cease working on a theory that he accepts just because he sees no way of further improving it. But the point I want to make now is that if, as I hold, it is impossible for a methodology to give good advice to scientists about what they should and should not work on, then a methodology that equates 'accept' and 'reject' with 'work on' and 'cease working on', if it comes up with any rules of acceptance and rejection at all, will inevitably come up with ones that are either bad or empty. This is borne out by the methodology of scientific research programmes.

157

As its name suggests, this takes the units of methodological appraisal to be not theories but research programmes (henceforth RPs). Now RPs progress, stagnate, or degenerate; and in the case of rival RPs, one may *overtake* the other. So it might seem that one rule should be this: if RP_1 is now degenerating and, moreover, has been overtaken by RP_2, which is progressing, then RP_1 should be rejected in favour of RP_2. And there are places where Lakatos wrote in this vein. For instance:

> I give criteria of progress and stagnation within a programme and also rules for the 'elimination' of whole research programmes. . . . If a research programme progressively explains more than a rival, it 'supersedes' it, and the rival can be eliminated (or, if you wish, 'shelved'). (1971, p. 112)

Again:

> The idea of competing scientific research programmes leads us to the problem: *how are research programmes eliminated?* . . . *Can there be any objective reason to reject a programme* . . . ? Our answer, in outline, is that such an objective reason is provided by a rival research programme which explains the previous success of its rival and supersedes it by a further display of *heuristic power.* (1970, p. 69)

Later, he crisply summarised this idea thus: 'But when should . . . a whole research programme be rejected? I claim, only if there is a better one to replace it' (1974, p. 150).

All that seems clear enough. But it was vitiated by his surprising insistence that one normally *cannot tell*, at the time, which of two RPs is doing better: 'It is very difficult to decide . . . when a research programme has degenerated hopelessly or when one of two rival programmes has achieved a decisive advantage over the other' (1971, p. 113). This suggests that Lakatos's rejection rule should read: '*if* you could tell, which you normally cannot, that RP_2 is doing better than RP_1, then you should reject RP_1'. But it turns out that this formulation is still too strong. For he insisted that an RP that has degenerated and been overtaken by a rival may always 'stage a comeback' (1971, p. 113)—provided, of course, that people are allowed to go on working on it: 'There is never anything inevitable about the triumph of a programme. Also, there is never anything inevitable about its defeat. Thus pigheadedness . . . has more "rational" scope' (*ibid.*). The meaning of this last sentence was indicated later: 'One may rationally stick to a degenerating programme until it is overtaken by a rival *and even after*'

(p. 117). To accommodate this further qualification we may reformulate his "rejection rule" thus: 'If you could tell, which you normally cannot, that RP_2 is doing better than RP_1, then you may reject RP_1 or, if you prefer, continue to accept RP_1.' The promised rules of elimination have fizzled out.

So it is hardly surprising that Feyerabend hailed this methodology as in accord with his own supreme maxim, *Anything goes*, or as an 'anarchism in disguise' (*1975*, p. 181). He warmly endorsed what Lakatos had said about the acceptance of inconsistent or falsified theories: 'I also agree with two suggestions which form an essential part of Lakatos's theory of science.... Neither blatant internal inconsistencies, ... nor massive conflict with experimental results should prevent us from retaining and elaborating a point of view that pleases us for some reason or other' (p. 183).

Feyerabend gave a strange new twist to Lakatos's misequation of 'reject' with 'cease working on':

> The methodology of research programmes thus differs radically from inductivism [and] falsificationism.... Inductivism demands that theories that lack empirical support *be removed*. Falsificationism demands that theories that lack excess empirical content over their predecessors *be removed*. Everyone demands that inconsistent theories ... *be removed*. The methodology of research programmes neither *does* contain such demands nor *can* it contain them. (P. 186)

According to Feyerabend, then, practically everyone except Lakatos and himself holds a view that implies that, once Russell had discovered those contradictions in it, mathematics should have been *removed*. We need not pursue this pis aller further.

4.52 *Scepticism and the Semantic Idea of Truth*

Let us now return to (A*). This leaves scientists free to work on whatever they like, but debars them from adopting theories that they recognise to be false. There is no doubt that (A*) can be combined without incoherence with (B): I will call their combination (B*). This says that if a theory T_j is deeper and more unified, predictively powerful and exact than any rival theory, then T_j should be accepted as the best theory in its field *provided* that no positive reason has been found for supposing T_j to be false. If (A*) is indeed the maximum feasible core that can be retrieved from (A), then we have good reason to suppose that (B*) is the optimum aim for science. But doubts may still be raised about the feasibility of (A*). It retains from the original (A) the idea that science should aspire after truth; and we remember Poincaré's question whether this 'most legitimate, most imperative aspiration' is not entirely vain. Probability-scepticism implies that there can be no working criteria for the application of the term 'true' to hypotheses in the ocean of uncertainty; and objectors may say that this reduces the idea of aspiring after truth to an empty piety. They might go further and declare that it nullifies the idea of truth itself. I will take this last objection first.

Probability-scepticism presupposes that the hypotheses to which it applies possess truth values even though we cannot determine what these are. As we have seen, a standard argument for probability-scepticism points out that, in cases where some evidence e is regarded as confirming a hypothesis h, there will typically be a set of mutually exclusive hypotheses, h, h', h'' ..., all standing in a similar relation to e; if the set is exhaustive, then one of these is *true* and all the others are false; and if the set is very large, then the probability that h is the true one is vanishingly small. Popper (1962, p. 373) coined the term *criterion philosophy* for a philosophy that asserts or, more often, takes for granted, that a key term that is not chaperoned by criteria for its application is empty. He pointed out that, since there can be no criterion of truth, a criterion philosophy must lead to disappointment and scepticism; and, I would add, to a scepticism far more despairing than probability-scepticism which, as I said, presupposes the existence of true propositions in denying that we can determine which these are.

4.53 *Sense with an Uncertain Reference*

The idea asserted or assumed by a criterion philosophy is widely accepted by nonphilosophers as well as philosophers. Indeed, it seems highly plausible. Consider the following two denoting phrases: 'The richest man in Greece' and 'The prettiest girl in England'. An accountant would point out that it is not so easy to put a figure on a man's fortune. What if he owns paintings that have never been valued, or shares in a company whose shares are not quoted? All the same, we feel, there are criteria that make it possible in principle to identify the richest man in Greece. But 'the prettiest girl in England' is quite another matter. Even rough, working criteria are lacking here. This phrase may perhaps arouse some psychic response in those who encounter it, and this may lull them into supposing that it has some meaning; but since it is in principle impossible that we could determine whom it denotes, it is an *empty* phrase.

Very well; but now consider 'The longest lived dinosaur'. On the assumption that there were dinosaurs, that this species is now extinct, and that the span of life (say, from the first crack of the egg to the last heart beat) of no two dinosaurs was exactly the same, this phrase denotes a single object, but we have no way of telling which; and the sentence 'The longest lived dinosaur was a female' has a truth value, though we will never know which. Or consider a more intriguing example. Call such pairs of prime numbers as 3 and 5, 17 and 19, or 29 and 31, 'twinned primes'. This expression has a perfectly clear meaning. Now form the denoting phrase 'The largest twinned primes'. Assume, rightly or wrongly, that it will never be known whether the series of twinned primes eventually dries up or carries on to infinity. On that assumption, there *may* be a pair of numbers denoted by this phrase, but we have no way of telling *which* pair, if any, they are. Yet the phrase has a perfectly clear sense. One recall's Frege's statement: 'In grasping a sense, one is not assured of a reference' (1892, p. 58).

Bearing in mind these two examples of terms that have a clear sense but an essentially uncertain reference, let us revert to the term 'true'. Let $C_1, \ldots C_K$ be the set of the complete sentences (this idea was explained in §2.21) of a suitable language L. In the light of Tarski's theory of truth, we can give a perfectly clear meaning to the metalanguage statement that one of the sentences in this exhaustive and mutually exclusive set is *true* and all the others are false. And we *know* this semantic fact quite independently of any optimistic or pessimistic epistemological view we may hold as to the possibility of identifying the true one. Probability-scepticism is a pessimistic epistemological position; but it has not the slightest tendency to undermine the semantic concept of truth.

The idea that there can be no general criterion for truth was anticipated by Kant in a striking passage. Kant said that when people ask the old question, 'What is truth?', they are not asking what the word means: 'That it is the agreement of knowledge with its object, is assumed as granted; the question asked is as to what is the general and sure criterion of the truth of any and every knowledge' (*1781/87*, A 58 = B 82). But *this* question, Kant went on, is absurd: if one man asks it and another tries to answer it they will present 'the ludicrous spectacle of one man milking a he-goat and the other holding a sieve underneath.' The question is absurd, according to Kant, because a 'sufficient and at the same time general criterion of truth cannot possibly be given . . . no general criterion can be demanded. Such a criterion would by its very nature be self-contradictory' (A 59 = B 83).

4.54 *The Nonempty Demand for Possible Truth*

I turn now to the objection that, even if probability-scepticism does not nullify the concept of truth, it does nullify the idea that science should aspire after truth; to urge that it should, while proclaiming probability-scepticism, it may be said, is like encouraging people to go gold prospecting but adding that they will never be able to tell whether they have found gold, or even whether what they have found has some probability of being gold. One part of the answer to this objection has already been given: if your 'most imperative aspiration' is to seek truth, then you will not tolerate inconsistencies within whatever system of hypotheses you accept. (Kant insisted that logic provides a negative criterion of truth, A 59 = B 84.) Descartes may have conformed with (B) but he certainly offended against (B*) when he: (i) explained reflection and refraction of light in terms of light particles that behave like tiny tennis balls, sometimes bouncing off a resistant surface with undiminished velocity and sometimes passing through a yielding surface with a reduced velocity; and (ii) explained the absence of any delay in the appearance to a terrestrial observer of an eclipse of the moon in terms of the instantaneous propagation of light (see Sabra, *1967*, pp. 57f).

My next point carries this negative requirement further. I will propose in chapter 7 that the empirical basis for science consists of level-1 statements, about physical things or events, which cannot be verified or justified by, but can be tested against, perceptual experience. Each of us possesses, at level-0, some transient little bits of certainty. Nothing can be solidly built on these but they can be used as negative controls on what level-1 statements we accept. Instead of speaking of an ocean of uncertainty around an island of certainty it might be better to speak of an ocean of uncertainty in which little rocks of certainty are constantly popping up and subsiding. We might say, with apologies to Michael Oakeshott (*1962*, p. 127), that in science men sail a *rocky* ocean without harbour or anchorage. And (A*) behoves us to try, in navigating this ocean, not to run into any of these rocks. Or to put it more prosaically: probability-scepticism obliges us to relinquish proposition (II), the experientialist thesis that *knowledge* of the external world must be derived from perceptual experience; but (A*) calls on us to replace it by proposition (II*) which was introduced in §3.5 and says that perceptual experience should exercise a negative control.

Thus (A*) imposes a twofold requirement on the whole system of scientific statements, from level-1 up to a level-4, adopted by a person at a given time: he should have tried his best to eliminate both any internal inconsistencies within it and any inconsistencies between the level-1 statements in it and those level-0 reports of his that he *knows*, at the time, to be true.

To anyone who accepts (A) as the main aim of science, probability-scepticism will of course seem a thoroughly negative and destructive doctrine. But someone who accepts the weaker (A*), together with the semantic conception of truth and the strong (B)-components of our much modified version of the Bacon-Descartes ideal, may hold that probability-scepticism can be positively fruitful. It has sometimes happened in the history of science that a theory has seemed so massively, so overwhelmingly, well confirmed that its truth can no longer be doubted. The geocentric hypothesis was viewed in this way, for many centuries. And so, of course, was Newtonian theory, though its reign was shorter. Popper has said:

> Like almost all of his contemporaries who were knowledgeable in this field, Kant believed in the *truth* of Newton's celestial mechanics. The almost universal belief that Newton's theory *must* be true was not only understandable but seemed to be well-founded. Never had there been a better theory, nor one more severely tested. . . .
>
> Here was a universally valid system of the world that described the laws of cosmic motion in the simplest and clearest way possible— and with absolute accuracy. Its principles were as simple and precise as geometry itself. (*1963*, p. 185)

Probability-scepticism says nothing that would in the least depreciate Newton's achievement except this: the claim that this theory is certainly true, or very probably true, and even the claim that the mass of supporting evidence for it has at least raised somewhat the probability that it is true, have to be discarded; for whenever a highly exact theory T seems to be brilliantly confirmed by evidence E, there is a huge, indeed infinite, set of possible alternatives T', T'' . . . to T, each having a relation to E similar to that which T enjoys. Now provided we retain the semantic conception of truth which, as I said, is presupposed by probability-scepticism, the argument can continue as follows: one of the theories within the set T, T', T'' . . . is true and the rest are false. The chance that the one true theory is T itself is vanishingly small. Therefore there is a strong chance that there exists, as yet unformulated, a theory that is better than T, namely one that retains everything in T that is true and replaces everything in T that is false by true propositions of at least equal strength.

Probability-scepticisim plus the semantic conception of truth cannot by themselves carry the argument further. For probability-scepticism immediately adds that if, perchance, people should hit upon the alternative to T, if there is one, that is in this sense better than T, getting right everything that T gets right or gets wrong, they could never know that it is in this way better than T. But if, having thus opened up the possibility of a superior alternative to T, we now bring in our aim (B*), in which the demand for probable truth plays no role, the argument can resume. If the scientific community were to regard T as actually verified, they would be completely deterred from seeking a rival, incompatible theory superior to it. If they regard it as very probably true, they will be rather strongly deterred. But if, in line with probability-scepticism, they regard its chance of being true as vanishingly small, they will not be deterred from trying to construct a rival theory which, though its chance of being true will likewise be vanishingly small, is nevertheless superior to T relative to (B*), a theory that is deeper and more unified, predictively powerful, and exact than T and, moreover, one that is at least possibly true in the light of currently available evidence. And perhaps an Einstein among them will succeed. Probability-scepticism, by destroying any complacency concerning the veridical status of a reigning theory, helps to underline the fact that in no field is it ever right to assume that the task of science has been well-nigh completed. The possibility of a revolutionary advance always lies ahead.

The claim made above that (A*), although it no doubt retrieves less than we would wish from (A), nevertheless adds something important to (B), so that (B*) is significantly more demanding than (B) alone, will be further defended in §8.2. I will there argue that (B*) does, whereas (B) alone would not, call for tests to be carried out on promising theories. It is the addition of (A*) to (B) that leads to the conclusion that the best theory in its field is the one that is best corroborated rather than the one that is most corroborable. Another important advantage of (B*) over (B) is that this aim for science now satisfies our fifth adequacy requirement, concerning truth.

It also satisfies the fourth requirement, concerning impartiality. That it satisfies the third, namely that it can serve as a guide in the choice of theories, will be shown in §8.3. As to the second and first requirements, for feasibility and coherence: this aim *appears* to be feasible and coherent. Has not science been triumphantly fulfilling it over the last four hundred years? However, we need to elucidate it and sharpen its formulation before we can be satisfied on this score; for we have so far been operating with merely intuitive ideas concerning the increasing depth, unity, predictive power and exactitude of scientific theories; and we will encounter some serious difficulties when we try to give them precision. But if these difficulties can be surmounted; and if the adequacy conditions laid down in §4.1 were right; and if the Bacon-Descartes ideal as initially formulated was as comprehensive and ambitious as anyone could ask; and if its subsequent revision into an aim that can be progressively fulfilled was reasonable and indeed called for by the impartiality requirement; and if (A*) retrieves all that is feasible from (A); then (B*) surely is a nonarbitrary cognitive aim for theoretical science, for it is the optimum aim.

5

The Optimum Aim
Elucidated

5.1 A Comparative Measure of Testable Content

What I am claiming to be the optimum aim for science has so far been depicted only in an informal and intuitive way. The key idea, in (B1), of one explanation being *deeper* than its predecessor is in urgent need of clarification; and so is the idea, in (B2), of one theory being more unified than its predecessor. Moreover, it has turned out, in recent years, to be difficult to give precision to the idea in (B3) of one theory having more predictive power or a greater testable content than another in cases where one revises the other. Since this idea will turn out to be involved in all the other (B)-components of (B*) I will begin with it. But before turning to it I will make two preliminary remarks about the following elucidations of these four (B)-components.

5.11 *An Antitrivialisation Principle*

I speak of *elucidating* (B*), rather than of imposing an interpretation of my own on its various components, because I think that very little freedom of interpretation remains if we are guided in our investigation of them by the following meta-principle: any philosophical account of scientific progress must be inadequate if it has the (no doubt unintended) implication that it is always *trivially easy* to make theoretical progress in science. We might call this an antitrivialisation principle. If, for instance, a theory of confirmation has the implication that, given a body of evidence, we can immediately construct, in a perfectly ad hoc way, the hypothesis that, of all possible hypotheses, will be the one that is "best" confirmed by that evidence, then this theory of confirmation infringes this principle. (We encountered an example of this in §2.52.) Or suppose that an attempted explication of 'theoretical depth', or of predictive power, has the implication that, given a good scientific hypothesis, we can immediately construct a "better" one by merely plugging some new theoretical predicates into it, or by tacking some arbitrary sentences onto it, then such an explication would run foul of this principle. For we know very well that genuine scientific advances are not achieved in any such trivially easy way, and we can be confident that any such easily manufactured "better" or "best" theory would be brushed aside as worthless by the scientific community.

My other preliminary remark is that I will be content, in the case of all these four (B)-components, to provide a pairwise measure that will normally enable us to compare only competing, or otherwise related, theories in the same field. It might be nice if we had across-the-board measures that would enable us to compare, say, Dalton's atomic theory with 'All ravens are black' and with Freudian theory, for depth, predictive power, etc., but that would be surplus to our requirements. Our aim (B*) lays down desiderata for theoretical progress in science, that is, for an existing theory T_i to be superseded by a better theory T_j. To elucidate this aim it will be enough if we can make pairwise comparisons of theories in the same field. Pairwise comparisons are of course transitive.

5.12 *Popper's Criteria for Empirical Content*

The need for a superseding theory to exceed its predecessor in testable content has been stressed by Karl Popper. More particularly, he requires a deeper theory to do this. And he added an important rider: whenever

> a new theory of a higher level of universality successfully explains some older theory *by correcting it*, then this is a sure sign that the new theory has penetrated deeper than the older ones. The demand that a new theory should contain the old one approximately, for appropriate values of the new theory, may be called (following Bohr) the *'principle of correspondence'*. (1957, p. 202)

(Bohr's principle of correspondence has been widely discussed in recent years, for instance by: Koertge, 1969; Post, 1971; Krajewski, *1977*; Szumilewicz, 1977; Yoshida, *1977*.) What typically happens when a new theory T_j satisfies this principle with respect to an earlier theory T_i is this: in T_i there is a parameter, say Φ_1, that is taken to be a constant; but in T_j this parameter becomes a variable that is dependent on some further variable, say Φ_2, which plays no role in T_i; and the value given to Φ_1 by T_j approaches the fixed value it is given by T_i as Φ_2 tends to some limit, say zero, or infinity, or the speed of light. For instance, where Newton took the mass m of a body to be constant, Einstein

167

distinguished its variable mass m from its fixed rest mass m_o and related them by the equation $m = m_o \sqrt{1 - v^2/c^2}$, where v is the body's velocity and c is the speed of light. In this case $m = m_o$ in the limiting case where $v = 0$; and we could say here that this T_j yields this T_i "as a limiting case". But it often happens that a T_j does not allow Φ_2 actually to reach the limiting value at which Φ_1 would attain the constant value it is given by T_i. An obvious example is Galileo's acceleration constant g for freely falling terrestrial bodies which, in Newton's theory, turns into a variable dependent on the distance between the centres of gravity of the freely falling body and of the earth. It is a consequence of Newton's theory that no freely falling body, or body the only force acting on which is gravity, falls with an absolutely constant acceleration, though it may of course fall a relatively small distance with a virtually constant acceleration.

So an adequate criterion for a theory T_j to have more empirical content than a theory T_i, which I will abbreviate to $\mathrm{Ct}(T_j) > \mathrm{Ct}(T_i)$, must capture the case where T_i is in the relation of correspondence to T_j; more generally, it must capture cases where T_j both goes beyond *and revises* T_i. Bearing this in mind, let us consider the criteria for $\mathrm{Ct}(T_j) > \mathrm{Ct}(T_i)$ that Popper has proposed.

I will henceforth abbreviate Popper's term 'potential falsifier' to PF; and I will call the first criterion proposed by him *the PF criterion*. This says that $\mathrm{Ct}\,(T_j) > \mathrm{Ct}(T_i)$ if every PF of T_i is a PF of T_j but not vice versa. This will hold in cases where T_j has PFs and T_i does not and in cases where the PFs of T_i are a proper subclass of those of T_j. However, as Popper pointed out (*1934*, p. 126), there are many cases where 'comparison by means of the subclass relation' will not work; in particular it will not work in cases where T_j revises the empirical content of T_i. So Popper introduced an additional criterion that may be called *the dimension criterion*. Let $P_1 a$, $P_2 a$ and $P_3 a$ be what he called 'relatively atomic' basic statements (for instance, they might report measurements of the volume, weight, and temperature of a body). Then we may say that the degree of composition of a PF consisting of $P_1 a \wedge \sim P_2 a$ is two while that of $P_1 a \wedge P_2 a \wedge \sim P_3 a$ is three. Let us call a PF that just suffices to contradict a theory a minimal PF of that theory. Then we may assess two theories having a common field of application for empirical content by comparing the degree of composition of minimal PFs of one with that of the other, empirical content varying inversely with degree of composition. For instance, if T_i were $\forall x(P_1 x \rightarrow P_2 x)$ and T_j were $\forall x((P_1 x \wedge P_2 x) \rightarrow P_3 x)$, then we would have $\mathrm{Ct}(T_i) >$

$Ct(T_j)$ by the dimension criterion; but the PF criterion would give no answer here because the subclass relation does not hold between the PFs of the two statements.

Unfortunately the dimension criterion gives an unwanted answer when two theories are in the relation of correspondence. Suppose T_i places a constant value on parameter Φ_1 whereas T_j makes Φ_1 a function of Φ_2, where Φ_2 is a variable that does not occur in T_i. Other things being equal, a minimal PF of T_j will be more composite than one of T_i since it will have to put a particular value on Φ_2 whereas a PF of T_i can leave Φ_2 unspecified.

We might take, as crude surrogates for two theories being in this relation, the hypothesis h_1 that all mice have long tails and the hypothesis h_2 that all rodents have long tails if male and short tails if female. The former puts a constant value on mouse tail length while the latter makes rodent tail length a function of the animal's sex. The hypothesis h_2 both goes beyond and revises h_1, which it matches with its theorem h_1' that all mice have long tails if male and short tails if female. It would seem that h_1' has as much empirical content as h_1: concerning male mice they say the same and concerning female mice one says that they have long tails and the other that they have short tails. So it seems that the dimension criterion is unfair to h_1' when it says that it has less empirical content than h_1 just because its minimal PFs are more composite than those of h_1.

In (*1972*, pp. 52–53) Popper proposed a new criterion, which is more comprehensive than the previous two and which seems intuitively satisfactory. I will call it *the question-answering criterion*. It says that $Ct(T_j)$ > $Ct(T_i)$ if T_j can answer with at least equal precision every (empirical) question that T_i can answer but not vice versa. However, this criterion has run into serious difficulties. One is this. Let T be some highly falsifiable universal scientific theory; and assume that, though very powerful, T does not entail all the falsifiable universal statements that can be formulated in the language in which it is expressed. And now consider its relation to the widest constituent that can be formulated in that language. Call this constituent C_K. (We met C_K in §2.23 and again in §2.53.) It has the form $\exists x Q_1 x \wedge \exists x Q_2 x \ldots \wedge \exists x Q_K x$, where K is the number of Q-predicates in the language. Our T is highly falsifiable and C_K is unfalsifiable, so we have $Ct(T) >$ $CT(C_K)$ by the PF criterion. However, we cannot say that T has more empirical content than C_K by the question-answering criterion; for although there will be empirical questions that T can answer and C_K

cannot, the converse is also true. Consider the question, 'Is g true?' where g is one of the falsifiable universal statements *not* entailed by T but expressible in the language of T. Then T can only answer 'Don't know' whereas C_K will answer 'No'; for C_K entails the negation of every (nonanalytic) universal statement in the language.

And there is worse to come. David Miller (1975) showed that the question-answering criterion is unable to compare *any* two theories T_i and T_j in cases where neither of them is complete (in the sense indicated in §2.21) and one *revises* the other. For if neither is complete, there will be at least one statement, say p, that is logically independent of T_j; and if T_j revises T_i then there will be consequences c_i of T_i whose negation is entailed by T_j. We now disjoin such a c_i with p and ask: is $c_i \lor p$ true? If c_i and p are both empirical statements, this is an empirical question; and while T_i answers 'Yes', T_j answers 'Don't know'. You may feel that this is only an irritating little logical point; but logical points have to be accepted, however irritating.

The idea is now quite widely accepted, especially among Popper's critics, that no way has been found of comparing the predictive power of competing scientific theories: 'the attempt to specify content measures for scientific theories is extremely problematic if not literally impossible' (Laudan, *1977*, p. 77); 'we have to this day no precise, useful accounts of [explanatory and predictive power] that are generally applicable and that serve to make distinctions among competing theories' (Glymour, *1980*, p. 43); 'We do not have any viable way of comparing theories as to content' (Newton-Smith, *1981*, pp. 55–56).

I tried to find a way round Miller's point in my (1978c). I there developed what I will here call *the counterpart criterion*. My idea was to define a relation of counterparthood between two statements in a way that guarantees that if c_i and c_j are counterparts, then $\mathrm{Ct}(c_i) \approx \mathrm{Ct}(c_j)$, that is, c_i and c_j have equal amounts of empirical content. If c_i and c_j diverge in what they say, they are *incongruent* counterparts. Then we can say, in cases where T_j revises T_i, that $\mathrm{Ct}(T_j) > \mathrm{Ct}(T_i)$ if T_j has an incongruent counterpart T_i' such that T_j strictly entails T_i', for we will have $\mathrm{Ct}(T_j) > \mathrm{Ct}(T_i')$ together with $\mathrm{Ct}(T_i') \approx \mathrm{Ct}(T_i)$.

Certain weaknesses in this criterion were pointed out by Oddie (1979) with whom I had many discussions. (The occurrence of his name in parentheses in what follows will signal an acknowledgment to him.) In the meanwhile I have arrived at a definition of counterparthood that is both simpler and less restrictive than the previous one; and I have come to see that, with the help of this idea, Popper's original PF criterion can be generalised in what seems a straightforward and intuitively satisfactory way, and one which enables it to handle the case where T_i and T_j are in the relation of correspondence or more generally the case where T_j both goes beyond and revises T_i. I will call it *the comparative testability criterion* or, for short, the CT criterion.

I turn now to the updated definition of counterparthood. The discussion will become a little technical in places and a reader who prefers to take the whole thing on trust can go to §5.15. But I may perhaps remark that this is one of those cases mentioned in the Preface where the technicalities, if successful, have a certain importance.

5.13   *Incongruent Counterparts*

My definition of counterparthood will proceed in stages. I will first introduce the idea within the context of a very simple qualitative language L_1 whose predicates consist just of atomic P-predicates and of the Q-predicates generated by these. (If P_i and P_j are atomic predicates, then assigning P_i to an individual is always consistent with assigning either P_j or $\sim P_j$ to that individual.) Afterwards I will extend the definition to quantitative languages.

And there will be another major restriction on the kind of predicates that occur in L_1. The underlying intuition behind the idea of incongruent counterparts is this. Suppose that two indicative sentences, say s_i and s_j, are exactly similar except that, at one or more places where a predicate P_i occurs in one of them, $\sim P_i$ occurs in the other; then subject to certain provisos to be introduced shortly, s_i and s_j are incongruent counterparts, and so also (Oddie) are any two sentences logically equivalent to, respectively, s_i and s_j. I call a P-predicate that occurs unnegated at one place in either s_i or s_j and negated at the corresponding place in the other a *disputed* predicate, or a predicate whose *sign is reversed* when we go from s_i to s_j. Now if sign reversal of a predicate P_i is to leave the amount of empirical content equal, then P_i and $\sim P_i$ should be equally precise, or of the same width; $\sim P_i a$ should tell us as much about a as does $P_i a$. If P_i were 'is crimson', for instance, it would fail this requirement; one is told more about the colour of, say, a flower if told that it is crimson than if told that it is not crimson.

I will try to meet this difficulty with the help of what I will call *dichotomising* predicates. Some philosophers hold that, while it is false to say that the number 17 is even or that Cleopatra is male, it is not false but absurd to say that 17 is male or that Cleopatra is even: such utterances involve a category mistake. Fred Sommers (1963) relied on this idea to define the *span* of a predicate as the range of things to which it can be applied truly or falsely but not absurdly. Thus the span of 'male' includes female animals but not numbers and that of 'even' includes odd numbers but not animals. A pair of predicates, say P_i and P_i', can be said to be a dichotomising pair if they satisfy the following conditions: (i) anything that is P_i is necessarily $\sim P_i'$ and vice versa; (ii) P_i and P_i' have the same span; (iii) the properties denoted respectively by P_i and P_i' do not admit of degrees; (iv) a thing within their span must be either P_i or P_i': there is no middle possibility. Thus 'hot' and 'cold' are not a dichotomising pair: they satisfy conditions (i) and (ii) but not (iii) and (iv) (a tepid cup of tea being neither hot nor cold).

In some cases the extension of a dichotomising predicate consists of half the things in its span; for instance, it does so in the case of 'even number'. And in some other cases it consists of roughly half of them. Presumably about half of all animals are male and about half of all spun coins that landed with one face up have landed heads up. At one time I relied on this equality, or rough equality, of extensions to justify the claim that the statements $P_i a$ and $P_i' a$ are equally informative, where P_i and P_i' are such a dichotomising pair. But David Miller persuaded me (personal communication) that statistical considerations are out of place here. What is needed for $P_i a$ and $P_i' a$ to be equally informative is not that the number of things that are P_i should be (approximately) equal to the number that are P_i'; it is rather than P_i and P_i' should be *equally precise*. A switch can be in one of these two states: on or off. I have no idea whether, at any given time, more of all the world's switches are usually on or off. But to say that a switch is on is to characterise it as precisely as to say that it is off. Of course, relative to background information, one of two equally precise descriptions may be more, perhaps much more, informative than the other. If I already know of a randomly selected integer that it is a prime number, I learn vastly more if told that it is even than if told that it is odd; and if I already know that a person called Smith is bald I would be startled to be told that Smith is female. But we are here concerned with the informativeness of statements considered on their own rather than relative to other statements. And if two statements give conflicting but equally precise descriptions of the same thing(s) then, considered on their own, they are, I shall assume, equally informative. (I am not, of course, using 'informative' as a success-word: information may be misinformation.)

So I am going to suppose that the *P*-predicates of our preliminary language L_1 are not only monadic but dichotomising. I will also suppose that such a predicate is applied only to things within its span: 'even' to integers, 'male' to animals, and so on. This means that, for every monadic predicate P_i in L_1, P_i and $\sim P_i$ are a dichotomising pair. Such a language would be very artificial and restricted; but we will stay with it for only a short (or not long) time, proceeding afterwards to a language with scales divided into equal intervals.

Let s_i and s_j be two indicative sentences, in our restricted language L_1, in which only P-predicates (as opposed to Q-predicates) occur; and suppose, to begin with, that each predicate occurring in s_i occurs only once. Then if s_i and s_j are identical except that some or all of the predicates in s_i have their signs reversed in s_j, and if s_i and s_j are not logically equivalent, then they are incongruent counterparts. For instance, s_i and s_j might be respectively P_1a and $\sim P_1a$, or $\exists x\,(P_1x\,\wedge\,P_2x)$ and $\exists x(P_1x\,\wedge\,\sim P_2x)$, or $\forall x(P_1x \rightarrow P_2x)$ and $\forall x(P_1x \rightarrow \sim P_2x)$. But $P_1a \leftrightarrow P_2a$ and $\sim P_1a \leftrightarrow \sim P_2a$, which are logically equivalent (Miller), are congruent counterparts.

A problem arises if the same predicate occurs more than once in s_i. Consider $\forall x((P_1x\,\wedge\,P_2x)\,\vee\,(P_1x\,\wedge\,P_3x))$ and $\forall x((P_1x\,\wedge\,P_2x)\,\vee\,(\sim P_1x\,\wedge\,P_3x))$. Here, P_1 is a disputed predicate that occurs twice in each sentence, its sign being reversed at one place but not at the other. Should we allow that these two sentences are incongruent counterparts? No; for they are not equally informative. One way in which their unequal informativeness reveals itself is the following. The predicates P_1, P_2 and P_3 generate eight Q-predicates (going from $P_1\,\wedge\,P_2\,\wedge\,P_3$ through to $\sim P_1\,\wedge\,\sim P_2\,\wedge\,\sim P_3$); and the first of the above two sentences excludes five of these Q-predicates while the second excludes four. Their failure to be incongruent counterparts also reveals itself in the following way. I will prove later that if c_i and c_j are incongruent counterparts, then every consequence of c_i has a counterpart (not necessarily incongruent) among the consequences of c_j and vice versa. Now one consequence of the first sentence above is $\forall x\,P_1x$; but this has no counterpart among the consequences of the second sentence, which entails neither $\forall x\,P_1x$ nor $\forall x \sim P_1x$.

I dealt with this in (1978c) by requiring that the sign of a disputed predicate be *systematically* reversed in the other sentence. Thus an incongruent counterpart of $\forall x((P_1x\,\wedge\,P_2x)\,\vee\,(\sim P_1x\,\wedge\,P_3x))$ would be $\forall x((\sim P_1x\,\wedge\,P_2x)\,\vee\,(P_1x\,\wedge\,P_3x))$. These two sentences do exclude the same number of Q-predicates and the consequences of the first are in one:one correspondence with those of the second.

But I have since found that this requirement is too strong. I said earlier that an adequate measure for empirical content must capture the case where T_i and T_j are in the relation of correspondence and I proposed, as a crude precursor of that relation, the relation between 'All mice have long tails' and a hypothesis about rodents that strictly entails 'All male mice have long tails and all female mice have short tails'. Suppose that in our L_1 the following two sentences have been formulated:

s_1: $\forall x(P_1 x \rightarrow P_3 x)$

s_2: $\forall x(((P_1 x \wedge P_2 x) \rightarrow P_3 x) \wedge ((P_1 x \wedge \sim P_2 x) \rightarrow \sim P_3 x))$

Now s_1 is logically equivalent to the following sentence:

s_3: $\forall x(((P_1 x \wedge P_2 x) \rightarrow P_3 x) \wedge ((P_1 x \wedge \sim P_2 x) \rightarrow P_3 x))$

(One could imagine a proponent of s_1 being stung into reformulating it in this padded out way after being accused by a proponent of s_2 of overlooking the P_2-factor. It would be as if he replied: 'I don't *overlook* it; I assert that it makes no difference.') Now in s_2 and s_3 the disputed P_3 occurs twice in each sentence but its sign is reversed only on its second occurrence. Yet the two sentences seem equally informative: they say the same about things that are $P_1 \wedge P_2$ and concerning things that $P_1 \wedge \sim P_2$, one says that they are all P_3 and the other that they are all $\sim P_3$.

In our L_1 every Q-predicate is as precise as every other one, because for every P_i it picks up either P_i or $\sim P_i$, and each P_i is a dichotomising monadic predicate with the same width as $\sim P_i$. I now introduce the idea of two statements including/excluding equal numbers of Q-predicates. Any general statement in L_1 in which only P-predicates occur is equivalent to one in which only Q-predicates occur, each of these being governed by an existential quantifier or else by a negative existential quantifier. And I say that s_i and s_j include/exclude equal numbers of Q-predicates if (i) every Q-predicate governed by an existential quantifier in s_i is paired with one similarly governed in s_j and vice versa, and (ii) ditto for ones governed by a negative existential quantifier. Let me illustrate this with an example modeled on the pair of sentences 'Mice exist and all mice have long tails' and 'Mice exist and all male mice have long tails and all female mice have short tails.' We might express these by

s_1: $\exists x P_1 x \wedge \forall x(P_1 x \rightarrow P_3 x)$

s_2: $\exists x P_1 x \land \forall x (P_1 x \to (P_2 x \leftrightarrow P_3 x))$.

Of the eight Q-predicates generated by these three P-predicates let Q_1 $= P_1 \land P_2 \land \sim P_3$ and $Q_3 = P_1 \land \sim P_2 \land P_3$ and $Q_5 = P_1 \land P_2 \land \sim P_3$ and $Q_7 = P_1 \land \sim P_2 \land \sim P_3$. Then we can restate s_1 and s_2 thus:

s_1: $\exists x (Q_1 x \lor Q_3 x) \land \sim \exists x (Q_5 x \lor Q_7 x)$

s_2: $\exists x (Q_1 x \lor Q_7 x) \land \sim \exists x (Q_5 x \lor Q_3 x)$

And we now see that s_1 and s_2 include/exclude equal numbers of Q-predicates. And since all Q-predicates are equally precise, it follows that s_1 and s_2 convey equal amounts of information.

I now say that, in our L_1, c_i and c_j are incongruent counterparts if they satisfy these two conditions:

1. they can be expressed by two sentences, whose predicates are all P-predicates, the only difference between them being that at one or more places the sign of one or more P-predicates is reversed; and when any predicates occurring inessentially are eliminated, they still have predicates in common;

2. they can be expressed by two sentences, whose predicates are all Q-predicates, the only difference between them being that one or more of the Q-predicates included (excluded) by one is different from that included (excluded) by the other.

(The second part of condition 1 was added to meet a point raised by Frank Jackson.)

It may be asked why I do not rely on condition 2 alone and say that c_i and c_j are counterparts if and only if they include/exclude equal numbers of Q-predicates. Well, in our L_1, any such pair of statements would indeed be equally precise; but they might otherwise be quite unrelated to one another. For instance, $\forall x (P_1 x \to P_2 x)$ and $\forall x (P_3 x \to P_4 x)$ would be turned into incongruent counterparts, whereas I want to restrict this relation to cases where c_i and c_j are not merely equally precise but also variants of one another or where one may be said to revise the other.

If c_i and c_j are incongruent counterparts, their respective consequence classes will match each other or be in one:one correspondence, in the sense that each consequence c_i' of c_i has a counterpart (congruent or incongruent) c_j' among the consequences of c_j and vice versa. I will now provide an informal proof for this. It will be a shade technical and a reader who prefers to take it on trust should go to §5.14.

The first step is to express the incongruent counterparts c_i and c_j in a way that displays the fact that they include/exclude equal numbers of Q-predicates. For instance, if c_i and c_j were respectively $\exists x P_1 x \wedge \forall x(P_1 x \rightarrow P_3 x)$ and $\exists x P_1 x \wedge \forall x(P_1 x \rightarrow (P_2 x \leftrightarrow P_3 x))$ we may, as before, express them by

$$c_i: \quad \exists x(Q_1 x \vee Q_3 x) \wedge \sim\exists x(Q_5 x \vee Q_7 x)$$

$$c_j: \quad \exists x(Q_1 x \vee Q_7 x) \wedge \sim\exists x(Q_5 x \vee Q_3 x).$$

Call a predicate that is included by c_i, or by c_j, but not by both, a *disputed* predicate, and likewise for a predicate that is excluded by one but not the other. In our example Q_3 and Q_7 are the disputed predicates.

The next step is to pair each disputed predicate that c_i includes with a disputed predicate that c_j includes (if there are two or more it does not matter which is married to which so long as there is no bigamy), and ditto for disputed predicates excluded by c_i or c_j. And let f be a function that sends each disputed predicate to the one with which it is paired. In our example, f sends Q_3 to Q_7 and Q_7 to Q_3. It is clear that $c_i = f(c_j)$, $c_j = f(c_i)$, and $c_i = f(f(c_i))$.

In a language tailored to c_i and c_j, a depth-1 constituent (see §2.22 above) will say of each of the Q-predicates generated by the P-predicates occurring in the original c_i and c_j that it is, or else that it is not, instantiated. Each constituent will entail either c_i or else $\sim c_i$, and c_i is equivalent to the disjunction of those that entail it, say $C_{i_1}, \ldots C_{i_n}$. (For instance, if c_i is $\exists x (Q_1 x \vee Q_3 x) \wedge \sim\exists x(Q_5 x \vee Q_7 x)$ then C_{i_1} might be $\exists x Q_1 x \wedge \exists x Q_2 x \wedge \exists x Q_3 x \wedge \exists x Q_4 x \wedge \sim\exists x Q_5 x \wedge \sim\exists x Q_6 x \wedge \sim\exists x Q_7 x \wedge \sim\exists x Q_8 x$.) Now consider $f(C_{i_1}), \ldots f(C_{i_n})$: these are clearly constituents and, moreover, the constituents in the d.n.f. of c_j. (For instance, since in our example f sends Q_3 to Q_7 and Q_7 to Q_3, $f(C_{i_1})$ is $\exists x Q_1 x \wedge \exists x Q_2 x \wedge \sim \exists x Q_3 x \wedge \exists x Q_4 x \wedge \sim \exists x Q_5 x \wedge \sim\exists x Q_6 x \wedge \exists x Q_7 x \wedge \sim \exists x Q_8 x$ which entails our c_j, namely $\exists x(Q_1 x \vee Q_7 x) \wedge \sim\exists x(Q_5 x \vee Q_3 x)$.) Let $c_i{}'$ be a consequence of c_i other than c_i itself. Then in the d.n.f. of $c_i{}'$ there will be one or more constituents not in the d.n.f. of c_i. For ease of exposition suppose that there is just one, $C_{i_{n+1}}$. Then $f(C_{i_{n+1}})$ is a constituent not in the d.n.f. of c_j and $f(C_{i_1}) \vee \ldots \vee f(C_{i_n}) \vee f(C_{i_{n+1}})$ is a consequence $c_j{}'$ of c_j other than c_j itself. Moreover $c_i{}'$ and $c_j{}'$ are isomorphic and include/exclude equal numbers of Q-predicates; they are therefore (congruent or incongruent) counterparts.

The converse also holds: if c_i is strictly entailed by $c_i{}''$, then $c_i{}''$ will have an incongruent counterpart c_i'', equivalent to $f(c_i'')$, that strictly entails c_j. For c_i'' in d.n.f. will be equivalent to the disjunction of some but not all of the constituents in the d.n.f. of c_i, say, $C_{i_1}, \ldots C_{i_{n-1}}$; and $f(C_{i_1}) \vee \ldots \vee f(C_{i_{n-1}})$ is a statement c_j'' that strictly entails $f(C_{i_n})$ $\vee \ldots \vee f(C_{i_{n-1}}) \vee f(C_{i_n})$, i.e. c_j. Eventually there will be, in this language, the maximumly (or a maximally) strong, consistent statement that entails c_i, namely the narrowest (or if there are more than one, a narrowest) constituent in its d.n.f., say C_{i_1}; and $f(C_{i_1})$ will be a constituent of the same width (see §2.23 above) that entails c_j.

5.14 *The Problem of Language Dependence*

The question arises whether the relation of counterparthood is language dependent: can it happen that two sentences that are incongruent counterparts in one language, say L_F, cease to be so when translated into another language, say L_G? David Miller, in a personal communication which I reported in (1978c, pp. 371–72), argued that it could. His argument was addressed to my earlier definition of counterparthood but, if valid, it also hits the updated version presented above. This is clearly an important question; it is also a rather technical one. So I will summarise my answer to it, which will occupy the present subsection, in advance. Very briefly, Miller gets his result by operating with two intertranslatable languages, say L_F and L_G, where the primitive predicates $F_1, F_2 \ldots$ in L_F are equivalent to molecular predicates in L_G and the primitive predicates $G_1, G_2 \ldots$ in L_G are equivalent to molecular predicates in L_F. My answer is twofold. First, if $G_1, G_2 \ldots$ really are the *primitive* predicates of L_G, then their equivalence to molecular predicates in L_F could never be established. But if we waive this objection, a second objection can be raised. Assume that L_F is an observational language whose primitive predicates are well suited for reflecting observable differences between the things being investigated; then the language L_G will be ill suited for this, being at once undersensitive and oversensitive to such observable differences; in which case we have reason to prefer L_F to L_G as the more "natural" observation language.

I now present his case. Let L_F contain just three primitive predicates, F_1, F_2 and F_3. Then the following are incongruent counterparts by both the previous and the present definitions:

c_1: $\forall x((F_1 x \wedge F_2 x) \vee (\sim F_1 x \wedge F_3 x))$

c_2: $\forall x((\sim F_1 x \wedge F_2 x) \vee (F_1 x \wedge F_3 x))$.

We now introduce predicates G_1, G_2 and G_3 as follows: G_2 is equivalent to $(F_1 \land F_2) \lor (\sim F_1 \land F_3)$ so that c_i could be rewritten as $\forall x G_2 x$, and G_3 is equivalent to $(\sim F_1 \land F_2) \lor (F_1 \land F_3)$ so that c_j could be rewritten as $\forall x G_3 x$; and G_1 is equivalent to F_1. And then we introduce predicates H_1, H_2 and H_3 whose relation to G_1, G_2 and G_3 is exactly analogous to that of G_1, G_2 and G_3 to F_1, F_2 and F_3. Thus $H_1 = G_1$, $H_2 = (G_1 \land G_2) \lor (\sim G_1 \land G_3)$, $H_3 = (\sim G_1 \land G_2) \lor (G_1 \lor G_3)$. A truth-table check will reveal that H_1, H_2 and H_3 are respectively equivalent to F_1, F_2 and F_3. So long as G_1, G_2, G_3 and H_1 H_2 H_3 are defined predicates within L_F no difficulty is created for our counterpart criterion. But now consider a language L_G in which G_1, G_2 and G_3 are the *primitive* predicates, H_1, H_2 and H_3 being defined as before. In this language we could express c_i and c_j by

$$c_i: \quad \forall x((H_1 x \land H_2 x) \lor (\sim H_1 x \land H_3 x))$$

$$c_j: \quad \forall x((\sim H_1 x \land H_2 x) \lor (H_1 x \land H_3 x)).$$

But our first requirement for incongruent counterparts is that they be logically equivalent to sentences s_i and s_j in which only primitive predicates occur, the only difference between s_i and s_j being that at least one of these predicates has its sign reversed in at least one place. But if we reformulate c_i and c_j using just the primitive predicates of L_G, they become respectively $\forall x G_2 x$ and $\forall x G_3 x$, and their counterparthood disappears.

Miller's inversion manoeuvre, as we may call this turning of undefined and defined predicates in L_F into, respectively, defined and undefined predicates of L_G, does not have such startling consequences for content comparisons as it appears to have for verisimilitude comparisons. *There it can reverse the ordering* (Miller, 1974a) whereas here it only renders statements incomparable that are comparable in the other language. For it to do this, however, the supposedly undefined predicates of L_G must be genuinely primitive predicates of an autonomous language. For if L_G were merely an extension of L_F, its "primitive" predicates being merely abbreviations for defined predicates of L_F, then all statements in L_G could be unpacked into statements containing only primitive predicates of L_F. But if G_1, G_2 and G_3 really are *primitive* predicates in L_G, how is their equivalence to the defined predicates G_1, G_2 and G_3 in L_F established? That they look alike tells us nothing; the French word 'laid' *looks* like the English word 'laid'. Is there supposed to be an $L_F - L_G$ dictionary? If so, how was it compiled? If its compiler introduced G_1, G_2 and G_3 with definitions involving F_1, F_2 and F_3, then L_G would be a mere extension of L_F. Or are we to suppose that its compiler discovered an empirical correlation between uses of, for example, G_2 in L_G and uses of $(F_1 \wedge F_2) \vee (\sim F_1 \wedge F_3)$ in L_F? But empirical correlations do not establish semantic equivalences, as we know from Quine.

But let us waive this difficulty and raise a different question. Let there be two observational languages that are intertranslatable in the sort of way we have been considering. Might one of them be preferred as more natural, or realistic, or basic, than the other? Consider the following game. There is a pack of cards; each card has one of the following eight figures on it:

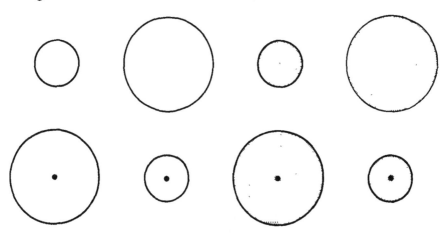

The pack is brand new and there are no discernible differences between any two cards displaying similar figures. There are two players, Mr A and Miss B, and a referee. Mr A has two buttons; one produces a bleep, the other a number of clicks according to how often it is pressed. Miss B has a bell. The referee repeatedly places a pair of cards face up on the table. Mr A is required to report how many observable differences, independent of their spatial position, there are between just the two cards on the table, pressing his bleep button if there are none and making one/two . . . clicks if there are one/two . . . such differences. Miss B can challenge A's report by ringing her bell; in which case A must either concede or else defend himself by actually pointing to the differences. If he defends himself, B must either concede or else point out an error in A's defence. If neither concedes, the referee's decision is final. Each time a report of A's goes unchallenged by B he wins \$1. If he is successfully challenged, he pays B \$20; if unsuccessfully challenged, he is paid $5 by B. If a decision goes to the referee, the player judged to be in the wrong is fined $10. After a hundred goes, the players exchange roles. We assume them to be sharp-sighted, alert, and keen to earn an honest dollar.

The referee keeps a written record of each game. For this purpose he is equipped with an observation language L_F with just three primitive predicates F_1, F_2 and F_3. An entry in his record might look like this:

	F_1	F_2	F_3
a_i	1	0	1
b_i	0	0	1

This says that the ith pair of cards consisted of one that is $F_1 \wedge \sim F_2 \wedge F_3$ and one that is $\sim F_1 \wedge \sim F_2 \wedge F_3$. I will assume that all his entries are correct. An entry containing 0 and 1 in the same column reports an observable difference between the cards. Thus the above report says, in effect, that there was just *one* observable difference between them (they differed with respect just to F_1).

Suppose that, after discarding cases where the bell rang, we find that there is a happy correlation between the number of clicks and the number of observable differences as reported in the language L_F. We might conclude that the primitive predicates in L_F are nicely suited to the observable characteristics of the figures on the cards.

Let us now turn to a language L_G having three primitive predicates G_1, G_2 and G_3. L_G is intertranslatable with L_F and the referee could equally have kept his record in this language. Suppose, however, that when the previous observation report is translated into L_G it becomes:

	G_1	G_2	G_3
a_i	1	0	1
b_i	0	1	0

This suggests that there were three observable differences between the cards on the ith occasion; yet there was only one click. Or suppose that on the jth occasion the report as expressed in L_F was this:

	F_1	F_2	F_3
a_j	1	0	1
b_j	0	1	0

This says, in effect, that there were three observable differences between the cards; and on this occasion there were indeed three clicks. But suppose that when translated into L_G this becomes:

	G_1	G_2	G_3
a_j	1	0	1
b_j	0	0	1

This suggests that there was only one observable difference between the cards on this occasion. If such things happened, I think that we could reasonably complain that there is not a happy fit between the primitive predicates of L_G and the observable characteristics of the figures on the cards; this language seems both undersensitive and oversensitive to observable differences, sometimes suggesting that there was one when three were observed by the players, and sometimes that there were three when one was observed.

Yet this is exactly what would happen if L_G were constructed from L_F in the same way that Miller's (1974a) *'hot-Minnesotan-Arizonan'* weather language was constructed from Tichy's *'hot-rainy-windy'* weather language; that is, we have $G_1 = F_1$, $G_2 = F_1 \leftrightarrow F_2$, and $G_3 = F_1 \leftrightarrow F_3$. (A language that exhibits a more extreme kind of undersensivity-*cum*-oversensitivity has been constructed by Oddie, 1981, pp. 252f.)

I learnt from David Armstrong (1975, pp. 151f.) the following important lesson: from the fact that a certain predicate can be truly assigned to certain things it does not follow that those things possess a property answering to that predicate. Let Mx mean: 'x is made of india rubber if and only if x speaks Chinese fluently'. Then M can be truly predicated of the Empire State Building, the Lisbon earthquake, and Socrates; but it does not follow that these three entities all have a property in common, namely a property denoted by M. The properties of things are not at the beck and call of our man-made predicates. Rather, we want our predicates to be well adjusted to those properties with which we are concerned, say in an experimental investigation. I do not assert that every Miller-type predicate is bound to be maladjusted. As he pointed out, 'homosexual' is a predicate of the $P_1 \leftrightarrow P_2$ type; and experimenters interested in certain kinds of symmetry may likewise need essentially molecular predicates. But I do say that, relative to a particular investigation, it may be possible, as it was in our card game, to establish that one observation language is more "natural" than another in the sense that its primitive predicates are better adjusted to the observable properties in question. And I propose that if one is more "natural" than the other, and if two statements can be exhibited as counterparts when expressed in one language but not when expressed in the other, then we should accept the verdict yielded by the more "natural" language.

5.15 *The Comparative Testability Criterion*

Equipped with the idea of incongruent counterparts I now turn to the generalisation of Popper's PF criterion. The idea of the empirical (or testable or predictive) content of a theory was introduced near the beginning of §4.4 but it will be as well to spell it out again here. Popper's idea was, of course, that the empirical content of a theory is correlated with its class of PFs, a PF being a conjunction of singular observation statements that entails the denial of the theory. Let T be a set of axioms. Then to each PF of T there corresponds a statement, which I earlier called a singular predictive implication or SPI for short, that is entailed by T and is simply the negation of the PF. If $F_1a \wedge \sim F_2a$ is a PF of T, then its negation, $\sim(F_1a \wedge \sim F_2a)$ or $F_1a \rightarrow F_2a$, is a SPI of T. Only observational predicates occur in PFs and SPIs.

From the set of all the consequences of T the subset of those that are SPIs will be denoted by $CT(T)$. Thus $CT(T)$ mirrors T's class of PFs. If T entails no SPIs we write $CT(T) = \emptyset$; otherwise $CT(T) > \emptyset$. Let T_i and T_j be two sets of axioms. Then $CT(T_i) = CT(T_j)$ means that the SPIs of T_i are identical with those of T_j, while $CT(T_i) \subset CT(T_j)$ means that the SPIs of T_i are a proper subset of those of T_j. It should be noted that we may have $CT(T_i) = CT(T_j)$ although T_i and T_j are not logically equivalent, and $CT(T_i) \subset CT(T_j)$ although T_j does not entail T_i (as would be the case if T_i were 'All mice have long tails and angels exist' and T_j were 'All rodents have long tails'). By $CT(T_i) \approx$

$CT(T_j)$ I will mean that the SPIs of T_i and of T_j are in one:one correspondence or can be paired off with one another. This will of course be so if either

    (1a) $CT(T_i) = CT(T_j) = \emptyset$
or (1b) $CT(T_i) = CT(T_j) > \emptyset.$

It will also be so if

    (1c) T_i and T_j are incongruent counterparts.

For in that case every consequence c_i of T_i will have a counterpart c_j among the consequences of T_j and vice versa; hence for any SPI that is a consequence of T_i there will correspond a SPI of T_j of equal precision, and vice versa.

I turn now to the generalisation of the PF criterion. By $CT(T_j) >$ $CT(T_i)$ I will mean that the SPIs of T_i are in one:one correspondence or can be paired off with those in a proper subset of the SPIs of T_j or, to put it another way, that every PF of T_i is paired with a PF of T_j but not vice versa. This will of course be so if

    (2a) $CT(T_i) = \emptyset$ and $CT(T_j) > \emptyset$
or (2b) $CT(T_i) \subset CT(T_j).$

It will also be so if

    (2c) T_i has an incongruent counterpart T_i' and $CT(T') \subset CT(T_j).$

For in that case for any SPI that is a consequence of T_i there will correspond an equally precise SPI that is a consequence of T_j but not vice versa.

We could collapse the three conditions for $CT(T_i) \approx CT(T_j)$ and the three conditions for $CT(T_j) > CT(T_i)$ each into a single condition as follows:

    (1) $CT(T_i) \approx CT(T_j)$ if and only if every PF of a (congruent or incongruent) counterpart of T_i is a PF of T_j and vice versa;
    (2) $CT(T_j) > CT(T_i)$ if and only if every PF of a (congruent or incongruent) counterpart of T_i is a PF of T_j but not vice versa.

But for some purposes it is desirable, especially in the case of (2), to keep these conditions distinct; in what follows I will sometimes justify a claim of the form $CT(T_j) > CT(T_i)$ by referring to a specific condition. Moreover, each of the three conditions in (2) has a distinctive character: (2a) expresses Popper's original principle of demarcation between scientific and nonscientific statements; (2b) expresses his PF criterion; and (2c) is my generalisation thereof.

This CT criterion has the important advantage over both the PF criterion and my previous counterpart criterion that it, unlike them, can compare such statements as:

s_1: All mice have long tails

s_2: All male rodents have long tails and all female rodents have short tails,

where s_2 goes beyond and revises s_1 but without systematic sign reversal. It also has some further advantages over the counterpart criterion. (A reader who prefers to skip these should go to §5.16.) I mentioned, in connection with the PF criterion, the case where the widest constituent C_K, which has no PFs, is up for comparison with a testable T. But the counterpart criterion is equally unable to effect a comparison here (Oddie). To establish $\mathrm{Ct}(T) > \mathrm{Ct}(C_K)$ it would need there to be an incongruent counterpart of C_K that is strictly entailed by T; but one of several unique features of C_K is that it is the one nonanalytic statement in L_1 that has *no* incongruent counterpart. As I mentioned in §2.23, two constituents are incongruent counterparts if they are of the same width and hence include/exclude equal numbers of Q-predicates. But this widest constituent is unique in including all Q-predicates and excluding none. However, by condition (2a) of the CT criterion we have $\mathrm{CT}(T) > \mathrm{CT}(C_K)$.

Another advantage is this. By the counterpart criterion we have $\mathrm{Ct}(T_j) > \mathrm{Ct}(T_i)$ if T_j strictly entails an incongruent counterpart T_i' of T_i. But that would obtain if T_j were merely T_i' plus 'Angels exist' or whatever; whereas we have $\mathrm{CT}(T_j) \approx \mathrm{CT}(T_i)$ by the CT criterion if T_j has no excess *testable* content over T_i'. Conversely, if T_j has excess testable content over T_i' but failed to entail T_i' merely because the latter contained 'Angels exist' or whatever, then T_i and T_j would be incomparable by the counterpart criterion, but not by the CT criterion; for the latter relies essentially on the proper subset relation between sets of SPIs (or sets of PFs) rather than on the relation of strict entailment.

The last advantage to be mentioned here of the CT over the counterpart criterion is that the former, but not the latter, can exhibit the fact, insisted upon in §2.53, that if C_1 is a narrower constituent than C_2, then $\mathrm{CT}(C_1) > \mathrm{CT}(C_2)$. Let C_1 be a narrowest constituent that says

C_1: $\exists x Q_1 x \wedge \forall x Q_1 x$

and let C_2 be a wider one that says

$$C_2: \quad \exists x Q_2 x \ \wedge \ \exists x Q_3 x \ \wedge \ \forall x (Q_2 x \ \vee \ Q_3 x).$$

To establish $\mathrm{Ct}(C_1) > \mathrm{Ct}(C_2)$ by the counterpart criterion, one would need to show that C_1 strictly entails an incongruent counterpart of C_2. But that is impossible. The incongruent counterparts of C_2 are just those constituents with the same width as it and no constituent can be entailed by another one. But to establish $\mathrm{CT}(C_1) > \mathrm{CT}(C_2)$ one needs to show that there is an incongruent counterpart C_2' of C_2 such that $\mathrm{CT}(C_2') \subset \mathrm{CT}(C_1)$. Now $\mathrm{CT}(C_1)$ denotes the SPIs of C_1; and the SPIs of $\exists x Q_1 x \ \wedge \ \forall x Q_1 x$ are identical with those of $\forall x Q_1 x$ alone: the addition of the existential clause does not generate any new SPIs. So $\mathrm{CT}(C_1) = \mathrm{CT}(\forall x Q_1 x)$. One incongruent counterpart C_2' of C_2 is $\exists x Q_1 x \ \wedge \ \exists x Q_3 x \ \wedge \ \forall x (Q_1 x \ \vee \ Q_3 x)$. And for the same reason $\mathrm{CT}(C_2') = \mathrm{CT}(\forall x (Q_1 x \ \vee \ Q_3 x))$. It is now obvious that $\mathrm{CT}(C_2') \subset \mathrm{CT}(C_1)$ and hence that $\mathrm{CT}(C_1) > \mathrm{CT}(C_2)$.

5.16 *More and Much More*

One last point concerning notation. One sometimes wants to say that a theory has *much more* testable content than some other statement, for instance, that Newtonian theory has much more testable content than 'The moon orbits the earth.' But our apparatus, as it stands, does not entitle us to say this. We have no metric for testable content. We can, in suitable circumstances, say that one set of SPIs is larger than another, but not by how much.

I think that a metric for content might be constructed from ideas already presented in this book along the following lines. A language tailored to the statements up for comparison would generate some definite number of constituents. The basic idea would be that a statement says more, other things being equal, the more constituents it excludes; however, a statement also says more in excluding a wider constituent than in excluding a narrower one. Now if initial probabilities were distributed over the constituents in the way indicated in § 2.23, these would vary with their width. Thus we might measure the relative content of a statement s_i simply by adding up the initial probabilities of the constituents excluded by it. And if we found that the ratio of this measure of s_i's content to that of some statement s_j exceeded some preassigned value, we might go on to say that s_i has *much more* content than s_j or that $\mathrm{Ct}(s_i) \gg \mathrm{Ct}(s_j)$.

I hasten to add that I shall not attempt such a construction. Our need for a metric is not strong enough to justify the labour. But I am emboldened by the above speculation to introduce the notation $CT(T)$ $\gg CT(T')$ as a *façon de parler* in cases where T has, intuitively, far more testable content than T'. No undue reliance will be placed on this undefined notion.

5.17 *Extension to Quantitative Languages*

We have now to extend the CT criterion to quantitative theories. At a pinch we could regard the dichotomising predicates of our qualitative language L_1 as each splitting a "scale" into two equal intervals. Instead of 'male' and 'female', for instance, we could introduce a variable ('sex') which takes the values 1 or 0 (no male chauvinism is intended in giving the value 1 to males and 0 to females). More generally, we might replace P_i and $\sim P_i$ by $\pi = 1$ and $\pi = 0$. Let us now turn to a quantitative language L_2 that has genuine scales, for instance for angle, length, weight, temperature, etc. Let each scale have a smallest interval, e.g. one second of arc. This should not be greater than a "just noticeable difference" or *minimum sensibile*. The scale is divided into indefinitely many equal intervals, each denoted by a number. Let variables α, β, γ . . . be associated respectively with each scale. Such a variable takes as values the discrete numbers denoting intervals on the scale. The equivalent in L_2 of sign reversal in L_1 is the replacement of one such value by a different but equally precise value. Rather as a Q-predicate in L_1 assigns the values 1 or 0 to every P-predicate in L_1, so a Q-predicate in L_2 assigns a precise value to each variable in L_2. Thus if L_2 has just the variables α, β and γ, then $\alpha = 1 \wedge \beta = 1000 \wedge \gamma = 10$ would be a Q-predicate in L_2.

We may think of an experimental law as consisting of a rubric ('For all light rays . . . ', 'For all gases . . . ') governing a functional equation of such variables. The equation can be put in the form $f(\alpha, \beta \ . \ . \ .) = 0$ in cases where, as in the ideal gas law, the value of any one variable can be got from a specification of the others. In cases where there is only one-way derivability, say of the value of α from values of β and γ, it can be put in the form $\alpha = f(\beta, \gamma)$. In cases where two or more variables in the equation relate to the same scale, as in the Snell-Descartes law of refraction, the same Greek letter with different subscripts can be used, say $f(\alpha_1, \alpha_2) = 0$ or $\alpha_1 = f(\alpha_2, \alpha_3)$. (Since the variables may take only discrete values, f must send fractions and decimals to the nearest whole number.)

We can now say that two experimental laws l_i and l_j are incongruent counterparts if they satisfy these two conditions:

1. they can be expressed by two equations involving the very same variables but governed by different functions; and when any variables occurring inessentially are eliminated, they still have variables in common;

2. for all finite ranges of the values of the variables occurring in the equations, the Q-predicates included/excluded by l_i are in one:one correspondence with those included/excluded by l_j.

The second requirement means that the equations are equally precise.

5.18   *The Correspondence Relation*

I said earlier that it is particularly important that our CT measure should capture the case where a theory T_j is in the relation of correspondence to T_i. In such a case T_j introduces variables that do not occur in T_i. A familiar example of this is the relation of the van der Waals gas law, $(P + a/V^2)(V - b) = rT$, to the ideal gas law, $PV = rT$, where a/V^2 allows for attraction between the molecules and b for their volume. We saw earlier that 'All mice have long tails' can be brought into the relation of counterparthood with 'All male mice have long tails and all female mice have short tails' by reformulating it in a padded-out but equivalent way, as 'All male mice have long tails and all female mice have long tails'. And the same can be done with the ideal gas law. For instance, we could rewrite it as $PV + a/a - b/b = rT$. This can be regarded as asserting, against van der Waals, that neither attraction between molecules nor their volume makes any difference. It is sometimes said that Newton's and Einstein's theories are incommensurable because, where Newton just had *mass*, Einstein distinguished mass (m) from rest-mass (m_o) and related them by $m = m_o/\sqrt{1 - v^2/c^2}$. But we could restate Newton's assumption concerning the invariance of mass with respect to velocity by the padded-out equation $m = m_o + v/v - c/c$. This says that velocity makes no difference. In this complicating way we may prepare a law for comparison with a more discriminating successor. Starting with, say, the laws $\alpha = f_i(\beta,\gamma)$ and $\alpha = f_j(\beta,\gamma,\delta)$, we replace the former by $\alpha = f_i'(\beta,\gamma,\delta)$ where f_i', unlike f_j, is insensitive to changes in the value of δ. If for any triple of finite values of β, γ and δ equally precise values for α are determined by f_i' and by f_j, then the laws are incongruent counterparts. And we can now say that if T_i and T_j are quantitative theories, then $CT(T_j) >$

$CT(T_i)$ if there is a T_i' such that T_i and T_i' are counterparts in the above extended sense, and $CT(T_i') \subset CT(T_j)$. This completes my elucidation of component (B3) of our optimum aim.

5.2 Explanatory Depth

I turn now to the elucidation of (B1), the demand for progressively deeper explanations of empirical phenomena. The idea of theoretical depth has considerable importance in Popper's philosophy of science. 'If at all possible,' he has said, 'we are after deep theories' (*1972*, p. 55). But he was pessimistic about the possibility of any sharp characterisation of this idea:

> I believe that this word 'deeper' defies any attempt at exhaustive logical analysis, but that it is nevertheless a guide to our intuitions. . . . The 'depth' of a scientific theory seems to be most closely related to its simplicity and so to the wealth of its content. . . . Two ingredients seem to be required: a rich content, and a certain coherence or compactness (or 'organicity'). . . . It is this latter ingredient which, although it is intuitively fairly clear, is so difficult to analyse. . . . I do not think that we can do much more than refer here to an intuitive idea, nor that we need do much more. (1957, p. 197)

I am more optimistic than Popper was about the possibility of analysing these interconnected ideas, of depth, and of the compactness or unity of a theory. The analysis of the latter will be postponed until § 5.3. I proceed now to the analysis of the former. I will consider three kinds of case, the first and second of which can, I believe, be handled without too much difficulty, while the third, which is the most important, is more challenging. In the first case (a), empirical generalisations or exact experimental laws, at level-2 or level-3, are explained by a level-4 theory that is deeper than them. In the second case (b), a level-4 theory T_i is explained by a deeper theory T_j which turns T_i into a theorem of itself. In the third case (c), a theory T_i is superseded by a deeper theory T_j that conflicts with it in various ways. I begin with case (a).

5.21 *Gilbert's Magnetic Theory*

It will help to keep this case separate from cases (b) and (c) if we suppose that, in a field where there had previously existed a certain amount of rough and ready empirical knowledge about certain kinds of phenomena, a genuinely scientific theory now appears for the first time. And it will help to focus our ideas if we have an example before us. William Gilbert's (*1600*) theory fits the bill. The phenomena with which it was concerned were of course the loadstone (I follow his translator's spelling) and the mariner's compass. The former had been known since antiquity and the latter for at least three centuries. There had been previous "theories" concerning them but these, unlike Gilbert's theory, seem to have been highly ad hoc and not to have yielded any novel predictions. (There was an effluvium "theory", according to which atoms go out from a loadstone, get entangled with the atoms of a piece of iron, and then seek to return home, dragging the latter with them. And there was a "theory" that endowed celestial matter near the pole star with special properties that attract compass needles.)

One novel consequence of Gilbert's theory was the following (*1600*, pp. 201–203): if two iron needles, suspended by silk threads, close together but not quite touching, are gently lowered until they almost touch a loadstone with one of its poles immediately beneath them, then the needles will draw apart. Let us represent this generalisation by $\forall x$ ($Fx \rightarrow Gx$), where Fx describes the initial conditions of the experiment and Gx its outcome (the drawing apart of the needles).

I will now outline very briefly the theory with which Gilbert explained this and related phenomena. A loadstone, according to Gilbert, is endowed with a very special kind of energy; it transmits instantaneous rays of magnetic force in all directions (pp. 123, 150) and thereby spreads an invisible 'sphere of influence' (p. 161) around itself. Magnetic force is a physical reality but it is *not corporeal* (p. 189): it passes without any interruption through thick, dense barriers of nonmagnetic matter. Moreover, there is no diminution of magnetic power with the passage of time, as presumably there would be if it were due to the emission of effluvia involving a continuous expenditure of matter and energy: 'if with one loadstone you magnetize one thousand compass needles for mariners' use, that loadstone not less powerfully attracts iron than it did before' (p. 62).

The magnetic forces emanating from a loadstone have two foci or 'primary termini' (p. 23), a north and a south pole. Like poles repel, and unlike poles attract, each other. When a loadstone attracts a piece of hitherto unmagnetised iron so that the latter becomes 'united with perfect union' to itself, the iron is thereby 'powerfully altered and converted, and absolutely metamorphosed into a perfect magnet' (p. 110).

We can now see what Gilbert's explanation for the drawing apart of the needles will be. As the needles proceed further into the loadstone's sphere of influence, they become increasingly affected by its magnetic energy until, by the time that they are nearly touching it, they have been metamorphosed into two small magnets, each with its south (north) pole at the lower and north (south) pole at the upper end. Thus they repel each other at both ends and so move part.

Leaving this notion undefined for the moment, we can say that Gilbert's theory has a *theoretical ontology*: it postulates the existence of something that is physically real, but immaterial and invisible, namely magnetic energy and immaterial rays of magnetic force. And with the help of this theoretical ontology it both explained various empirical generalisations already accepted among people familiar with loadstones and compass needles and predicted experimental generalisations, such as the one about the drawing apart of the needles, of a novel kind. (A definition of empirical novelty will be given in § 8.23.) Gilbert's theory satisfied Whewell's requirement that a deep hypothesis, one that gets hold of nature's 'alphabet' as he put it, must enable 'us to explain and determine cases of a *kind different* from those which we contemplated in the formation of our hypothesis' (*1840*, ii, p. 65). Let G denote the corpus of already accepted empirical generalisations that are now, for the first time, explained by a genuinely deep theory T. (We may expect T to effect some revising and sharpening of the generalisations in G.) Then for T to satisfy Whewell's requirement we must have $CT(T) > CT(G)$.

One argument for this requirement is that it is called for by the antitrivialisation principle. For suppose that we allow that, for T to be a deep theory, it is enough that T has a theoretical ontology and does not matter if $CT(T) = CT(G)$. Then it would be trivially easy, having fixed G, to construct a "deep" T. Let $F_1, F_2 \ldots$ be observational predicates and $\theta_1, \theta_2 \ldots$ be theoretical predicates. For ease of exposition let G consist just of the empirical generalisation $\forall x\, (F_1 x \to F_2 x)$. Then it is trivially easy first to devise a theoretical ontology, say $\exists x\, \theta_1 x \wedge \forall x\, (\theta_1 x \to \theta_2 x)$, and then to elaborate this into a "theory" that explains exactly G; for instance, we could elaborate it into

$$\exists x\, \theta_1 x \wedge \forall x((F_1 x \to \theta_1 x) \wedge (\theta_1 x \to \theta_2 x) \wedge (\theta_2 x \to F_2 x)).$$

(The effluvium "theory" of loadstones mentioned earlier seems to have had some such structure as this.)

A second argument for Whewell's requirement is this. When a genuinely deep theory is eventually introduced into an area hitherto occupied by empirical generalisations, it will, if true, get hold of some rather pervasive structural features underlying diverse phenomena, that the theory will thereby be able to connect in unexpected ways. For instance, Gilbert's theory connected the way steel filings, strewn on paper above a loadstone, orient themselves with variations as in the dip of a compass needle at different latitudes, given a fundamental assumption of his, namely that the earth is a great loadstone. The structural explanations of already known phenomena that a deep theory provides should also lead to the discovery of novel phenomena.

We must now turn to the idea that it is largely in virtue of its theoretical core that the empirical coverage of a deep theory is wider than the body of already accepted generalisations in its field. Before we can identify such a core, we need a distinction to have been drawn between observational and theoretical predicates, and I will begin with this question.

5.22 *"Observational" and "Theoretical" Predicates*

There are two considerations that delimit the unavoidable arbitrariness of any classification of predicates into those that are "observational" and those that are "theoretical". One is that some predicates are assuredly observational and others are assuredly theoretical. Gilbert reported that the colour of some loadstones is 'dark blood-red' (p. 18); and that is assuredly an observational predicate. On the other hand, his concept of an immaterial magnetic force or energy is assuredly theoretical. The other consideration is that between these two limits there is a spectrum of predicates, all partly observational and partly theoretical, but some of which are assuredly more observational or less theoretical, than others. Thus 'iron' is less purely observational than 'grey', and perhaps more observational than 'loadstone', and assuredly more observational than 'magnetic energy'. Again, 'tree' is more observational than 'planet', which is still somewhat observational, unlike 'the centre of gravity of the solar system', which denotes something essentially unobservable.

My view here is very much in line with that of Newton-Smith (*1981*, pp. 22f.). He speaks of

a rough spectrum of terms determined by the following principles:

1. The more observational a term is, the easier it is to decide with confidence whether or not it applies.
2. The more observational a term is, the less will be the reliance on instruments in determining its application.

3. The more observational a term is, the easier it is to grasp its meaning without having to grasp a scientific theory. (pp. 26–27)

Mary Hesse had previously written in a somewhat similar way of some predicates (namely, the more observational and less theoretical ones) being 'better entrenched' and 'learnable and applicable in a *pragmatically* simpler and quicker manner' than others (*1974*, pp. 22–23).

Fortunately, it will turn out that, for the purposes of comparing two theories, whether for depth, testable content, unity or exactitude, it will not matter much just where we make the "observational"/"theoretical" cut, provided that our classification satisfies the twofold requirement (i) that predicates that are assuredly observational, or assuredly theoretical, are classed as such, and (ii) that there is no gerrymandering of the boundary in the sense that, although a predicate P_i is more observational than P_j, P_i is classified as "theoretical" and P_j as "observational". The main role that the "observational"/"theoretical" distinction will play is in the elucidation of the idea of one theory being deeper than another. And the reason why it will not matter much just where we make the cut is this: if one theory dominates another theory, or a body of accepted generalisations, with respect to depth then it also dominates it with respect to width, or empirical coverage. Thus if we shift the line so that, say, some predicates previously classed as "theoretical" become "observational", then the theoretical content of the dominant theory will be somewhat reduced and its empirical content increased, but it should retain its dominance in both respects.

Some empiricists draw the line between "observational" and "theoretical" very close to the observational end of the spectrum in the hope that the empirical basis can be kept pure and uncontaminated by dubiety. I will argue in chapter 7 that while there can be no verification of level-1 statements, there can be rational acceptance of them; and in §7.5 I will add that a rather sophisticated level-1 statement, say about the position of a planet or the resistance of a coil or the existence of interference patterns, is no less open to rational acceptance than one about a pointer reading. So we are under no philosophical obligation to be very austere in what we class as "observational" predicates; rather, we are free to pursue a liberal policy in line with scientific practice and to classify predicates in which there is a considerable admixture of theoreticity as "observational". Consider 'loadstone'. I have little doubt that if Gilbert had had to choose between classing this as "observational" or "theoretical", he would have opted for the former. Of course, you cannot tell simply by looking that a piece of rock is a loadstone, any more than you can tell simply by looking that a shiny yellow object is made of gold. Tests have to be made. Gilbert wrote: 'And by good luck at last the loadstone was found, as seems probable, by iron-smelters or by miners in veins of iron ore. On being treated by the metallurgists, it quickly exhibited that strong powerful attraction of iron—no latent or obscure property, but one easily seen of all' (pp. 1–2). In what follows I will assume that scientists working in a certain field have agreed, concerning the predicates occurring in the laws and theories with which they are concerned, which are to be counted as "observational", and that they have drawn the line in a reasonably liberal way and in accordance with the twofold requirement mentioned above. Henceforth I will use F (or F_1, F_2 ...) for "observational" predicates and θ (or θ_1, θ_2 ...) for "theoretical" predicates.

5.23 Theoretical Cores

Einstein used to speak about the 'fundamental assumptions' of a theory that also contains many subsidiary assumptions (e.g. 1920, p. 124). Lakatos used the term 'hard core'. I have sometimes spoken of the 'theoretical ontology' of a theory and also of its 'metaphysical core'. All these labels denote very much the same thing, and pretty well all students of science (other than empiricists of the kind whose deflationary policies we considered in §4.4) would agree that this thing is in some ways the most distinctive and important part of a theory, or what lies at its heart. The question now is how to identify such a theoretical core, as I will call it. A rather natural answer suggests itself, namely that we should pick out those of the theory's premises in which only theoretical predicates occur. The trouble with this answer is that the same theory can be axiomatised in very different ways, and with a larger or smaller number of axioms. Indeed, one can of course always conjoin its several premises into *one* composite axiom; and since this one axiom would be bound, in the case of a testable theory, to contain some observational predicates, a theory thus axiomatised would no longer have a theoretical core, by this answer.

In view of this I proposed, in (1975), to identify the theoretical core of a theory in an indirect way. We may think of a Ramsey-sentence T_R (see §4.41) of a theory T as representing the latter's empirical content. So as a first step we might try locating its purely theoretical content among those consequences of T that are not consequences of T_R. But then a second step is needed; for some of these consequences will be testable. Indeed, T itself is a consequence of T but not of T_R. So my proposal was that we should identify T's theoretical core with the set of those of its consequences that are neither consequences of T_R nor testable. But this proposal has the disadvantage that T's theoretical core, thus defined, cannot be summarised in a few propositions, since it consists of a set of statements that is not axiomatisable. (The theoretical content of a theory, on this (1975) criterion, is analogous to its falsity content; the latter likewise consists of the set of its consequences *minus* a proper subset of them, namely, those that are true, and is not closed under deduction.)

Now it would be possible to go back to the first answer, namely that T's theoretical core consists of those of its premises in which only theoretical predicates occur, if we had a way of ensuring that, starting with one big composite axiom for T, we could break this up or decompose it into a number of axioms each of which is, in some sense, a "natural" propositional unit that ought not to be further decomposed. The difficulty with this latter idea is that, from a logical point of view, such a process of decomposition can always be carried further: having arrived at an axiom-set $B_1, \ldots B_n$ where each B_i seems intuitively to be a "natural" propositional unit, we could always decompose a B_i further, for instance into $B_i \vee p$ and $B_i \vee \sim p$, where p is an arbitrary statement. However, in §5.33 I will present certain rules with which a "natural" axiomatisation of a theory should comply, the last of which calls for as much decomposition as possible *provided* that the previous rules are not violated; and one of the intentions behind them is precisely to stop the decomposition process becoming degenerate.

At this stage I will anticipate by taking these rules as given and supposing that the theories with which we are concerned have been axiomatised in accordance with them. On this supposition each of the axioms of a theory will fall into one or other of the following three categories: those in which occur (i) only theoretical predicates, (ii) both theoretical and observational predicates, (iii) only observational predicates. If an axiom-set T has axioms in category (i), I will call their conjunction, to be denoted by T_H, the theoretical or hard core, or fundamental assumptions, of T. I like to think that Einstein would have endorsed this; for he often insisted that the further science advances, 'the more the *basic* concepts and axioms distance themselves from what is directly observable' (1949, p. 27, my italics). I will denote by A the conjunction of all the remaining axioms, the auxiliary or subsidiary assumptions of T. Some of these will be bridging laws in category (ii), axioms that relate the theoretical predicates in T_H to observational predicates; and a bridging law, which in the simplest case has the form $\forall x(F_1 x \rightarrow \theta_1 x)$ or $\forall x(\theta_2 x \rightarrow F_2 x)$, is normally untestable. However, we can expect that A by itself, when not conjoined with T_H, will have some testable content. For instance, if T were Newtonian theory then A would include assumptions about the mean distances of the planets from the sun, their relative masses, etc. In such a case we have $CT(A) > \varnothing$. On the other hand, T_H by itself, when not conjoined with A, will of course be untestable since it lacks predicates of the kind that occur in PFs: we are bound to have $CT(T_H) = \varnothing$; but we should also have $CT(T_H \wedge A) > CT(A)$ and even, if I may be permitted the expression (see §5.16), $CT(T_H \wedge A) \gg CT(A)$.

I take it for granted that a theory, realistically interpreted, typically asserts the existence of those unobservable entities that it invokes in its explanations of phenomena; Gilbert's theory asserts the existence of magnetic energy, Fresnel's of light-waves, Maxwell's of electromagnetic energy. After saying that his talk of a 'kind of motion' and a 'kind of strain' in the medium should be considered merely as illustrative, Maxwell significantly added: 'In speaking of the Energy of the field, however, I wish to be understood literally' (quoted by Hesse, *1961*, p. 209). I know that Newton was cagey about the existence of gravitational attraction. But according to his biographer Westfall, this was due more to his anxiety not to alarm the reader than to genuine doubt as to its reality:

> Newton began to worry about the reception of the concept of attraction. Initially, he had intended to state his position straightforwardly. In the early versions of Books I and II, he had spoken of attractions without apology and described the gravity of cosmic bodies as a force that arises from the universal nature of matter. (*1980*, p. 462)
>
> As he worried about the reception of his work, he began to hedge. The line in the preface about bodies being impelled by causes unknown was only part of a more extensive camouflage in which he insisted that his mathematical demonstrations did not entail any assertions about the ontological status of forces. (P. 464)

Thus a T_H will include statements of the form $\exists x\, \theta x$, which postulate the existence of those unobservable entities to which T appeals in its

explanations of phenomena, as well as universal statements characterising those entities.

We cannot demand that the fundamental assumptions T_H of a theory
T play an *indispensable* role in generating its testable content; for we
know from Ramsey's device (§4.41 above) that all T's theoretical predicates could be replaced by existentially quantified predicate variables,
leaving its testable content intact. But we can demand that T_H should
make a *vital contribution* to its testable content (in rather the same way
that a person's kidneys make a vital contribution to his bodily well-
being, even though he might survive if they were removed and he were
placed on a kidney machine). And we can express this as the demand
that the testable content of T's fundamental assumptions reinforced by
its subsidiary assumptions should be much greater than any testable
content that its subsidiary assumptions alone may have.

We can now summarise our requirements for a case (a) increase of
depth as follows. Let T be the axioms of the theory in question and let
G denote those of the empirical generalisations explained by T that
were generally accepted within the scientific community before T was
introduced. Then T is deeper than G if:

(i) $CT(T) > CT(G)$;
(ii) T consists of a theoretical core T_H and auxiliary assumptions
      A;
(iii) $CT(T_H) = \varnothing$ but $CT(T_H \wedge A) \gg CT(A)$.

5.24   *Gilbert and Kepler*

It is instructive, in the light of the foregoing, to compare Gilbert's
achievement with Kepler's. Kepler's worldview was quite as metaphysical as Gilbert's. Indeed, there are important similarities and affin-
ities between them. Kepler spurned the hitherto prevailing idea of a
merely descriptive astronomy that would accurately map the motions
of heavenly bodies without providing any physical explanation of them.
He wanted to discover their ultimate causes. His Second (equal areas)
Law provided a key component: this law showed, in contrast to Copernicus, who had located the centre of the planetary orbits *near* the
sun, that *the sun itself* is right at a focus of every planetary orbit. As
Gerald Holton says (*1973*, p. 80), Kepler's was the first truly heliocentric
system. So he took those ultimate causes to be invisible forces, rather
similar to Gilbert's magnetic rays, radiating from the sun: 'all the
manifold movements are carried out by means of a single, quite simple

magnetic force' (1605). The planets are driven round by a force emanating from the sun.

But now comes the big disappointment. Whereas the lower-level empirical laws in Gilbert's system are directly related to its central ontology, Kepler never succeeded in connecting his laws to his ontology. He assumed that the driving force from the sun diminishes in the same way as light; and he further supposed that light diminishes *linearly* with distance. Thus if the velocity of a planet varies with the force exerted on it, then on Kepler's assumptions a planet twice as far from the sun as another one, in completing one orbit, will travel twice the distance at half the speed, and its period will be 4 times that of the nearer planet. If we correct Kepler's mistake about light, and assume that the intensity of the sun's force varies inversely with the square of the distance, then the outer planet's period will be 8 times that of the inner one. Yet by Kepler's own Third Law the ratio is neither 4 nor 8 but $\sqrt{8}$.

Mill absurdly underrated Kepler's achievement when he argued that Kepler had produced 'a mere description' of the paths of the planets (*1843*, III, ii, 3). But it has to be admitted that Kepler's theoretical ontology, unlike Gilbert's, was not organically related to his laws; even if it could be squared with the latter, which seems doubtful, it failed to make a vital contribution, or indeed any contribution, to the testable content of his system.

5.25   *The Fresnel-Maxwell Example*

In case (b), a theory T_i, which is already deeper than the empirical generalisations that it explains, is subsumed *without revision* under a deeper theory T_j. I think that this has happened rather rarely in the history of science. One example, which Popper (1957, pp. 202–203) gave with acknowledgments to Einstein, was the incorporation of Fresnel's wave theory of light into Maxwell's electromagnetic wave theory; and I will use this example to illustrate what I shall say about greater depth in a case of this kind.

Suppose that T_j strictly entails T_i, and that T_i and T_j each has a theoretical core, which I will label respectively T_{iH} and T_{jH}. What conditions should be satisfied for us to regard T_j as deeper than T_i?

Suppose for the moment that T_j had no more testable content than T_i, or $\mathrm{CT}(T_j) = \mathrm{CT}(T_i)$. Since T_j strictly entails T_i, this would mean that T_j has excess content over T_i but that this excess content is merely metaphysical and fails to enhance its testable content; in short, *idle* metaphysical content. To allow *that* to count as greater depth would infringe our antitrivialisation principle; given a suitable existing theory T_i one could always manufacture a "deeper" one merely by tacking onto it 'Angels exist' or whatever. So we must clearly require (i) that T_j has more testable content than T_i, or $CT(T_j) > CT(T_i)$.

Let condition (i) be satisfied and now suppose, more interestingly, that T_j's theoretical ontology is not richer than T_i's. It cannot be poorer either, since T_j entails T_i. So in this case they would have the *same* theoretical ontology: $T_{i_H} = T_{j_H}$. This would mean that the explanation for $CT(T_j) > CT(T_i)$ is that the *subsidiary* assumptions A_j of T_j are stronger than the subsidiary assumptions A_i of T_i, or that T_j can be got from T_i by augmenting the latter's subsidiary assumptions. But that would hardly justify us in declaring T_j to be *deeper* than T_i. In such a case I would say that T_i and T_j have the same depth and that T_j is *wider* (has more predictive power) than T_i merely in virtue of its stronger subsidiary assumptions. This suggests that we should further require (ii) that T_j has a richer theoretical ontology than T_i or that T_{j_H} strictly entails T_{i_H}.

Suppose conditions (i) and (ii) to be satisfied. We are still not home and dry. For we might have achieved this by starting with the previous case, where T_j is wider but not deeper than T_i, and then adding 'Angels exist' or whatever to its theoretical ontology. We need to require (iii) that the excess content of T_j's richer theoretical ontology *contributes* to its having excess testable content. How can we spell out this requirement? Since T_j entails T_i the subsidiary assumptions A_j of T_j must be either the same as, or else stronger than, the subsidiary assumptions A_i of T_i. Suppose that they are stronger (for if they are the same it must be T_{j_H} that is responsible for $CT(T_j) > CT(T_i)$). Then let us augment T_i with T_j's stronger subsidiary assumptions: instead of taking $T_{i_H} \wedge A_i$ we take $T_{i_H} \wedge A_j$ as the theory with which we will compare T_j. Requirement (iii) will be satisfied if we have $CT(T_{j_H} \wedge A_j) > CT(T_{i_H} \wedge A_j)$; for as they now stand, the only difference between the premises of the two theories is the excess content of T_{j_H} over T_{i_H}: so it must be the greater richness of T_j's theoretical core that gives it its greater testable content. Had we had $CT(T_j) > CT(T_i)$ but $CT(T_{j_H} \wedge A_j) = CT(T_{i_H} \wedge A_j)$, we would have been back with the case where T_j is wider but not deeper than T_i.

I think that the foregoing is exemplified by the relation of Maxwell's electromagnetic wave theory to Fresnel's wave theory of light. Hertz's dramatic corroborations, in 1887–1888, of Maxwell's theory, involving the transmission and reception of electromagnetic signals, testify to its excess testable content; and there can be no doubt that this excess content was generated with the help of the theory's fundamental assumptions, involving the idea of electromagnetic disturbances spreading through an electromagnetic medium, or ether. And Maxwell himself suggested, to put it in our terminology, that the theoretical core of Fresnel's wave theory of light is properly contained in that of the electromagnetic theory:

> All that has been said with respect to the radiations which affect our eyes, and which we call light, applies also to those radiations which do not produce a luminous impression on our eyes. (*1890*, ii, p. 766)

> The properties of the electromagnetic medium are therefore as far as we have gone similar to those of the luminiferous medium. . . . If [the velocity with which an electromagnetic disturbance would be propagated through the medium] should be equal to the velocity of light, we would have strong reason to believe that the two media, occupying as they do the same space, are really identical. (P. 771)

To sum up my examination of the relatively easy case (b): given that T_i already has some depth (and hence has a theoretical core T_{i_H}) and that T_j strictly entails T_i, then T_j is deeper than T_i if

(i) $Ct(T_j) > CT(T_i)$;
(ii) T_{j_H} strictly entails T_{i_H};
(iii) $CT(T_{j_H} \wedge A_j) > CT(T_{i_H} \wedge A_j)$.

Let us now turn to the more difficult case (c).

5.26 *Revisionary Reductions*

There is a good reason why it happens so rarely in the history of science that a theory T_i is taken over by, and subsumed under, a deeper theory T_j without its content being disturbed and revised in significant ways. A scientific theory usually imposes a principle of conservation on any physical stuff or entities that it treats as fundamental. For instance, the caloric theory of heat treated caloric, or heat substance, as uncreatable and indestructible, and in Dalton's atomism, atoms were of course treated as uncreatable and indestructible:

> We might as well attempt to introduce a new planet into the solar system, or to annihilate one already in existence as to create or destroy a particle of hydrogen. All the changes we can produce, consist in

separating particles that are in a state of cohesion or combination, and joining those that were previously at a distance. (*1808*, p. 212)

But other things that, according to the theory, are built up from fundamental units will not, in general, be assumed to obey a conservation principle: such compound entities are usually regarded as decomposable into elements that can be recombined in other ways. Thus if T_j goes to a deeper ontological level than T_i, it is rather likely that something that T_i had treated as uncreatable and indestructible will be treated as creatable and destructible by T_j; and this important theoretical disagreement may have large implications.

There was considerable discussion some years ago of the question, 'When is a theory T_i superseded by a *deeper* theory T_j?' in the slightly different form, 'When is a theory T_i *reduced* to a theory T_j?' One early and important contribution was Popper's (1957); another was Nagel's (*1961*, chap. 11).

The problem of the nature of a scientific reduction seemed, at least to some investigators, to pose the following dilemma: (1) there surely cannot be reduction without deduction; (2) but (a) there may be little overlap between the vocabularies of the theories in question, and (b) there may even be theoretical conflict and empirical disagreement between them. Some writers, notably Nagel, seemed to suppose that the only problem was the apparent discrepancy between (1) and (2a), and that this could be eliminated without much difficulty; links between the new theoretical concepts of T_j and the older concepts employed in T_i could be established by correspondence rules, which might be analytic or conventional, or else synthetic and contingent, bridging laws. Once this sort of hiatus between T_j and T_i has been bridged in that way, T_i would, it was claimed, turn out to be deducible from T_j. Thus Nagel wrote:

A reduction is effected when the experimental laws of the secondary science [our T_i] (and if it has an adequate theory, its theory as well) are shown to be the logical consequences of the theoretical assumptions (inclusive of the coordinating definitions) of the primary science [our T_j]. (*1961*, p. 352)

With the help of these additional assumptions all the laws of [T_i] . . . must be logically derivable from the theoretical premises and their associated coordinating definitions in [T_j]. Let us call this the "condition of derivability". (P. 354)

Mario Bunge (in *1967*, i, p. 508) likewise required a deeper T_j to *entail* a less deep T_i. He relaxed this (in 1968) to the requirement that T_j entails 'a large part' of T_i but not conversely, adding: 'In particular, if T_j entails the whole of T_i, then T_i can be said to be reduced to T_j' (p. 129; I have altered his T and T' to my T_j and T_i). I think it is fair to say that Nagel and Bunge wanted to turn the difficult case (c) back into the relatively easy case (b).

Kemeny and Oppenheim (1956) had previously taken an important step in the right direction by requiring T_j to yield deductively, not T_i itself, but the predictive content of T_i. This provides for the very real possibility that there is radical disagreement between the theoretical ontologies of T_i and T_j. (The Newtonian ontology of absolute space, absolute time, indestructible corpuscles, action across empty space, is of course in radical disagreement with GTR, though there is near continuity between them at the empirical level.) But it fails to provide for the possibility that there is some disagreement between the predictive consequences of the two theories, an objection which the authors themselves noticed (p. 17). Clark Glymour suggested that what T_j should do with respect to T_i is to show under what conditions T_i *would* be true and to contrast those conditions with conditions that actually obtain (1970, p. 341). Using our notation, his idea might be put like this:– Assume that T_j conflicts with T_i. Let T_j be equivalent to $T_{J_H} \wedge A_j$ where T_{J_H} is T_j's theoretical core. Then there should be an A_j' that differs from A_j and such that $T_{J_H} \wedge A_j'$ entails T_i. When we compare A_j' with A_j we will usually find that T_i would hold only under certain limiting or counterfactual conditions.

The snag with this proposal is that there may be *no A_j'* that could be *consistently* joined with T_{J_H} to yield T_i. This point had been emphasized by Popper (in 1957). Suppose we try to derive Kepler's laws from Newtonian theory; we will not quite get them so long as the planets are assumed to have a nonzero mass; and if we set their mass at zero, Newton's laws will no longer apply to them (p. 201). So Popper required that what T_j should explain with respect to T_i is why T_i, though false, had nevertheless been as successful as it was; and T_j does this by showing that the empirical content of T_i comes very close to what is the truth according to T_j. Another way in which Popper expressed this idea was to say that, in the case of a scientific reduction where there is a more or less radical conflict between the theoretical cores of T_i and T_j, T_j shows that the actual phenomenal world *simulates* a world of the kind depicted by T_i (*1972*, pp. 266–270); according to T_j the reality behind the phenomena is very unlike what T_i declares it to be; nevertheless the phenomenal world behaves *almost as if* the reality behind it *were* as T_i declares it to be.

This idea is well illustrated by the relation of the kinetic theory of energy to the caloric theory of heat. According to the theoretical ontology of the latter theory, the reality behind phenomena involving freezing, melting, vaporisation, and changes in temperature, is movements of a heat substance distinct from ponderable matter. This stuff, which obeys a conservation law, is in Dalton's words, 'an elastic fluid of great subtility, the particles of which repel one another, but are attracted by all other bodies' (*1808*, p. 1). This mutual repulsion means that the fluid is always tending to disperse itself until a uniform distribution is reached. Thus if a hot body and a cold body are juxtaposed, caloric will flow out of the former into the latter until both are at the same temperature. If caloric goes on flowing out of liquid matter, the latter's temperature will continue to fall until it reaches freezing point; then it will stop falling and the body will lose only latent heat until all the matter has solidified; then its temperature will start falling again.

The kinetic theory is a theory of heat according to which there is no such stuff as *heat*. What is conserved through all heat changes is not caloric but *energy*. Heat can be created (as by friction) when energy in one form is converted into energy in another form. Behind heat phenomena there are swarms of tiny molecules moving randomly, often with enormous velocities, and collectively possessing a mean kinetic energy. When a hot and a cold body are juxtaposed, collections of molecules with, respectively, a comparatively high and a comparatively low mean kinetic energy interact; there is a molecular bombardment. The molecules near the surface of the cooler body become more excited and collide more frequently with those further in, and so on. The mean kinetic energy of its molecules is raised and that of the hotter body's is lowered, until a uniform level is reached. So it is as if heat flowed out of one into the other.

Kenneth Schaffner (1967) took up Popper's view of a scientific reduction: he required a deeper theory T_j to be reinforced with bridging assumptions in such a way that it now entails a corrected version T_i' of the reduced theory T_i, a version that 'produces numerical predictions which are "very close" to' those of T_i, so that T_j explains why T_i, though false, 'worked as well as it did' (p. 144). Apropos this idea that the deeper theory should entail laws that approximate those of the reduced theory, Glymour commented: 'In some sense this is doubtless correct, but it leaves a great deal unsaid. When, for example, do one set of laws approximate another set of laws?' (1970, p. 341). I believe that an answer to this question can be given with the help of the idea, introduced in §5.17, of two experimental laws being incongruent counterparts. We can say that the law l_j approximates the law l_i if l_i and l_j are incongruent counterparts and, moreover, the function f_j in l_j gives numerical values that are close to those given by f_i in l_i. And we can add that two predicted values are close to each other if either they are not discernibly different or else fine measurements are needed to detect the difference.

We can lay down conditions for a case (c) increase of depth in analogy with case (a) as follows. Let G_i be the set of all the experimental laws and generalisations entailed by T_i. Then for T_j to be deeper than T_i, there must be a set G_i' of experimental laws and generalisations entailed by T_j such that

(i) $CT(G_i') \approx CT(G_i)$;
(ii) T_j is deeper than G_i' in the sense laid down for case (a) at the end of §5.23.

Condition (i) will obtain if every member of G_i has a counterpart in G_i' and vice versa.

5.3   Theoretical Unity and Exactitude

Having elucidated components (B1) and (B3) of our conjecturalist version of the Bacon-Descartes ideal, let us now turn to (B2). This expresses the idea that as science penetrates to deeper levels, so its laws and theories should become less compartmentalised and more unified, with the eventual possibility that the whole of physical nature is brought under one comprehensive theory. Something like (B2) inspired Einstein's theoretical work to the end of his life.

5.31   *When Does a Set of Axioms Constitute a Scientific Theory?*

So one question we need to answer is: when is one theory more unified than its predecessor(s)? But there is a prior question that is more urgent: when does a conjunction of statements constitute one unitary theory? Hitherto, in this book, the idea of a scientific *theory* has been very much taken for granted, as though theories present themselves as readily identifiable and, as it were, natural units. Now we may think of the content of a theory as encapsulated in its axioms; and its axioms must be *logically independent* of each other. (They must not be mutually inconsistent, and if one were entailed by the others it would not be an axiom but a theorem.) Then why should not the conjunction of *any* logically independent statements constitute a theory? For instance, why should not 'All cows are herbivorous' plus 'All crocodiles are thick-skinned' plus 'All crows are black' constitute a zoological theory no less than the conjunction of Newton's logically independent axioms constitutes a cosmological theory? And why should not the conjunction of Newton's axioms with, say, those of Bohm-Bawerk's *Capital and Interest* constitute one theory superior to either of its two halves taken separately?

It is rather remarkable that, although scientific theories are taken as the basic units by many philosophies and nearly all histories of science, there is no extant criterion, so far as I am aware, for distinguishing between a theory and an assemblage of propositions which, while it may have much testable content, remains a rag-bag collection. We saw earlier (at the beginning of §5.2) that Popper in (1957) was inclined to despair of the possibility of an analysis of the idea of the compactness or organicity of a scientific theory; and at around that time Oppenheim and Putnam wrote in a similarly pessimistic vein: 'Unity of Science in ᵗhe strongest sense is realized if the laws of science are not only reduced to the laws of some one discipline, but the laws of that discipline are in some intuitive sense "unified" or "connected". It is difficult to see how this last requirement can be made precise' (1958, p. 4). Robert Causey (*1977*, pp. 111–121) put forward a number of '*necessary* conditions for a theory to be unified', adding: 'I doubt that any set of general logical and semantical conditions would be sufficient for unity' (p. 120). While I have not grasped the purport of all his ten conditions, there is one, the seventh, that is surely right: it says that in each axiom there must be at least one predicate that also occurs in another axiom. Nagel had called a theory that satisfies this condition "compendent" (*1961*, p. 346). This condition knocks out our cows-crocodiles-and-crows "theory" (but not a "theory" whose axioms are 'All ravens are black' and 'Some swans are black').

So there is a crying need for a criterion, or for conditions that are necessary and sufficient, for theoryhood. If we can obtain that, we can proceed to the question of one unified theory being *more* unified (or perhaps, more unifying) than its predecessor(s).

5.32 *The Organic Fertility Requirement*

These are tricky issues and an investigation of them is bound to be somewhat technical. One might suggest that one thing that is wrong with 'All ravens are black' and 'Some swans are black', considered as a pair of axioms, is that, although they have a predicate in common, this does not enable their conjunction to generate fresh consequences. The trouble with that suggestion is that *any* conjunction, say $p \land q$, of two independent statements has consequences, for instance $p \leftrightarrow q$, that are not consequences of either statement by itself. However, we can get round that difficulty by confining ourselves to the testable or predictive contents of the statements in question. As a condition for theoryhood I will propose what I call the *organic fertility requirement*, which I will now formulate in an unguarded way. (Later, we will guard it against certain tricks.) Let T be a set of axioms, and assume that T is testable. It will be remembered that we are equating the testable content of such a T, or $CT(T)$, with the set of SPIs (singular predictive implications) entailed by it. Since T is testable we have $CT(T) > \varnothing$. Now provided that T consists of two or more axioms, we could partition it into T' and T'', where T' is one proper subset of its axioms and T'' is the remainder. Thus $T' \neq T \neq T''$ and $T' \land T'' = T$. The idea behind the organic fertility requirement is that if T is a genuine theory, then however we partition its axioms we shall always find that their conjunction is organically fertile in the sense that the whole has more testable content than the sum of the testable contents of its two parts. Let $CT(T') \cup CT(T'')$ denote a set of SPIs consisting of the union of the SPIs of T' and the SPIs of T''. Then an unguarded statement of the organic fertility requirement would say that T is a unified theory if, for any such partition of it into a T' and a T'', we always have $CT(T) > CT(T') \cup CT(T'')$ and never $CT(T) = CT(T') \cup CT(T'')$. A special case of a failure of this requirement would be one where $CT(T) = CT(T')$ and $CT(T'') = \varnothing$.

I hasten to add that $CT(T'') = \varnothing$ is not in the least objectionable by itself. It is to be expected that a scientific theory that is at all deep contains several axioms that are untestable in isolation, and T'' might be one or more of these. But $CT(T'') = \varnothing$ *together with* $CT(T) = CT(T')$ means that T'' is a piece of *idle* metaphysics within T. What we ask of a separately untestable axiom, say T'', of T is that it should be empirically fecund in the sense that, when conjoined with the other axioms T' of T, it generates a whole array of predictive consequences that are not consequences of T' alone, so that although we have $CT(T'') = \varnothing$ we also have $CT(T' \land T'') > CT(T')$.

However, this essentially simple underlying idea faces various technical difficulties connected with the fact that one theory can be axiomatised in very different ways. I will try to meet these difficulties by drawing up a set of rules to disallow "unnatural" axiomatisations. A reader who is willing to take all this on trust should turn to the end of §5.33.

5.33 *Rules for "Natural" Axiomatisations*

The organic fertility requirement, or OFR as we may call it for short, would obviously be useless if all theories, however good, automatically fail it. Now the same theory can of course be axiomatised in very different ways; and unless we impose restrictions on the way in which a theory may be axiomatised when it is being tested for unity, or for the organic fertility of its axioms, all theories will fail OFR. It might be supposed that OFR would likewise be useless if it were always possible to reaxiomatise a rag-bag "theory" in such a way that it now passes OFR. But this is not so (a point I owe to Heinz Post). The idea is that OFR will serve as a *test* for organicity: for a "theory" to fail the test establishes that it is not a genuine, unified theory; but passing the test, on a particular partition of T into T' and T'', does not establish that T is unified: perhaps it would fail the test on another partition. (In order positively to establish that it is unified, one would need to show that it passes on every possible partition; whether that could be shown we will consider later.)

So we need to reinforce OFR with a set of conventional rules that will disallow what I will call OFR-defeating reaxiomatisations, that is, reaxiomatisations of theories that, if permitted, would ensure that any theory automatically fails OFR. And my primary objective in what follows is to provide a set of rules that will do just that. But I would like at the same time to achieve a little more than just that. Glymour has said that 'the positivists had no account of what, if anything, makes one system of axioms more "natural" than another, . . . and today we are no better off in this regard' (*1980*, p. 39). And I would like my rules not merely to exclude OFR-defeating reaxiomatisations but to exclude them as being, in one way or another, more or less "unnatural", say because they are needlessly complex or lack perspicuity. I would prefer my rules to seem a reasonable rather than an arbitrary piece of philosophical legislation. I am far from confident that I shall succeed in this secondary objective. I shall not be very surprised if a critic comes up with an obviously "unnatural" kind of reaxiomatisation that slips by the rules to be presented. Nor shall I be too disappointed *unless* it is at the same time an OFR-defeating reaxiomatisation. How I would respond to this latter kind of counter-example I will consider later.

Concerning the language in which the theories are formulated, I will assume that a conventional cut between "observational" and "theoretical" predicates has been agreed upon, that it conforms with what was said in §5.22, and that the primitive predicates that belong to the observational vocabulary of this language are well co-ordinated with observable properties and are not at once undersensitive and oversensitive to observable differences in ways of the kind indicated towards the end of §5.14. I will also assume that this language has an underlying logic and that any mathematics involved has been axiomatised separately.

One more preliminary concerns what is to count as a PF. We have hitherto been operating, a shade casually, with the straightforward idea that a PF is a conjunction of (positive or negative) level-1 statements. Thus if F_1 and F_2 are observational predicates, then $F_1a \land \sim F_2a$ is a PF of any theory that has $\forall x \ (F_1x \rightarrow F_2x)$ as a consequence. But in the present context we have to be more careful. If we allow that a disjunction of two PFs is a PF (I owe this point to Worrall) we will turn the conjunction of Newton's and Bohm-Bawerk's theories into a unified theory; for let e_1 be PF of the former (NM) and e_2 of the latter (BB); then CT(NM $\land$ BB) > CT(NM) $\cup$ CT(BB) since NM $\land$ BB would have a PF, namely $e_1 \lor e_2$, that is not a PF of either NM or BB. And the same will happen if we allow certain kinds of molecular predicates to occur in PFs. Let $F_1a \land \sim F_2a$ and $F_3a \land \sim F_4a$ be PFs of, respectively, NM and BB. Introduce predicates G_1 and G_2 such that $G_1 = F_1 \land F_3$ and $G_2 = \sim F_2 \lor \sim F_4$; then $G_1a \land G_2a$ would be a PF of NM $\land$ BB but not of NM or of BB.

It is obvious that the class of PFs cannot be closed under *all* Boolean operations. If it were closed under deduction, for instance, we would have to accept ever-weaker consequences of a PF as PFs. In Popper's original characterisation (*1934*, §28) of basic statements, his underlying idea was that the strengthening operation of conjunction preserves basic statement status, but weakening operations generally do not. Thus if F_1a is a basic statement, so is $F_1a \land \sim F_2a$. But the negation of the latter is not a basic statement. For negation is here a weakening operation; it turns the above conjunction into the disjunction $\sim F_1a \lor F_2a$. And in (*1963*, p. 386) he explicitly excluded disjunctions of basic statements from the class of basic statements. In line with this, I propose that a statement is to count as a PF if and only if it is equivalent to some conjunction of atomic observation statements, the latter having the form *Fa*, or *~Fa*, where *F* is a primitive observational predicate of the kind specified earlier. A corollary of our stipulation that the disjunction of two PFs is not a PF is that the conjunction of two or more SPIs is not another SPI, rather as the conjunction of two or more pebbles is not another pebble. If $e_1 \rightarrow e_2$ is one SPI and $e_3 \rightarrow e_4$ is another, $(e_1 \rightarrow e_2) \wedge (e_3 \rightarrow e_4)$ is not a third SPI. Thus NM $\wedge$ BB does not have SPIs that are not SPIs of either NM or BB alone.

In the hope of keeping them nonarbitrary, my axiomatisation rules will be governed by three, I trust uncontroversial, adequacy requirements for a "good" or "natural" or "happy" axiomatisation of a scientific theory. Any two alternative axiomatisations of a given system of statements must of course be logically equivalent; each must have just these as its consequence class. And this means that preferences between alternative axiomatisations cannot be based on semantic considerations but only on considerations that are essentially pragmatic. In addition to the obvious requirement of the independence of the axioms there is a requirement of economy: an axiom set for a powerful scientific system may need to be rather rich but all needless complexity should be avoided. Another is a requirement of perspicuity; an axiomatisation should make it comparatively easy to trace a theorem back to those propositional units that just suffice, collectively, to entail it. To be able to pinpoint as sharply as possible which axioms are responsible for a theorem is likely to be particularly important in the case of a falsified theorem. The perspicuity requirement calls for finite axiomatisations where possible and I will assume that, its mathematics having been separately axiomatised, the factual content of a scientific theory always can be finitely axiomatised.

I now present five rules that a "natural" axiom set should satisfy, explaining and justifying them afterwards.

1. *Independence requirement:* each axiom in the axiom set must be logically independent of the conjunction of the others.
2. *Nonredundancy requirement:* no predicate or individual constant may occur inessentially in the axiom set.
3. *Segregation requirement:* if axioms containing only theoretical predicates can be separately stated, without violating other rules, they should be.
4. *Wajsberg's requirement:* an axiom is impermissible if it contains a (proper) component that is a theorem of the axiom set, *or becomes*

one when its variables are bound by the quantifiers that bind them in the axiom.

5. *Decomposition requirement:* if the axiom set can be replaced by an equivalent one that is more numerous (though still finite) without violating the preceding rules, it should be.

Besides helping to exclude "unnatural" axiomatisations, the first four rules also have the function of preventing the decomposition called for by rule 5 from becoming degenerate.

Rule 1 is uncontroversial. An obvious corollary of it is that tautological axioms are impermissible. (Remember that the underlying logic and the mathematics of the scientific theories are assumed to have been taken care of separately.) Rule 2 is also uncontroversial, and obviously justified by the economy requirement. A predicate, or individual constant, occurs inessentially in an axiom set if there is an equivalent axiom set in which neither that predicate (individual constant) nor any substitute for it occurs. Rule 2 helps to prevent the decomposition called for by rule 5 from becoming degenerate by forbidding the decomposition of a "natural" axiom, say $\forall x(P_1 x \rightarrow P_2 x)$ into, say, $\forall x((P_1 x \wedge P_3 x) \rightarrow P_2 x)$ and $\forall x((P_1 x \wedge \sim P_3 x) \rightarrow P_2 x)$ or into $\forall x((x \neq a) \rightarrow (P_1 x \rightarrow P_2 x))$ and $P_1 a \rightarrow P_2 a$. Rule 3 is called for by the perspicuity requirement on the assumption, here taken for granted, that an axiomatisation that enables us to go at once to a theory's fundamental assumptions is more perspicuous than one that has fundamental and auxiliary assumptions all intermixed.

In order to do justice to the people involved I would like to explain how I arrived at rule 4. At one time I had been toying with two alternative rules. One was that no *axiom* should reappear as a component of another axiom; but this, while it excluded many OFR-defeating reaxiomatisations, was too weak: in particular it would not prevent the decomposition of a "natural" axiom B into $B \vee p$ and $B \vee \sim p$. So I inclined to the stronger rule that no *proposition* should recur as a component of two or more axioms. But this seemed too strong. Then Elie Zahar suggested a rule that seemed just right, namely that no *theorem* of the axiom set should occur as a component of any axiom. I called this 'Zahar's requirement' and it is preserved in the unitalicised part of the present rule 4. It dealt very satisfactorily with various OFR-defeating reaxiomatisations formulated within the propositional calculus. However, Clark Glymour had drawn my attention to the fact that some of these reaxiomatisations could be replicated within the predicate calculus; and to deal with them I introduced an additional rule, since discarded. Later, David Miller read a draft of this chapter. He pointed out that what I was calling 'Zahar's requirement' had been anticipated by M. Wajsberg, whom Miller had previously reported in these words:

> An axiom is *organic with respect to a system A* if it contains no segment which is a theorem of A 'or becomes one as soon as its variables have been bound with an appropriate quantifier'. (1974b, p. 187)

Miller was here relying on two reports, one by Lukasiewicz and Tarski (1930, p. 45) who are the authorities for the first part of the quotation, and one by Sobocinski (1955–56, p. 60) who is the authority for the second part, inside quotation marks, and which is taken up in the italicised part of rule 4. This latter, which is a strengthening of what I had called 'Zahar's requirement', enables the rule to deal with those OFR-defeating reaxiomatisations within the predicate calculus due to Glymour. It seems clear that Wajsberg was using whatever Polish term is here translated as 'organic' in some such sense as 'unitary' or 'non-molecular'. It is an open question whether a conjunction of axioms each of which is 'organic' in this sense will exhibit *organic fertility* in my sense.

I will begin by considering rule 4 in its application to axiomatisations within the propositional calculus. Suppose that a theorem of an axiom set occurs as a (proper) component in one or more of the axioms. Then we will not increase the content of the axiom set if, as a first step, we state this theorem as a separate axiom; and having done that we will not decrease content if, as a second step, we replace the theorem, wherever it had occurred as a component of an axiom, by a tautology (I am indebted here to Zahar); after which we can eliminate all such tautological components and ask whether this separately stated axiom is still a theorem of the conjunction of the other, thus modified, axioms. If it is, we can eliminate it altogether from the axiom set, in line with the economy requirement; if it is not, then we have eliminated it from the other axioms and given it its rightful place as an independent axiom, in line with the perspicuity requirement. Here are two illustrations of this rule at work. (i) We start with the "unnatural" axiom set B_1, $c_1 \leftrightarrow B_2$, $c_2 \leftrightarrow B_3$, where B_1 entails c_1 and B_2 entails c_2. We first state c_1 and c_2 as axioms, replacing them where they had occurred by the tautology t, yielding c_1, c_2, B_1, $t \leftrightarrow B_2$, $t \leftrightarrow B_3$. We then eliminate tautological components, leaving c_1, c_2, B_1, B_2, B_3. Finally, we discard c_1 and c_2 in accordance with rule 1, since they are theorems of the other axioms, leaving B_1, B_2, B_3. (ii) We start with the "unnatural" axiom set $B_1 \lor p$, $B_1 \lor \sim p$, $B_1 \leftrightarrow B_2$, B_3, in which B_1 is a theorem of the set and a component of its first three axioms. Exporting it out of them yields B_1, $t \lor p$, $t \lor \sim p$, $t \leftrightarrow B_2$, B_3. Discarding tautological axioms by rule 1 leaves B_1, $t \leftrightarrow B_2$, B_3, and discarding tautological components leaves B_1, B_2, B_3. A needlessly complex axiom set has been simplified, and all redundancies removed, by the application of rule 4 (in co-operation with rule 1).

An obvious corollary of rule 4 is that no *axiom* may reappear as a component of another axiom. And this suffices to eliminate many OFR-defeating reaxiomatisations, one of which is the following (due to Worrall). Let T be a powerful theory with a theoretical core. Then T is equivalent to $T_R \land (T_R \to T)$ where T_R is a Ramsey-sentence for T and $T_R \to T$ is what Tuomela (*1973*, p. 59) calls a Carnap-sentence. Then we have $CT(T_R) = CT(T)$ and $CT(T_R \to T) = \varnothing$, contrary to OFR. But this reaxiomatisation is forbidden, since the Ramsey-sentence that constitutes the first axiom reappears as a component of the Carnap-sentence that constitutes the second axiom.

Another "unnatural" reaxiomatisation excluded by rule 4 is the following. Let B_1, B_2, ... B_n be a "natural" axiom set for what Tichy would call a lousy theory, every one of whose axioms has been separately falsified. A follower of Feyerabend, who holds, as we saw in §4.51, that 'massive conflict with experimental results should [not] prevent us from retaining' a theory, might reaxiomatise it as B_1, $B_1 \leftrightarrow B_2$, ... $B_1 \leftrightarrow B_n$. He could then say that, with the trifling exception of B_1, all the axioms of this theory are *verified*!

Rule 4 is also needed to prevent the decomposition called for by rule 5 from becoming degenerate. As we have already seen, it forbids the "unnatural" decomposition of a B_i into $B_i \lor p$ and $B_i \lor \sim p$. Again, if c_1, c_2, ... c_n are consequences of a "natural" axiom set B_1, ... B_n, rule 4 forbids the decomposition of the latter into, say, c_1 and $c_1 \to B_1$, c_2 and $c_2 \to B_2$, ... c_n and $c_n \to B_n$.

I turn now to the application of rule 4 to OFR-defeating reaxiomatisations within the predicate calculus. Glymour's example (personal communication) might be presented as follows. Let a theory T have a theorem G that captures all the empirical content of T, but without being logically equivalent to T. In other words, T strictly entails G but G entails every SPI entailed by T and $CT(G) = CT(T)$. A trivial example of this would be one where T consists of the three axioms $\forall x(\theta_1 x \to \theta_2 x)$, $\forall x(F_1 x \to \theta_1 x)$, and $\forall x(\theta_2 x \to F_2 x)$, and G

is $\forall x(F_1 x \rightarrow F_2 x)$, θ_1 and θ_2 being theoretical, and F_1 and F_2 obser-vational, predicates. But we can reaxiomatise T using G as one axiom and as the other an axiom which I will label $(G \rightarrow T)$ and which reads as follows:

$$(G \rightarrow T): \forall x((F_1 x \rightarrow F_2 x) \rightarrow ((\theta_1 x \rightarrow \theta_2 x) \wedge (F_1 x \rightarrow \theta_1 x)$$
$$\wedge (\theta_2 x \rightarrow F_2 x))).$$

This pattern of reaxiomatisation replicates one within the propositional calculus that we considered earlier, namely T_R and $T_R \rightarrow T$, where the Ramsey-sentence T_R reproduces all the empirical content of T. But the present one, unlike the previous one, is not excluded by the first part of rule 4; the above axiom does not contain a (proper) component that is a theorem of T, because only a proposition can be a theorem and this axiom does not contain a proper component that expresses a prop-osition. But this reaxiomatisation is excluded by the italicised part of Wajsberg's requirement because the above axiom $(G \rightarrow T)$ contains various components (such as $F_1 x \rightarrow F_2 x$ or $\theta_1 x \rightarrow \theta_1 x$) that become theorems of T when their variables are bound by the quantifier that binds them in this axiom, namely $\forall x$. It should be mentioned that this reaxiomatisation is also excluded by rule 3 for failing to state separately the "fundamental assumption" of our trivial "theory", namely $\forall x(\theta_1 x \rightarrow \theta_2 x)$.

Finally, we have rule 5, which seems uncontroversial. One often wants to pinpoint as narrowly as possible those axioms in an axiom set that are specifically responsible for a particular theorem, especially in cases where the theorem has been falsified; and this means that one wants large, composite axioms to be broken up into small propositional units, indeed into smallest propositional units, provided that these units remain "natural" or "organic" in Wajsberg's sense and are not dismembered artificially. I do not think that it could ever be proved that a given axiomatisation satisfies all the above rules. Consider rule 2: it has often happened in the history of science that a theoretical term that had at first been regarded as primitive is subsequently shown to be definable by other terms of the theory; and a defined term is of course eliminable in that it could be everywhere replaced by its *definiens*. Or consider rule 5: I doubt whether it could ever be proved of a given axiom set that there does not exist an equivalent but more numerous axiom set satisfying rules 1–4. In view of this, I say that a given axiomatisation is permissible so long as it is not known not to comply with the above rules; it is to be presumed innocent unless proved guilty. I hasten to add that I am not seeking to ban "impermissible" axiomatisations: anyone is free to axiomatise something in any unusual way he pleases. It is only when we apply the organic fertility requirement to an axiom set, to discover whether it constitutes a unified scientific theory, that "impermissible" axiomatisations are indeed impermissible.

I now introduce the idea of a *permissible partition* of an axiom set T into two subsets T' and T''. This is very simple. If $B_1, \ldots B_n$ are the axioms of a permissible axiom set T and if T' is a proper (and nonempty) subset of these B_i's and T'' consists of the remainder, then T' and T'' is a permissible partition of T. And the organic fertility requirement, negatively expressed, now says:

> OFR: An axiom set T is not a unified scientific theory if there is a permissible partition of it into T' and T'' such that $CT(T) = CT(T') \cup CT(T'')$.

Could it be established that an axiom set T *is* a unified theory? A permissible axiom set is one that is not known not to satisfy our rules. So one that is permissible today may become impermissible tomorrow, after the discovery of a violation. So there can be no absolute proof that a theory is unified; perhaps it will fail OFR on tomorrow's reaxiomatisation. However, relative to today's axiomatisation it is in principle possible to establish that it is unified. The number of permissible partitions of it, though it may be large, is finite and it would, in principle, be possible to check that it passes OFR on each of them.

In conclusion let us consider the (I trust hypothetical) eventuality that a critic comes up with an OFR-defeating type of reaxiomatisation that slips by rules 1–5. Then I would hope that some reasonable modification of the rules, in the spirit of our adequacy requirements, would rectify matters. But suppose that it looks as though no such modification will do the trick: his is a recalcitrant counter-example. In that event I would retreat to the following fallback position: any type of reaxiomatisation that has the result that *any* theory axiomatised in this way *automatically* fails OFR is *eo ipso* an impermissible axiomatisation. For we know that the decisive difference between a theory such as Newton's, and a rag-bag "theory" such as our one about cows, crocodiles, and crows, is that the premises of the former, on any reasonable axiomatisation, do mesh together to generate new testable content (more specifically, new SPIs) in a way that those of the latter do not. And a way of axiomatising the former that obliterates this vital difference cannot be perspicuous.

5.34 *"Kuhn-Loss"*

In recent years a thesis has gained widespread acceptance among historians and philosophers of science that has the damaging implication that the essentially Popperian account of theoretical progress being presented in this book is inapplicable to the actual history of science. According to our account, one thing that is required for a new theory T_j to be a clear-cut advance over a hitherto prevailing theory T_i is that we have $CT(T_j) > CT(T_i)$. But the "Kuhn-loss" thesis, as it is often called, denies (at least on some interpretations) that a scientific revolution ever results in this sort of clear-cut superiority. Whether Kuhn himself originated this thesis, or even held it in quite this form, I am not sure. I have heard it reported (by Grünbaum) that Philipp Frank put forward something very like it. Anyway, the underlying idea is that in a scientific revolution there are gains *but also losses*, so that T_i and T_j are incomparable for content.

There are versions of this thesis that are entirely innocuous and, indeed, fully acceptable. For instance, if no distinction is drawn between theoretical content and empirical content, then it will indeed typically happen that there are losses of content in a scientific revolution; it has been stressed *ad nauseam* in the course of this book that it typically happens that much of the theoretical content of the earlier T_i will be repudiated by T_j. When Feyerabend gave as an example of Kuhn-loss, the disappearance of the problem of the specific weight of phlogiston after the chemical revolution (*1975*, p. 177), he exposed himself to Worrall's retort: 'But, *of course*, losses in *theoretical* content occur in revolutions, the interesting question is whether losses in *empirical* content occur' (1978, p. 70, n 49).

And the thesis can still take an innocuous form when it is restricted to empirical content. It has also been stressed ad nauseam in this book that T_j will, typically, *revise* the empirical content of T_i, if only to a rather slight extent, so that not all of the empirical consequences of T_i will be consequences of T_j. To put it in our notation, in most cases we will *not* have $CT(T_i) \subset CT(T_j)$. However, where such revision occurs, the empirical consequences of T_i that are "lost" may be matched by the gain of counterpart consequences of T_j, and we may still have $CT(T_i) < CT(T_j)$.

However, Feyerabend (1974, *1975*, pp. 177f.) had an interesting argument to the effect that partisans of a new theory T_j disguise the existence of Kuhn-loss and create the illusion of an increase in content by comparing it, not with its real historical predecessor, but with a suitably emasculated and tailored subtheory T_i, extracted from its real predecessor, such that $CT(T_j) > CT(T_i)$ does indeed hold. For instance, they compare Copernicanism, not with its real predecessor, which was Aristotle's geocentric cosmology, but merely with Ptolemy's astronomy.

Someone who opposes Feyerabend here might retort that it is he who creates the illusion of Kuhn-loss by comparing T_j, not with its real predecessor, but with a suitably expanded system T_i such that $CT(T_j) > CT(T_i)$ does not hold. (For instance, he includes Aristotle's theory of perception as a component of his geocentric cosmology; and it must be conceded that Copernicus had no counterpart to that.) It seems clear that a debate over this sort of claim and counter-claim will remain inconclusive in the absence of any criterion or test for theoryhood. Before the advent of the new theory there existed a loosely organised body of theory from which one party can select a large, and the other a relatively small, subsystem with which to compare the new theory for empirical content. But if both parties were persuaded to accept our criterion, such a debate need no longer be inconclusive. Both parties would agree that before the advent of the new T_j, there was a subsystem T_i for which $CT(T_i) < CT(T_j)$ holds. Assume that this T_i and this T_j are both unified theories by our criterion. The proponent of Kuhn-loss would add that T_i was only a part of a larger theory, say T_k; and assume that $CT(T_j) > CT(T_k)$ does not hold. Then the crucial question would be: is T_k a unified theory or only an assemblage of heterogeneous material?

5.35 *Increasing Unity*

I now turn to the main question before us in the elucidation of (B2), namely, when is a theoretical system *more* unified than its predecessor? We are already equipped to deal with one important case. Let S_i and S_j be theoretical systems consisting of the conjunction of a number of theories, where S_j has superseded S_i. And assume that S_j has at least as much testable content as S_i or $CT(S_j) \geq CT(S_i)$. Then we can say that the progression from S_i to S_j involved greater unification if the number of unified theories in S_j is less than that in S_i. In particular, if S_j consists just of *one* theory, whereas S_i consists of two or more, then there is increased unity. An important historical exemplification of this pattern is the case where S_j consists of Newtonian theory and S_i of Galileo's and Kepler's laws. I assume that the conjunction of the latter fails OFR or that $CT(G \wedge K) = CT(G) \cup CT(K)$. And we know that

$CT(N) \geq CT(G \wedge K)$. So this was a case where *one* theory is doing at least as much empirical work as had been done by two separate ones.

We might be tempted to add that if S_i and S_j contain the *same* number of unified theories, but S_j has *more* testable content than S_i or $CT(S_j) > CT(S_i)$, then there is again an increase in unification in the progression from S_i to S_j, and that this holds, in particular, where each of them consists of just *one* theory. But that would be a little too brisk. Let us now turn to cases where what is superseded by a later theory T_j is just one theory T_i that has not failed our theoryhood test. We may take, as our point of departure here, the popular saying, 'Progress in science consists in subsuming more and more under less and less.' This seems to express, in a rough-and-ready way, the idea that science becomes increasingly unified as it progresses. But how should we interpret this saying? We can easily interpret the first part: more is subsumed under T_j than T_i if $CT(T_j) > CT(T_i)$. But what about the second part? Suppose that $I_1, I_2, \ldots I_m$ and $J_1, J_2, \ldots J_n$ are permissible axiom sets (in the sense explained in §5.33) for T_i and T_j respectively. Then we might say that T_j subsumes more, and under less, than T_i if

(1) $CT(J_1 \wedge J_2 \ldots \wedge J_n) > CT(I_1 \wedge I_2 \ldots \wedge I_m)$;
(2) $CT(J_1) \cup CT(J_2) \ldots \cup CT(J_n) < CT(I_1) \cup CT(I_2) \ldots \cup CT(I_m)$.

However, that would be a shade precarious; for remember that a permissible axiom set is not one that is known to comply with rule 5 but only one that is not known not to comply with it; and it might turn out later that T_i can be properly decomposed into a more numerous axiom set, and this might upset the inequality in (2).

If, as we are supposing, the axiom set $I_1, \ldots I_m$ complies with our axiomatisation rules, and in particular with rule 3, the segregation requirement, and if T_i has a theoretical core, then a proper subset of those axioms will constitute this theoretical core T_{iH}, and the remainder will constitute its subsidiary assumptions A_i. And it is very unlikely that any subsequent replacement of this present axiom set for T_i by a more numerous (but permissible) one would disturb the partition of T_i into T_{iH} and A_i; and similarly for the partition of T_j into T_{jH} and A_j. And operating with these partitions will have the further advantage of keeping the present analysis of increasing theoretical unity in step with the previous analysis of increasing theoretical depth. Suppose that if we partition T_i and T_j into $T_{iH} \wedge A_i$ and $T_{jH} \wedge A_j$ the following strict inequalities held:

216

(1) $CT(T_{JH} \wedge A_j) > CT(T_{iH} \wedge A_i)$;
(2) $CT(T_{JH}) \cup CT(A_j) < CT(T_{iH}) \cup CT(A_i)$.

We might interpret these as saying that, in the case of T_j, (1) more is subsumed, and (2) it is subsumed under less, than in the case of T_i. Another way we could express the significance of the above inequalities is by saying that T_j's fundamental assumptions are more organically fertile than T_i's, since they have stronger implications at the empirical level when combined with weaker auxiliary assumptions.

Since for any T we have $T_H \wedge A = T$ and $CT(T_H) = \varnothing$, we can simplify the above to

(1) $CT(T_j) > CT(T_i)$;
(2) $CT(A_j) < CT(A_i)$.

This will serve as a sufficient condition for T_j to be more unified than T_i; but it is clearly too demanding to be a necessary condition as well. We should obviously relax (2) to

(2') $CT(A_j) \leq CT(A_i)$.

Another way in which we could express the same idea is this. Let $CT(T) - CT(A)$ be the set of those consequences of T that are (i) SPIs and (ii) not consequences just of T's subsidiary assumptions alone. (I will later call this the *corroborable content* of T.) Then a sufficient condition for T_j to be more unified than T_i is

(3) $CT(T_j) - CT(A_j) > CT(T_i) - CT(A_i)$.

Can we relax the requirement that the subsidiary assumptions of T_j should not be stronger than those of T_i? If we have $CT(T_j) > CT(T_i)$ together with $CT(A_j) > CT(A_i)$, we may suspect that T_j has the greater testable content merely because it has the stronger subsidiary assumptions. On the other hand, if we do not relax that requirement, it seems unlikely that we can capture the important case where T_j is in the relation of correspondence, in Bohr's sense, to T_i. As we saw in §5.12, it typically happens in that case that there is a parameter, say Φ_1, that T_i treats as a constant, whereas T_j treats it as a variable dependent on some other variable Φ_2 ignored by T_i. If values of Φ_2 are not directly measurable, then T_j will need bridging laws linking Φ_2 to measurable magnitudes, which will mean that the subsidiary assumptions of T_j are in some respects more complex than those of T_i.

I think that it is possible to relax (2) further but without betraying the underlying idea that for T_j to be more unified, or perhaps one should say more unifying, than T_i, its theoretical core or fundamental assumptions must play the decisive role in enabling it to explain more empirical laws. As my point of departure here I will take what was said in the previous section about the case, exemplified by the incorporation of Fresnel's theory into Maxwell's, where the deeper T_j strictly entails the less deep T_i. We there allowed for the possibility that T_j has stronger subsidiary assumptions by requiring that T_j should still dominate T_i when the latter is fitted out with the former's subsidiary assumptions, or

$$\mathrm{CT}(T_{j_H} \wedge A_j) > \mathrm{CT}(T_{i_H} \wedge A_j).$$

For the only difference now between the premises of *these* two theories is over their fundamental assumptions, and it is obvious that those of T_j are more fecund or organically fertile.

We cannot employ this method of comparison in cases where T_j revises T_i, as it will do if it is in the relation of correspondence to T_i, but we can employ one rather like it. We could look for an $A_j{}'$ such that (i) $A_j{}'$ entails A_i; (ii) $\mathrm{CT}(A_j{}') \approx \mathrm{CT}(A_j)$; and (iii) $A_j{}'$ is as good a partner, from an organic fertility point of view, for T_i that we can find, given (i) and (ii). And if we had

$$\mathrm{CT}(T_{j_H} \wedge A_j) > \mathrm{CT}(T_{i_H} \wedge A_j{}')$$

we could again say that the fundamental assumptions of T_j are more fecund or organically fertile than those of T_i.

We may summarise the conditions for T_j to be more unified, or more unifying, than T_i as follows:

(1) there is a permissible partition of the axiom set T_i into fundamental assumptions T_{i_H} and subsidiary assumptions A_i; and likewise for T_j;

(2) $\mathrm{CT}(T_j) > \mathrm{CT}(T_i)$;

(3) either (a) $\mathrm{CT}(A_j) \leq \mathrm{CT}(A_i)$ or (b) there is an $A_j{}'$ such that
  (i) $A_j{}'$ entails A_i
  (ii) $\mathrm{CT}(A_j{}') \approx \mathrm{CT}(A_j)$
  (iii) $\mathrm{CT}(T_{j_H} \wedge A_j) > \mathrm{CT}(T_{i_H} \wedge A_j{}')$
  (iv) there is no $A_j{}''$ such that $A_j{}''$ entails A_i and $\mathrm{CT}(A_j{}'') \approx \mathrm{CT}(A_j)$ and $\mathrm{CT}(T_{i_H} \wedge A_j{}'') > \mathrm{CT}(T_{i_H} \wedge A_j{}')$.

As was to be expected, our conditions for greater unity are virtually identical with those for greater depth: if one theory goes deeper than a rival, then it will explain, in a unified way, more laws at the empirical level than its rival can explain. Components (B1) and (B2) of our aim (B*) are really alternative expressions of the same demand, which I will henceforth call (B1-2) or the demand for greater depth-cum-unity.

5.36 *Exactitude*

It remains to consider (B4), the demand for increasing exactitude. The discussion of this will be a shade technical. I will argue that, rather as (B1) and (B2) collapse into (B1-2), so (B3), the demand for increasing predictive power, and (B4) collapse into one demand, which I will call component (B3-4) of our version of the Bacon-Descartes ideal. This conclusion will be reached at the end of this subsection.

The exactitude of a *statement* must not be confused with that of the predicates occurring in it. One statement may be *less* exact than another although all the predicates occurring in it are *more* exact than their counterparts in the other. Let F_1' be a more exact (determinate, precise) predicate than F_1 in the sense that it is logically true that anything that is F_1' is F_1 but not vice versa. (For instance, F_1 and F_1' might be 'blue' and 'sky blue'.) And let F_2' and F_3' be similarly more exact than F_2 and F_3. Let the statements h and h' say respectively: $\forall x(F_1 x \rightarrow ((F_2 x \rightarrow \sim F_3 x))$ and $\forall x(F_1' x \rightarrow ((F_2' x \rightarrow \sim F_3' x))$. It is easily seen that h is a more exact statement than h'. If F' is more determinate than F then $\sim F'$ is less determinate than $\sim F$. And these two statements are respectively equivalent to $\forall x(\sim F_1 x \vee \sim F_2 x \vee \sim F_3 x)$ and $\forall x(\sim F_1' x \vee \sim F_2' x \vee \sim F_3' x)$.

Then how, if at all, can we compare the exactitude of statements? Well, universal statements, and it is with these that our aim for science is primarily concerned, can always be reformulated in such a way that they ascribe *one* predicate, which may be very complex, to *everything*. For instance, we could reformulate the above two hypotheses, h and h', as $\forall x\ Hx$ and $\forall x\ H'x$ where H and H' are respectively $\sim F_1 \vee \sim F_2 \vee \sim F_3$ and $\sim F_1' \vee \sim F_2' \vee \sim F_3'$. And if we could go on to say that the predicate ascribed to everything by h is more determinate than the one ascribed by h', we could conclude that h is a more exact description of the world than h'.

But under what conditions would we be in a position to compare two such predicates? To make the task more manageable let us concentrate on the comparative exactitude just of the *empirical* content of two rival theories, T_i and T_j. I think that no injustice will result from this. There are theories, for instance in economics, that are mathematically elegant but disappointingly infertile with respect to predictive consequences. I am inclined to regard barren precision of that kind as a kind of scholasticism rather than as genuine exactitude. On the other hand, a deep theory that yields exact experimental laws is bound to contain exact equations at the theoretical level since its theoretical ontology makes a vital contribution, in the sense recently elucidated, to its empirical content. So it is, I think, reasonable to judge precision at the theoretical level indirectly, by its fruits at the empirical level.

So let us assume that the empirical content of the universal theories T_i and T_j can be represented by, respectively, $\forall x\, H_i x$ and $\forall x H_j x$, where H_i and H_j may well be very complex predicates. There is one case where we could undoubtedly say that H_j is more exact (determinate, precise) than H_i, namely if it were logically true that anything that is H_j is H_i, but not vice versa. In that case T_j would strictly entail T_i. What about cases where T_j, far from entailing T_i, conflicts with it? Well, suppose that there were a predicate H_i' of which we could say: (i) H_i and H_i' are *equally* precise and (ii) H_j is *more* precise than H_i' because it is logically true that anything that is H_j is H_i' but not vice versa. In that case we could clearly conclude that H_j is more precise than H_i. But when could we say of an H_i' that it has the same precision as a given H_i? My answer is that we could say this if (and only if) $\forall x\, H_i x$ and $\forall x\, H_i' x$ are incongruent counterparts in the sense explained in §5.1. It will be remembered that the definition for incongruent counterparts expressed in a qualitative language requires them, in the case where they are universal statements, to exclude the same number of Q-predicates, while that for a quantitative language requires the Q-predicates excluded by one of them to be in one:one correspondence with those excluded by the other. Satisfying this condition is a guarantee of equal precision, whereas two comparable statements that might otherwise look like incongruent counterparts but fail to satisfy it, cannot be equally precise. So my conclusion is that the condition for T_j to have more overall precision at the empirical level than T_i is the same as the condition for T_j to have more empirical content than T_i.

Thus it turns out that our optimum aim for science is less diverse than it may have originally appeared. It consists of (A*), the requirement concerning the possible truth of an accepted system of scientific statements, plus two demands, (B1-2) and (B3-4), which could be epitomised as the demands for increasing *depth* and increasing *width*: science should try to penetrate to ever deeper levels with theories having an ever wider coverage at the empirical level.

5.37   *Simplicity Again*

I have claimed that (B*) is as ambitious and comprehensive as it can be without overreaching itself and becoming infeasible or incoherent or violating any of the other adequacy conditions laid down in §4.1. Now (B*) does not include the idea of verisimilitude, nor does it explicitly mention simplicity. And many philosophers would say that it ought to include one, or both, of these. I will consider its omission of verisimilitude in §8.1 and turn now to simplicity.

In §3.42 a distinction was drawn between the methodological thesis that a simpler hypothesis should be preferred, other things being equal, to a less simple competitor, and the epistemological thesis that it should be preferred because it is more likely to be true. The epistemological thesis had to be jettisoned, and the question of the justification of the methodological thesis was left over for later consideration. We must now ask whether our modified version of the Bacon-Descartes ideal provides a justification for it.

The idea of simplicity in science arises at two distinct levels. One can ask of two rival level-3 experimental law-statements, or of two rival level-4 theories, whether one is simpler than the other. It was primarily the former question that we addressed in §3.4. We there considered the admirably clear-cut and straightforward criterion for simplicity proposed by Jeffreys and Wrinch (1921) and independently by Popper (*1934*), namely the "paucity of parameters" criterion. And as we there saw, if law l_1 is simpler than l_2 by that criterion, then l_1 is more determinate, has more precision, and is more testable, than l_2. The demand for increasing simplicity in this sense is implicit in component (B3-4) of our aim for science.

Now consider the question of the simplicity of theories. It can rightly be claimed that Newtonian theory is, in an important sense, simpler than the conjunction of Galileo's and Kepler's laws. But this cannot mean that NM yields experimental laws that are simpler, in the "paucity-of-parameters" sense, than those of Galileo and Kepler, for it does not. What it surely does mean is that NM achieves a theoretical unification, by bringing phenomena that had been regarded as belonging to two different domains under one unified set of laws. And the demand for increasing simplicity in this sense is implicit in component (B1-2) of our aim. So the methodological simplicity thesis, having been deprived of its epistemological justification, is given an alternative justification by our version of the Bacon-Descartes ideal.

5.38 *Is Our Aim Coherent?*

The elucidation of (B*) is now complete. We have found that it really consists of just three components: (A*), or the demand for possible truth; (B1-2), or the demand for increasing depth-cum-unity; and (B3-4), or the demand for increasing testable content. Is it coherent? It could easily happen, of course that T_j is (B)-superior to T_i but (A*)-inferior just because it has been falsified and T_i has not. We solve this by imposing a lexicographic ordering: that is, for T_j to be (B*)-superior to T_i we require, first that T_j satisfies (A*), and second, that T_j is (B)-superior to T_i. If T_j fails to satisfy (A*), it is not a candidate for (B*)-superiority.

Could (B1-2) and (B3-4) pull in opposite directions? Well, there is a logical possibility of their doing so. We earlier (in §5.42) cited Kepler as an outstanding example of someone who produced a superb system of level-3 experimental laws but who did not succeed in tying them together into a unified theory with a theoretical core that makes a vital contribution to its empirical content.

Now we could imagine that Newton's NM, instead of being followed by Einstein's GTR, has been followed by a latter-day Kepler who had come up with an even more superb system, call it S, of experimental laws not tied together into a unified theory, and that the relation of his S to NM was very similar to that of just the empirical content of GTR to the empirical content of NM. Were that to happen, an adherent of (B*) would face a dilemma: S would be superior to NM with respect to (B3-4) but inferior with respect to (B1-2); indeed, S would have no depth and would not even constitute a *theory*.

But although this exists as a logical possibility, there is a consideration that suggests that it is virtually impossible. This consideration has to do with the way in which a new theory T_j revises the empirical content of a successful earlier theory T_i when it is in the relation of correspondence to the latter, as was GTR to NM. Such a revision has the character of a certain *distortion* of the experimental laws of T_i. This distortion is, for the most part, very slight; indeed, there may be very few places where the predictive consequences of T_i differ discernibly from their counterparts in T_j. On the other hand, the distortion is highly *systematic*: all T_i's experimental laws may undergo a certain revision. And this means that T_j will revise T_i (though perhaps only very slightly) in places where T_i has been *performing very well* and not just in places, if any, where it has been in some empirical difficulty. For instance, NM had been performing very well with regard to the observed behaviour of all the planets, with the sole exception of Mercury, whose perihelion, unlike that of the others, had been observed to precess very slowly. Now one way in which GTR systematically revised NM was by yielding the consequence that there is a precession of the perihelion of *all* planets. So NM was as incorrect in implying that Jupiter's perihelion does not precess as in implying that Mercury's does not. However, GTR adds that the amount of their precession is so small as to be practically undetectable in the case of planets other than Mercury.

Now there is nothing unexpected or mysterious about this sort of small but systematic distortion of the experimental laws of T_i at the behest of T_j, given that T_j has a distinctive theoretical core involving new fundamental assumptions, to use Einstein's term, or a 'new, and powerful, unifying idea', to use Popper's phrase. Consider the revision that the classical gas law underwent when incorporated into the kinetic theory of gases. According to the theoretical core of the latter, gases are composed of huge numbers of molecules. Now molecules come in different *sizes* and they exert a certain *mutual attraction*; and neither of these factors was taken into account by the classical gas law, $PV = rT$. So we can understand why this should have been revised into the more complicated law $(P + a/V^2)(V - b) = rT$, where b takes account of their size and a/V^2 of their mutual attraction. Again, it is entirely understandable that NM, with its central assumption of universal gravitational attraction, should lead to Kepler's laws being revised so as to take interplanetary attractions into account.

But now consider the problem confronting a scientist who is a hard-nosed, tough-minded empiricist, and who hopes to supersede a powerful existing theory T, which has performed very well though it has run into some empirical difficulties, with a system S of experimental laws devoid of any theoretical core. He will be guided by empirical and methodological considerations but not, of course, by any speculative, theoretical ideas. Empirical considerations may encourage him to distort certain of T's experimental laws to secure a better fit with experimental measurements, but hardly to spread such distortions right through the whole corpus of T's, for the most part, highly successful system of laws. As to methodological considerations: well, he may be motivated by an ideal of simplicity. But this is likely to push him in the wrong direction. As we have frequently noticed, if T is in the relation of correspondence to its predecessor(s), then the laws yielded by it will, typically, be *less* simple than their predecessors just because some parameters that had been treated as constants are now turned into dependent variables. (The comparative testability criterion developed in §§5.15–5.17 does not penalise laws for being in this way more discriminating, though no less determinate, than their predecessors.) For instance, the Newtonian counterparts of Galileo's and Kepler's laws are assuredly less simple than the latter. The kind of simplication that NM achieved, namely theoretical unification, is just what a mere assembly of experimental laws never can achieve.

The view for which I am arguing had been vigorously affirmed by Whewell. In the 1847 preface to the second edition of his (*1840*) he wrote:

> Perhaps one of the most prominent points of this work is the attempt to show the place which discussions concerning Ideas have had in the progress of science. The metaphysical aspect of each of the physical sciences ... is a necessary part of the inductive movement. ... Physical discoverers have differed from barren speculators, not by having *no* metaphysics in their heads, but by having *good* metaphysics while their adversaries had bad; and by binding their metaphysics to their physics, instead of keeping the two asunder. (Pp. ix–x)

If this view is correct, then our aim is coherent in the sense that, if T_i is superior to T_i with respect to (B1-2), then it will necessarily be superior with respect to (B3-4); and if T_j is superior with respect to (B3-4), then it will not in fact be inferior, and may well be superior, with respect to (B1-2). Whether our aim satisfies all the other adequacy requirements is a question to which we will return in chapter 8.

6

....

Deductivism and
Statistical Explanation

6.1 The Challenge from Microphysics

Our conjecturalist version of the Bacon-Descartes ideal endorses proposition (III), the deductivist thesis. And this means that it requires the explanandum of a scientific explanation to be deducible from the explanans. Clearly, for premises *h* to constitute an explanans for an explanandum *e* there must be *some* kind of derivability of *e* from *h*; and according to proposition (III), there is only one valid kind of derivability, namely deducibility.

There are three ways in which a proposed explanation put forward in the actual course of science may fall short of the deductivist ideal. The first way is rather innocuous. It must be admitted that scientific explanations are seldom, in scientific practice, spelt out with sufficient rigour to satisfy a formal logician. Even in mathematics an informal outline of a proof is often offered in lieu of the proof itself. But this means, not that the explanandum is not deduc*ible* from the explanans, but only that it has not been deduc*ed* in a formally rigorous way. The premises may be sufficiently full and exact to entail the conclusion, even though the derivation has been carried out in a semi-intuitive way. As to the second: C. G. Hempel (*1965*, pp. 238, 423–24) introduced the term *explanation-sketch* to denote the case where, on the one hand, the premises, as they now stand, are *not* sufficient to entail the explanandum, but where, on the other hand, it is clearly indicated what sort of additional premises are needed for the explanandum to be deducible from the explanans; here, for example, an initial condition to be ascertained, there a background assumption to be articulated. As Hempel put it, an explanation-sketch 'needs to be filled out by more specific statements; but it points into the direction where these statements are to be found' (p. 238). Finally, certain statements may be claimed to explain an explanandum although the latter is admittedly not deducible from them and no indication is given as to how they might be filled out to render it deducible from them.

I hold that there is only *one* serious challenge to the deductivist ideal for scientific explanations, but that it is *very* serious. Many of the theories in contemporary physics that conform admirably with the demands of (B*) for depth and predictive power have an essentially *statistical* or probabilistic character; and it is widely held that such theories, although they explain much, cannot provide *deductive* explanations for what they explain. If we accepted this, we would face a painful dilemma. We *might* resolutely retain proposition (III) as part of our optimum aim for science and declare that statistical theories fail to provide genuine explanations. (Wolfgang Stegmüller seems to have reached this negative conclusion. After laying down 'three minimal requirements of adequacy' for scientific explanations, 1980, p. 38, he proceeded 'to the conclusion that there is no concept of statistical explanation which satisfies the three requirements' p. 48.) And since (B*) calls for deep, etc. *explanations*, we would have to conclude that statistical theories are not what is really wanted. This would be a serious violation of the fourth of the adequacy requirements, laid down in §4.1, for any proposed aim for science, namely the impartiality requirement. This requires our aim, among other things, to be impartial between determinism and indeterminism. If physical nature is, ultimately, indeterministic, then statistical theories will be the best that science can achieve as it penetrates to deeper levels.

The other alternative, if we accepted that statistical theories do not provide deductive explanations, would be to concede that an explanans may be related to its explanandum in some nondeductive way, or that there is a kind of derivation of conclusions from premises that is not deductive but is valid nevertheless. This would mean reneging on proposition (III). We saw in §1.4 that throwing out proposition (III) tends to encourage a general lowering of intellectual standards; it would be a sad irony if such a lowering were necessitated by the need to accommodate some of the finest achievements of contemporary physics.

This is the challenge that I will try to meet in this chapter. Before I turn to it, however, I will briefly examine what I consider to be a much less serious challenge to the deductivist ideal of scientific explanation. It results from the confluence of two distinct tendencies in recent philosophy, namely inductivism and ordinary language philosophy.

6.11 *Nondeductive "Explanation"*

Humean scepticism results from the thesis that any alleged piece of knowledge is *knowledge* only if it has been validly derived from experience. I quoted earlier Russell's statement that Humean scepticism could be avoided if we were allowed *just one* exception to that thesis. But an inductivist is likely to feel uneasy at the idea of a single departure from principles that are supposed otherwise to reign universally: it looks so very ad hoc. He is likely to prefer some systematic modification of those principles. More particularly, if he holds, as presumably he does, that deductivism sets impossibly high standards with respect to establishing scientific laws on a basis of observed facts, he is likely to extend this view to the employment of scientific laws in explanations of observed facts. After all, if evidence *e* could provide "inductive proof" for a hypothesis *h* not entailed by *e*, then by parity of reasoning it would seem that a hypothesis *h* could explain an explanandum *e* not entailed by *h*.

Support for such a nondeductivist view of explanation is also provided by ordinary language philosophy, which holds that how a term is ordinarily used provides a norm for its correct use. And it is a fact that in everyday discourse a proposition *h* often is accepted as a satisfactory explanation of an explanandum *e* although *e* could not have been predicted beforehand on the basis of *h* because *h* does not entail *e*.

One philosopher who has combined these two negative reactions, from inductivism and from ordinary language, to the deductivist ideal of explanation, is Stephen Barker. Apropos the former he declared: 'But inductive proof is only one of the types of nondemonstrative reasoning that are employed both in scientific inquiries and in everyday thinking. The reasoning typically involved in explanation is another' (1961, p. 256). And apropos the latter he declared: 'Thus, for instance, if the patient shows all the symptoms of pneumonia, sickens, and dies, I can then explain his death—I know what killed him—but I could not have definitely predicted in advance that he was going to die; for usually pneumonia fails to be fatal' (p. 271). Let us look into this last claim. Let *h* say that Jones went down with pneumonia on a certain date, and *e* say that a month later Jones was dead. Let *b* be background knowledge that includes the statistical fact that pneumonia is not usually fatal, though it is sometimes. And let *e'* say, falsely, that a month later Jones had recovered from this attack of pneumonia. Then we presumably have:

$$p\,(e',\,h\cdot b) > p\,(e,\,h\cdot b)$$

and this suggests that if *h* can "explain" *e* given that *e* turned out to be true, then *h* could have "explained" *e'* at least as well had *e'* turned out to be true. But *e'* entails $\sim e$. Thus *h*, as well as "explaining" the

227

occurrence of the event depicted by e, could equally have "explained" its *non*occurrence. I hold that such a dual-purpose "explanation" that will serve whichever way things go, does not provide a genuine explanation of the way things actually went. On the deductivist view of explanation, of course, it cannot happen that a (consistent) explanans h could serve to explain either e or e' where e' entails $\sim e$. What a deductivist can say about the present case is that this h, though insufficient by itself to explain this e, is statistically relevant, to use Wesley Salmon's (*1971*) term, to e; for we have:

$$p\,(e,\,h \cdot b) > p\,(e,\,\overline{h} \cdot b)$$

On the deductivist view, one can explain no more after the event than one could have predicted, with the same initial conditions, beforehand. What one could have predicted, given just the information that Jones has recently gone down with pneumonia, is that the chances that he will be dead within a month have significantly increased; and this is the most that h can explain.

6.12 *The Question of Detachment*

I turn now to the serious challenge. It might be restated thus:–
Our aim for science must allow for the possibility that indeterminism is true; but if indeterminism is true, there are aspects or levels of physical reality a true description of which would involve laws of an irreducibly statistical or probabilistic nature; and such laws cannot provide deductive explanations; they provide either nondeductive explanations or no explanations at all.

Since my examination of this claim will inevitably be somewhat technical I will state my conclusions in advance. I will concede that a statistical theory *interpreted according to determinism* cannot provide deductive explanations; but then, I will argue, such a theory so interpreted cannot provide *explanations*. Concerning statistical theories explicitly interpreted in an indeterministic way, as asserting the existence of objective chances "out there" in nature, I will concede that they cannot provide deductive explanations of single events; but, I will argue, this is to be expected because the occurrence at a particular time of a chance event simply defies all explanation. However, they can deductively explain statistical patterns and regularities exhibited by very large numbers of chance events, which is all that they should be expected to explain. (There is much agreement, but also some disagreement, between the position defended in this chapter and that of Peter Railtc: in his (1978), which will be discussed in §6.33.)

The hallmark of a deductive explanation is that its conclusion, the explanandum, can be *detached.* Suppose that our explanandum is once more that Jones has died; but this time the explanans consists of the major premiss 'Anyone who swallows a gram or more of arsenic is dead within six hours' and the minor premiss 'Jones swallowed two grams of arsenic at 0200 last Monday'. Let us introduce the following abbreviations:

$F(x,t)$	$=$	x swallowed a gram or more of arsenic at time t
$G(x,t)$	$=$	x is dead by time t
Δt	$=$	six hours
t_o	$=$	0200 last Monday
a	$=$	Jones

We may now represent our deductive explanation by the following schema:

$$\frac{\forall x \forall t \; [F(x,t) \rightarrow G(x,t + \Delta t)]}{G(a,t_o + \Delta t)} \quad F(a,t_o)$$

Had we been in possession of these premises very shortly after t_o, we could have derived and detached the conclusion as a prediction that we could have asserted *categorically*, without needing to hedge it by a ceteris paribus clause to safeguard it against the arrival of new and disturbing information. No augmentation of these premises that leaves them consistent could yield a conclusion that overrules the present prediction/explanandum.

6.13 *Hempel's Schema*

Now suppose that the above major premiss, which has the character of an exceptionless law, is replaced by a statistical law that says that the statistical probability that anyone who has swallowed a gram or more of arsenic is dead within six hours is .999, the minor premiss and the explanandum being as before. Would we now have an explanation of Jones's death? Hempel, at the time of his (1965), would have answered that we do indeed have here an explanation of this explanandum, though an 'inductive-statistical' one rather than a deductive one; and he would have represented it by something like the following schema:

$$\forall x \; \forall t \; (P \; [G(x, t + \Delta t), F \; (x, t)] \; = \; .999)$$
$$F \; (a, t_o)$$
$$\overline{\overline{}} \; [.999]$$
$$G(a, t_o + \Delta t)$$

The double horizontal line with its numerical index, which replaces the single line in the previous schema, is used by Hempel to indicate that the premises confer not full certainty but only 'near-certainty' on the conclusion (p. 394).

Hempel pointed out a decisive difference between the previous schema and the present one: the premises of the latter *can* be augmented in ways that would leave them consistent and overrule the present conclusion:

> For a proposed probabilistic explanation with true explanans which confers near-certainty upon a particular event, there will often exist a rival argument of the same probabilistic form and with equally true premises which confers near-certainty upon the nonoccurrence of the same event. . . . *This predicament has no analogue in the case of a deductive explanation.* (pp. 394–395)

For instance, we might expand our minor premiss to say that a stomach pump was used on Jones ten minutes after he swallowed the arsenic; and perhaps there is a statistical law that says that, in the case of anyone who has swallowed a gram or more of arsenic *and* had his stomach pumped out within ten minutes of doing so, the statistical probability that he is dead within six hours is only .1. Using $H \; (x, t)$ as an abbreviation for 'x had his stomach pumped out at time t' and δt for 'ten minutes' we could represent these augmented premises and the conclusion to which they give 'inductive-statistical' support thus:

$$\forall x \forall t \; (P[G(x, t + \Delta t), F(x, t) \cdot H \; (x, t + \delta t)] \; = \; .1)$$
$$F(a, t_o) \cdot H \; (a, t_o + \delta t)$$

$$\overline{\overline{}} \; [.9]$$
$$\sim G(a, t_o + \Delta t)$$

The previous conclusion was not detachable: it has here been overruled by one supported by premises entirely consistent with the previous premises.

Hempel's solution for this problem was to introduce what he called 'the requirement of maximal specificity' (p. 399): this says that in a statistical explanation of the form

$$
\begin{array}{c}
P\,(G,\,F) = r \\
\underline{\underline{Fa}} \\
Ga
\end{array}
\quad [\text{r}]
$$

the premises must be such that, if augmented by the set of all currently accepted scientific statements, the value of r would remain unchanged. We will consider this requirement later.

6.2 Statistical Explanation within Determinist and Indeterminist Settings

In the meanwhile let us ask how a determinist would view a Hempel-type statistical explanation that is assumed to satisfy the requirement of maximal specificity. Assume that its minor premiss describes initial conditions obtaining at or before t_o, while its conclusion predicts an occurrence at the later time $t_o + \Delta t$, and that the value of r is less than 1. A determinist would insist that, if its premises are true, then they could in principle be augmented in such a way that: (i) they remain true; (ii) the minor premiss still does not refer to any conditions obtaining after t_o; (iii) the value of r goes to either 1 or 0; for at t_o there must have already existed sufficient causal preconditions to determine the occurrence, or else the nonoccurrence, at $t_o + \Delta t$ of the event described by the conclusion below the double line. He would add that, since the explanation satisfies the requirement of maximal specificity, we could not in practice augment its premises in this way, at least at the present time; but this means that our present scientific knowledge in this area is essentially incomplete. Whether a determinist could allow that events can be *explained* by statistical laws we will consider later. Let us now turn to indeterminism.

6.21 *The Disintegration Law*

An example of an indeterministic law will be helpful and I will take over one used by Hempel, namely the disintegration law for radon atoms. A disintegration law says that atoms of a certain kind have a certain half-life. This means that the probability that an atom will disintegrate during its next half-life is $\frac{1}{2}$. The half-life of radon atoms is 3.82 days. Letting $F(x, t)$ and $G(x, t + \Delta t)$ mean respectively 'x is a radon atom at time t' and 'x has disintegrated by time $t + \Delta t$', where Δt is 3.82 days, we could formulate the disintegration law for radon atoms as

$$\forall x \; \forall t \; (P[G(x, t + \Delta t), F(x, t)] = \tfrac{1}{2}).$$

According to Max Born, whose authority I will accept, the distintegration law makes this probability value unconditional and autonomous in a rather striking way: it is 'quite impossible to affect the disintegration by any means whatever, whether by high or low temperatures, electric or magnetic fields, or any other influences' (*1963*, p. 237). This suggests that it is a matter of pure chance whether a particular radon atom disintegrates during the next 3.82 days or not, for nothing can make it or even encourage it to do so, or stop it or even discourage it from doing so. On this strongly indeterministic interpretation, the probability value of $\frac{1}{2}$ in this law reflects, not gaps in our knowledge, but irreducible indeterminacies "out there" in Nature. A Laplacean Demon, equipped with an exact knowledge of the state, at some instant, of every object in the universe plus an entire knowledge of all the laws of nature, could not possibly predict whether a particular atom will disintegrate during its next half-life, according to the disintegration law. As Jeans put it: 'Thus an atom had always the same chance of disintegrating, whatever its past history or present state might be. Here was a natural law of a kind hitherto unknown to science' (*1947*, p. 312).

There turns out to be an unexpected affinity between a statistical or probabilistic law of this kind and an exceptionless law involving what is sometimes called nomic necessity. Suppose one held that it is a sheer physical necessity that anyone who swallows a gram or more of arsenic inevitably dies within six hours: neither stomach pumps nor anything else can possibly save him. For most purposes this "law" could be expressed, as we expressed it earlier, by $\forall x \forall t \; [F(x,t) \rightarrow G(x,t + \Delta t)]$ where $F(x,t) = $ 'x swallowed a gram or more of arsenic at time t', $G(x,t + \Delta t) = $ 'x is dead by time $t + \Delta t$' and $\Delta t = $ 'six hours'. But this does not bring out this "law's" element of nomic necessity: it is consistent as it stands with the possibility that, for instance, an antidote to arsenic could be developed but never will be because it would be prohibitively expensive. I will now propose a formulation that is intended to bring out this nomic element. It will involve some additional technicalities; but I think that these will help us subsequently to do justice to essentially indeterministic laws such as the disintegration law.

The idea we want to catch is that, given just that x has swallowed a gram or more of arsenic at t, then *no matter what* the other circumstances at t may be, x is bound to be dead by $t + \Delta t$. Let us introduce the predicate variables ϕ, ψ, ... to range over properties, rather as the individual variables x, y, ... range over individuals. We might try to bring out the element of nomic necessity in our "law" by reformulating it thus:

$$\forall \phi \ \forall \psi \ \forall t \ \forall x \ \forall y \ (x \neq y) \ [(F(x, t) \wedge \phi \ (x, t) \wedge \psi \ (y, t)) \rightarrow \\ G \ (x, t + \Delta t)]$$

This says that if x is F at t (has swallowed a gram or more of arsenic), then no matter what other properties x may have at t and no matter what properties anything else may have at t, x is G at $t + \Delta t$ (dead within six hours).

6.22 *Future-Independent Properties*

However, to avoid inconsistencies we must impose two restrictions on the scopes of ϕ and ψ. I begin with ϕ. This must not range over any property H such that it is logically or conceptually impossible that something is both F and H at the same time. For instance, if $H(x, t)$ were 'x swallows nothing at t' it would be logically impossible that something is both F at t (swallows at least a gram of arsenic at t) and H at t; and if $H(x, t)$ were 'x is dead at t' or 'x is a marble statue at t' it would be conceptually impossible that something is both F and H at t. We may combine these two kinds of incompatibility, logical and conceptual, under Leibniz's term 'incompossibility': ϕ must not range over properties incompossible with $F(x, t)$. And we must likewise require that ψ does not range over any property H such that $H(y, t)$ is incompossible with $F(x, t)$ where $x \neq y$. No doubt the scope of ψ will thereby be less restricted than that of ϕ, but it will be somewhat restricted: for example, if $H(y, t)$ meant that y has destroyed every last gram of arsenic by t, then it would be incompossible with $F(x, t)$.

We must further require ϕ not to range over any property H such that $H(x, t)$ is incompossible with $G(x, t + \Delta t)$. (I am indebted to Oddie here.) Consider a property that most of us usually believe ourselves to possess, namely the property of having at least twelve more hours of life ahead of us. If this property were within the scope of ϕ, our "law" would say that anyone who swallows a gram of arsenic is dead within six hours even if he lives for at least twelve hours after swallowing it. And we must likewise require ψ not to range over any property H such that $H(y, t)$ in conjunction with $F(x, t)$ is incompossible with

$G(x, t + \Delta t)$. An example would be where $H(y, t)$ means: 'x is a miracle worker who cures anyone who swallows arsenic at t.'

For the possession by x of a property H at t to be incompossible with x's possession of the property G at $t + \Delta t$, the property H must have what I will call a *future-dependent* character. At the time I am writing this, call it t_o, Prince Charles is the heir apparent to the English throne and Susan is pregnant with her first child. Suppose we now say that at t_o Charles is a King-to-be and Susan is a mother-to-be. These two properties are future-dependent: whether Charles and Susan do in fact now possess them depends on how things will work out. If Charles will in due course be crowned, and Susan will give birth to a healthy baby, then they do now possess these properties; otherwise not. By contrast, the property of being heir apparent at t_o and the property of being pregnant at t_o are *future-independent*: nothing whatever that happens subsequently could deprive Charles of the one or Susan of the other. I say that a property H is future-independent if, given that an individual a was H at t_o, the future having subsequently unfolded as it did, then a would still have been H at t_o no matter how differently the future might have unfolded and even if the universe had been annihilated immediately after t_o. And I stipulate that ϕ and ψ are to range over future-independent properties only. Since the formulation of our "law" is already rather cumbersome, I will not try to incorporate these two restrictions into it, but will take it to be governed by the qualification: 'where $\phi(x,t)$ and $\psi(y,t)$ are compossible with $F(x,t)$ and ϕ and ψ do not range over future-dependent properties'.

It might be supposed that this idea of a future-independent property is in conflict with what was said in §3.12 about level-1 statements being corrigible because they impute *dispositional* properties to things, and hence carry predictive implications that may get refuted by future developments. But that is not so. Suppose that a fragile porcelain vase is acquired at t_o by a wealthy art collector who subsequently, at t_1, has it treated with a special resin which makes it very strong, though leaving its outward appearance unchanged. Then at any time during its existence up to t_1 this vase possessed the future-independent property of being fragile at that time: if it had then been dropped it would have broken. And at any time during its existence after t_1 it possessed the future-independent property of being nonfragile at that time: if it had then been dropped it would not have broken. This vase would never have possessed the future-dependent property that most vases possess of being perpetually fragile. A statement imputing a dispositional property to an

234

object at a certain time is corrigible because the object may subsequently behave in a way that indicates that it did not have that property *at that time*; it is not refuted if the object is subsequently altered in a way that deprives it of that property. Suppose that, shortly before the attempt on his life, President Reagan had had a medical checkup and been pronounced completely fit. That pronouncement would have been corrigible, and would have been refuted if he had gone down with a coronary attack a few hours later; it is not refuted by the fact that he went down with a bullet in his lung a few hours later.

6.3   The Objective Probability of a Future Event

Let us now consider the affinity between laws involving nomic necessity and the kind of strongly indeterministic law we were considering. I gave as a first formulation of the disintegration law for radon atoms $\forall x \forall t (P[G(x,t + \Delta t), F(x,t)] = \frac{1}{2})$, where Δt is the half-life of a radon atom, namely 3.82 days. But this formulation leaves open the possibility that if, in addition to knowing that x is a radon atom, we *also* knew that, say, it had been subjected to very high temperatures or was in a strong electro-magnetic field, then our estimate of the chance of its decaying during the next 3.82 days would alter. Yet according to Born, any such possibility should be excluded. We can exclude it by strengthening our formulation to read:

$$\forall \phi \forall \psi \forall t \forall x \forall y (x \neq y)(P[G(x,t + \Delta t), F(x,t) \cdot \phi(x,t) \cdot \psi(y,t)] = \frac{1}{2}.$$

This says that if x is an undecayed radon atom at time t, then no matter what other properties x may have at t and no matter what properties anything else may have at t, the probability that x will disintegrate during the next 3.82 days is always $\frac{1}{2}$. (Remember that ϕ and ψ range over future-independent properties only and ones that are compossible with $F(x,t)$.)

I will now introduce the idea of the objective probability, at a given time t_o, that a certain event will occur at, or will have occurred by, the later time $t_o + \Delta t$. Laplace envisaged an infinite intelligence ascertaining the exact state of every thing in nature by a single, instantaneous intuition. Let us denote such an ideally complete knowledge of all initial conditions coexisting at t_o, or of the state of the world at t_o, by W_{t_o}. Assume that no future-dependent predicates enter into W_{t_o}. Laplace's Demon was also endowed with a complete knowledge of all the laws of nature, which we may denote by K. And Laplace's idea was that, given W_{t_o} plus K, then to 'an intelligence sufficiently vast to submit these data to analysis ... , nothing would be uncertain and the future, like the past, would be present to its eyes' (*1812*, p. 4). But if the disintegration law is true, there are many future possibilities to which $K \wedge W_{t_o}$ gives a probability less than 1 and greater than 0. However, we can use Laplace's idealised concepts, which I am denoting by K and W_{t_o}, to define the *objective probability* at t_o that a particular event will occur at, or by, $t_o + \Delta t$. Let $G(a,t_o + \Delta t)$ be the statement that the event does so occur. If this statement has a determinate *logical* probability relative to $K \wedge W_{t_o}$, then that gives us the objective probability at t_o that the event will occur at, or by, $t_o + \Delta t$. This objective probability, which I will represent by $P_{t_o}[G(a,t_o + \Delta t)]$, is the best possible estimate that even a Laplacean Demon could make at t_o concerning its occurrence. So let us lay down:

$$P_{t_o}[G(a,t_o + \Delta t)] = {}^{\mathrm{df}} p\,[G\,(a,t_o + \Delta t),\, K \cdot W_{t_o}].$$

The objective probability of $G(a,t_o + \Delta t)$ at all times from $t_o + \Delta t$ on will of course be 1 or 0 according to whether the event does or does not occur by $t_o + \Delta t$. In rather the same way, the probability that a penny will land heads up, given that it is about to be spun, may be half, but the probability that it has landed heads up, given that it was spun and landed heads up, is 1.

It might seem that both K and W_{t_o} are hopelessly unwieldy concepts. But the happy fact is that we can still operate with them though almost entirely ignorant of their content, if we have an indeterministic law such as the disintegration law, say for radon atoms, plus an appropriate initial condition, say that a is a radon atom at t_o. We need make no other assumptions about the content of K and W_{t_o} than that K includes this law and that W_{t_o} includes this initial condition. For our law tells us that, provided only that a is F (an undecayed radon atom) at t_o, then no matter what (future-independent) properties (compossible with Fa) either a itself or anything else may have at t_o, the probability that a is G (has disintegrated) at $t_o + \Delta t$ is $\frac{1}{2}$. If K contains our law and W_{t_o} contains our initial condition, then we may regard $K \wedge W_{t_o}$ as a tremendous augmentation of these two premises; but since our law assures us that *any* (consistent) augmentation of them will leave our conclusion intact, it does not matter if we are largely, or even wholly, ignorant of the ways in which $K \wedge W_{t_o}$ augments them. We have here a deductive schema with a detachable conclusion, looking like this:

$$\forall\phi\forall\psi\forall t\forall x\forall y(x \neq y)(P[G(x,t + \Delta t), F(x,t)\cdot\phi(x,t)\cdot\psi(y,t)] = r)$$
$$\frac{F(a,t_o)}{P_{t_o}[G(a,t_o + \Delta t)] = r}$$

6.31 *A Comparison with Hempel's Schema*

For Hempel in (*1965*), the corresponding schema would look like this:

$$\forall x\forall t\ (P[G(x,t + \Delta t), F(x,t)] = r$$
$$\frac{F(a,t_o)}{G(a,t_o + \Delta t)} \quad [r]$$

The differences between our schemas are these.

1. His is simpler than mine, but then it is incomplete in that it needs to be supplemented by a meta-statement saying that it satisfies the requirement of maximal specificity, whereas mine is self-sufficient.

2. In his schema the conclusion is only inductively supported by the premises whereas in mine it is deduced from them and is detachable.

3. In his schema the value of r needs to be 'fairly close to 1', whereas in mine it can take all values between 0 and 1.

4. In his schema the statistical law that serves as the major premiss is open to either a deterministic or an indeterministic interpretation, whereas that in mine has an essentially indeterministic character.

5. In his schema the conclusion says that a certain event *will* have occurred within a certain time interval, whereas in mine it states the objective probability that it will have occurred.

Let us now consider in whose favour each of these five differences tells.

1. Hempel's requirement of maximal specificity has a close resemblance to the requirement of total relevant evidence discussed in §2.42, where it was argued that a person X could never verify that E contains all and only those evidence-statements that are both relevant to a hypothesis h and now known to X. It would seem that an analogous argument could be deployed against Hempel's requirement. For X to know that a given statistical explanation satisfies this requirement, he would need to have checked it against (in Hempel's words) 'all statements [currently] asserted or accepted by empirical science' (p. 395). Who could possibly do that? But if this has not been done, it is always possible that its premises could be augmented in a way that would both alter the value of r and be endorsed by current empirical science. If the

premises of a statistical explanation that complies with my schema are a part of currently accepted empirical science, one could, but would not need to, show that the requirement of maximal specificity is satisfied; if they comply with Hempel's schema, one would need to, but could not, show that this requirement is satisfied.

2. I need hardly dwell on the preferability, at least from the standpoint of this book, of the deductive structure, yielding a detachable conclusion, of my schema over the 'inductive-statistical' structure, not yielding a detachable conclusion, of Hempel's. A main purpose in this section is to uphold the ideal of deductive explanation in science, and more particularly to ward off the threat to that ideal seemingly posed by statistical theories.

3. A schema that can work, as mine can, with all values of r is obviously more generally applicable than one that is restricted, as Hempel's is, to values of r 'fairly close to 1'.

4. It is a prima facie advantage of Hempel's schema that it can work with any kind of statistical law (provided it gives a high value to r) irrespective of whether a deterministic or an indeterministic interpretation is put on it, whereas mine is restricted to indeterministic laws. However, I will now argue, determinism implies that no objective fact can really be *explained* by a merely statistical hypothesis. Hovering above any statistical "explanation" there is, according to determinism, a superior cognitive structure, perhaps invisible to mortal eyes, consisting of exceptionless laws, full and exact descriptions of initial conditions, and categorical predictions/explananda deductively derived therefrom; and in this ideal structure *probabilities play no role at all*; the probability values that figure essentially in statistical "explanations", so-called, *reflect gaps in our knowledge*. As Joseph Butler put it, a probability estimate, however nicely calculated,

> in its very nature, affords but an imperfect kind of information; and is to be considered as relative only to beings of limited capacities. For nothing which is the possible object of knowledge, whether past, present, or future, can be probable to an infinite Intelligence; since it cannot but be discerned absolutely as it is in itself—certainly true, or certainly false. (*1736*, p. 73)

And Laplace, the great probability theorist, shared this view of probabilities:

> The curve described by a simple molecule . . . is regulated in a manner just as certain as the planetary orbits; the only difference between them is that which comes from *our ignorance*.

Probability is relative, in part *to this ignorance*, in part to our knowledge. (*1812*, p. 6, my italics)

How can any fact in a deterministic world be *explained* by premises that betray *our ignorance*, or the gappy and imperfect state of our knowledge?

So while agreeing that my schema is more restrictive than Hempel's, I claim that its restriction to indeterministic laws is not gratuitous but reflects the fact that determinism is incompatible with the idea of a genuine statistical *explanation*.

5. It is a prima facie advantage of Hempel's schema that its conclusion asserts the occurrence of an event whereas in mine it asserts only the probability of the occurrence. His, it might be said, explains and predicts actual events, whereas mine deals with a metaphysical structure of objective probabilities hovering above actual events. This issue appears differently according to whether the conclusion concerns the outcome of a single trial or of a large number of trials. Let us consider the former first.

This difference between Hempel's schema and mine is closely connected with the previous two. If genuine statistical explanations involve indeterministic laws concerning chance events, to which they may give any probability between 0 and 1 exclusive, then one thing they cannot explain is why this particular chance event happened to occur. And this is so irrespective of whether the probability is low, middling, or high. Suppose that a photon is to be fired at a screen. Our premises tell us that it will land somewhere within a certain circular area, though just where is a matter of chance. We divide this area into, say, 100 concentric rings in such a way that, according to our premises, the photon has an equal probability of hitting any one of these rings. In the event it hits ring 57. Now instead of describing this as 'hitting ring 57', we might describe it as 'not hitting ring 56'; and our premises will endow these two descriptions with probabilities of, respectively, .01 and .99. But we can no more explain why it did not hit ring 56 than we can explain why it did hit ring 57; before the event, the probability that it would not hit any one ring *equaled* the probability that it would not hit any other one.

6.32 *Salmon on Very Low Probabilities*

I find it rather astonishing that Salmon, while highly critical of Hempel's restriction of r to values fairly close to 1, should at the same time repeatedly insist that a statistical explanation, applied to a single event, explains the actual occurrence of that event even if the value of r is very low (*1971*, pp. 9, 39, 58, 63, 76, 108). Salmon discussed one kind of case where this claim has a certain plausibility: namely, one where the premises do significantly *raise* the probability of the occurrence, even though they raise it only to a low level. He took over from Michael Scriven (1959) the example where the explanandum is that Jones is suffering from paresis. (Poor Jones is always in a bad way.) The explanans is that Jones has latent syphilis (minor premiss) and that, while only a small percentage of people with latent syphilis develop paresis, no one develops the latter who does not have the former (major premiss). However, Salmon did not restrict his claim to cases where the explanans does at least raise the probability of the explanandum. He even extended it to cases where it *lowers* it and to a *very low* level. Suppose that at t_o we had trapped one uranium atom, with an enormously long half-life, and one polonium atom, with a very short half-life of Δt. Assume that the probability that a uranium atom disintegrates during Δt is 10^{-35}. Suppose that one of our two atoms does in fact disintegrate by $t_o + \Delta t$, though we do not yet know which. Denote this disintegration by e and let A be the class of all atoms that disintegrate. According to Salmon, to provide a statistical explanation for e is to locate e in a homogeneous reference class for A. B is a homogeneous reference class for A if there is a statistical probability $P(A, B) = r$ such that no place selection, in the sense of Richard von Mises (*1928*, pp. 24f.), within B will alter the value of r. If B is the class of all uranium atoms and C that of all polonium atoms, then both B and C are homogeneous reference classes for A, according to the disintegration law as interpreted by Born: a selection of those members of B or C that have been subjected to high temperatures or that are in strong electromagnetic fields, for instance, or any other such selection, would leave the probabilities unchanged.

The probability that at least one of our two atoms will disintegrate by $t_o + \Delta t$ was $\frac{1}{2} + 10^{-35}$. We now investigate and find to our astonishment that it is not the polonium but the uranium atom that disintegrated. We can now locate e in a homogeneous reference class, namely B, and this gives us $P(e, B) = 10^{-35}$; and with this, according to Salmon, e is *explained*! He disarmingly added: 'I must confess to a feeling of queasiness in saying that an event is explained when we have shown that according to all relevant factors, its occurrence is overwhelmingly improbable' (p. 12).

I say that if, as we are supposing, the disintegration of an atom is a causally undetermined, chance event, then we could not explain a particular occurrence of such an event even with premises that assigned it a high probability, let alone with ones that assign it a vanishingly small probability.

6.33 Railton's Deductive-Nomological Model

In (1978) Railton had proposed a view of statistical explanation that has much in common with the view defended here. He too held that it is only genuinely indeterministic processes that are open to statistical explanation, and that such explanations involve 'irremediably probabilistic laws' (p. 207). If a law, say of the form $P(A, B) = \frac{1}{2}$, truly describes a statistical regularity, then for this to be a probabilistic law it must not be the case that there is a hidden factor, say C, such that whenever B and C are both present, A occurs, and whenever B but not C is present, A does not occur, C being scattered among about half of all initial conditions involving B. There must be 'no hidden variables' that would account for the statistical regularity in a deterministic way (p. 213). Apart from one vital difference, his schema for statistical explanations is very like the one presented here: a major premiss consisting of a nomological, probabilistic law, say $P(A, B) = $ r; a minor premiss saying that condition B is satisfied in a particular instance; and *deduced* from these a conclusion saying that there is an objective probability r that an A-type event will occur in this instance. Railton points out that, in contrast with Hempel's schema, there is no restriction of r to values close to 1 and no need for any requirement of maximal specificity.

But now comes the big difference. Suppose that the event in question, call it e, does in fact occur, though perhaps the probability of its occurring was low. Then the schema as so far presented, Railton says, 'leave[s] out a crucial part of the story: did the chance fact obtain?' (p. 217). Railton, like Salmon, still hankers after the idea that such a statistical explanation can explain the occurrence of an individual chance event. In his example, the event is the decay of an alpha particle, and had a low probability. He writes:

> Still, does [the explanans] explain why the decay took place? It does not explain why the decay *had* to take place, nor does it explain why the decay *could be expected* to take place. And a good thing, too: there is no *had to* or *could be expected to* about the decay to explain — it is not only a chance event, but a very improbable one. [The ex-

241

planans] does explain why the decay *improbably* took place, which is how it did. (P. 216)

How does it explain this? Railton introduces into his schema a 'parenthetic addendum' to fill the gap. This simply says that the event in question did (if it in fact did) or did not (if it in fact did not) take place. Let an explanandum *e* say that a particular alpha-particle did decay during a certain time interval. Just where, in our statistical explanation of this event, are we to insert this "parenthetic addendum": in the premises or in the conclusion? Were we to insert it into the premises, their conclusion would no longer be that the probability of *e* is r (where r is, in the present case, very low). If to the premises (a) 'If a fair coin is spun, the probability that it lands heads up is $\frac{1}{2}$'; and (b) 'This is fair coin and it has been spun', we add the "parenthetic addendum" (c) 'This coin landed heads up', the conclusion will now be that the probability that it landed heads up is not $\frac{1}{2}$ but 1. The alternative, and the one chosen by Railton, is that we add it to the conclusion. But this destroys the *deductive* character of his schema. Its conclusion will now be of the form: '$P(e) =$ r; moreover *e*' (alternatively: 'moreover $\sim e$'). It is as if from 'All men are mortal and Socrates is a man' one were to derive 'Socrates is mortal; moreover he died at the age of seventy'. On a genuinely deductivist view of explanation, one can explain, given an appropriate set of initial conditions, no more after the event than one could have predicted, given the same initial conditions, beforehand. As Railton himself points out (p. 224), his account violates this requirement: before the event the "parenthetic addendum" is unavailable and we can predict only the probability of the event. But according to Railton, we can be wise after the event and explain its actual occurrence. I say that nothing is *explained* by adding this parenthetic addendum to the explanandum.

6.34 *Microphysical Explanations of Macro-events*

I now turn to the question of deductive explanations for those macro-processes that are made up, according to microphysical theories, of enormous numbers of random micro-events. Let us continue with Hempel's example, where the explanandum is that a sample of 10 milligrams of radon was reduced, after 7.64 days, or two half-lives of a radon atom, to within 2.5 ± 0.1 milligrams. The major premise of the explanans is the disintegration law for radon atoms, and a minor premiss is 'that the number of atoms in 10 milligrams is enormously large (in excess of 10^{19})' (*1965*, p. 392). Let us denote the atoms composing the sample at the start of the 7.64 day period by $a_1, a_2, \ldots a_n$ where $n > 10^{19}$. Then it is as if each of n coins will be spun twice during this period and withdrawn if it lands heads at least once. In accordance with my schema, we can deduce, for each a_i ($i = 1, 2, \ldots$ n), that the objective probability, at the outset, that a_i disintegrates during this period is $\frac{3}{4}$. Let m be whatever is the number of a_i's that do in fact disintegrate. The question now is whether we can further deduce the approximate value of m/n, the proportion of disintegrating atoms.

Let us say that the probability that the value of m/n lies within the interval $r \pm \delta$ is $1 - \epsilon$; the Central Limit theorem enables us to calculate the value of ϵ for given values of n and δ. In the present case, where $n > 10^{19}$ and $\delta = 4\%$, $\epsilon < 10^{-(10^{15})}$. (I owe this computation to my colleague Susannah Brown.) When n is enormous, the prong of the familiar bell-shaped probability distribution curve has become so narrow, and its two tails have become so flat, that the curve is virtually indistinguishable from a vertical line at r (here $\frac{3}{4}$) with horizontal tails.

At this point, anyone, whether a deductivist, inductivist, or whatever, concerned with the nature of microphysics, faces a crucial question: what *physical* meaning should be given to vanishingly small values of ϵ such as $10^{-(10^{15})}$? Should we interpret our disintegration law as allowing that the macro-outcome *might* fall outside the δ-interval? Should we interpret the laws of thermodynamics as allowing that patches of ice and wisps of steam *might* form spontaneously in an ordinary bathtub because of hugely improbable distributions of the molecules? If we take this line, we in effect declare that, in the case of macro-processes that are the resultant of astronomically large numbers of chance micro-events, there can be no theory of such macro-processes that is at once *deep* (goes down to the micro-level) and *scientific*; the best that science could achieve here would be a combination of (a) certain level-3 experimental laws, of a deterministic nature, concerning the phenomenal characteristics of the macro-process, and (b) an indeterministic but ultimately metaphysical theory of the underlying micro-events, with a logical hiatus between (b) and (a). The macro-laws would exclude certain events (e.g. ice and steam spontaneously forming in the bathtub) that are permitted by the micro-theory. An experimental refutation of a macro-law would leave the micro-theory unscathed. Instead of a scientific *reduction* of macro-level to micro-level events, we would have an irreducibly two-level system: hard experimental laws with a cloud of soft metaphysical theory hovering above them.

6.35 *Cutting Off the Tails of the Distribution*

As I said, this is a dilemma that faces any philosopher concerned to do justice to microphysics, whatever his epistemological leanings. What is the alternative? Well, given a micro-theory that yields, at the macro-level, a probability distribution that is virtually indistinguishable from a vertical line with horizontal tails, the alternative is to impose an interpretation on the theory that has the effect, as Donald Gillies puts it, of 'cutting off the tails of the distribution' (*1973*, p. 173). In other words, given that the probability that a macro-outcome lies outside the δ-interval is less than $10^{-(10^{15})}$, we can say that the *physical* meaning of this vanishingly small probability is that the outcome does *not* lie outside the δ-interval.

Popper adopted this approach in (*1934*, pp. 198f.). Is there a non-arbitrary way of reinterpreting probabilistic hypotheses so as to render them falsifiable? I am an amateur in this area but Popper's (*1934*) answer to this question seems to me a good one and I will adopt it. We have three interrelated variables: δ, ϵ and n. Popper's first step was to fix δ. A measuring process for macro-outcomes will have an interval of imprecision, and Popper required δ to be contained within this interval. This renders predictions about macro-outcomes that are based on probabilistic hypotheses as stringently testable as ones based on deterministic hypotheses. With δ thus tied down, the next question is how large n must be for ϵ to be negligibly small? If the probability that the ratio m/n (say of disintegrating atoms in the sample) lies in the interval r $\pm \delta$ is $1 - \epsilon$, how large must n be for $1 - \epsilon$ to be physically equivalent to 1? Popper answered: n is large enough if it is so large that ϵ becomes *virtually insensitive to increases in the value of* δ: *in other words, a large increase in δ makes virtually no difference to ϵ.* What does 'virtually no difference' mean? There are cases where, although no sharp borderline can be drawn, one can say that something is definitely on one side of a vague border. A man who has lost 95 percent of the hairs on his head is assuredly bald although we cannot say at just what percentage of hair loss baldness commences. In rather the same way I suggest that if a large increase in δ decreases ϵ by less than $10^{-(10^{10})}$, then this increase makes virtually no difference to ϵ.

Let us try this out on our earlier example. Suppose that the limits of accuracy within which our sample of radon can be weighed are not less than $\pm \cdot 1\%$ and that we therefore fix δ at this value. Given that the original sample of 10 milligrams contained n atoms where n >

10^{19}, the probability that its weight, after two half-lives, will diverge from 2.5 milligrams by more than $.1\%$ is about $10^{-(10^{12})}$ (Susannah Brown). Increasing the value of δ from $.1\%$ to Hempel's value of 4% decreases the value of ϵ only to about $10^{-(10^{15})}$. Thus ϵ is virtually insensitive to changes in δ when $n > 10^{19}$.

An incidental advantage of this way of construing probabilistic theories in microphysics is that it brings them within the ambit of the measure for testable content that was developed in §5.1. If T_i and T_j are both probabilistic hypotheses, there is no special difficulty. We can proceed to a definition of counterparthood here in the following way. Suppose that T and T' are probabilistic hypotheses that say respectively: $\forall x[P(Gx, Fx) = r]$ and $\forall x(P(G'x, F'x) = r')$. Then T and T' are counterparts if (i) the corresponding universal laws, $\forall x(Fx \rightarrow Gx)$ and $\forall x(F'x \rightarrow G'x)$, are counterparts by the earlier definition, and (ii) r and r' are equally precise and are both greater than 0 and less than 1. And we can now say that $CT(T_j) > CT(T_i)$ if there is a counterpart T_i' of T_i such that every PF of T_i' is a PF of T_j but not vice versa.

In the case where T_j is a probabilistic theory in microphysics and T_i is a deterministic theory of classical physics, we can now proceed as follows. Let δ be an agreed interval, given the most precise of the current methods of measuring a certain magnitude, such that two predictions are experimentally indistinguishable if they do not diverge by more than δ. Assume that the classical T_i yields experimental laws for macroprocesses that are exact. Thus such a law might make a variable magnitude y a mathematically precise function of the independent variables x and z. Now there will be no loss of testable content, relative to current measuring methods, if we introduce an element of tolerance equal to δ into the laws of T_i. Thus a law of the form $y = f_i(x, z)$ can be weakened without loss of testable content to one of the form:

$$f_i(x, z) - \delta \leq y \leq f_i(x, z) + \delta.$$

Call the thus weakened law l_i, and suppose that the probabilistic theory T_j yields a statistical law l_j of the following form:

$$P[f_j(x, z) - \delta \leq y \leq f_j(x, z) + \delta] > 1 - \epsilon.$$

Then we may treat l_i and l_j as empirical counterparts if (i) the functions f_i and f_j are equally precise in the sense explained in §5.17; and (ii) according to T_j the number n of micro-events involved in an observable macro-event of the kind to which l_j applies is so enormous (say, $n > 10^{19}$) that the value of $1 - \epsilon$ is physically equivalent to 1, in the sense

245

recently explained. If every law of T_i has such a counterpart among the laws of T_j but not vice versa, then $CT(T_j) > CT(T_i)$.

There is a far-reaching analogy between the explanatory power of the deterministic theories of classical physics and the indeterministic theories of modern microphysics, according to the construal of the latter that I have been defending. Both explain empirical regularities by appealing to higher level structural laws that are taken as immutable and absolute. In the case of classical physics, such a law says that, given only that such-and-such conditions are satisfied, nothing whatever can prevent a certain outcome from following. In the case of microphysics, it says that, given only that such-and-such conditions are satisfied, nothing whatever can alter *the chance* that a certain outcome will follow. One could as well call the latter an "iron law" of chance as the former an "iron law" of nomic necessity. Both kinds of law, in conjunction with appropriate initial conditions, can explain empirical regularities and, indeed, singular macro-events (provided that the macro-event in question is the resultant of a huge aggregate of micro-events, and is not determined, as in the case of the electrocution or otherwise of Schrödinger's cat, by which way a single micro-event chances to go). The analogy breaks down, however, when we come down to individual events at the micro-level. According to classical physics, these too are completely causally determined and, at least in principle, explainable and predictable. I have argued that it is a mistake to call upon microphysics to *explain* an individual micro-event, such as the disintegration of a radon atom or the reflection of a photon; for if it really was a matter of chance which way it went, then the fact that it chanced to go this way rather than that simply defies explanation. What we *can* explain with the help of an appropriate indeterministic microphysical theory is why there was a precise objective probability that it would go this way.

I conclude that the deductivist ideal of scientific explanation is not, after all, in conflict with the fact that probabilities play an irreducible role in many of the most fundamental theories of modern physics.

7

The Empirical Basis

7.1 Do We Need an Empirical Basis?

Our aim for science (B*) calls for even deeper and more testable explanations; but explanations of what and testable against what? The natural answer is: empirical facts. But we are trying to elaborate a conception of science that is invulnerable to Humean scepticism; and, as we saw in §3.1, Humean scepticism hits level-1 statements, or singular statements that purport to describe empirical facts, as well as level-2 generalisations and other statements higher up in the "inductive ascent". So if we are to show that there can be rational adoption and rejection of higher level hypotheses under the guidance of (B*), we must first show that there can be rational adoption and rejection of level-1 statements.

The daring suggestion that science does not need acceptance of statements at this level, that science can proceed without an empirical basis, has been made by Joseph Agassi, who declared: 'We need speak neither of acceptance, nor of justification of acceptance, of any observation report' (1966, p. 114). And he added: 'The problem of the empirical basis is thereby disposed of' (p. 115). Would that that were so. But what are scientific theories now supposed to explain? Agassi answered: 'We merely have to demand that account be taken of the fact that some observation reports were made repeatedly, and that this fact be explained by some testable hypothesis' (p. 114). One recalls Mary Hesse's distinction, considered in §3.11, between an observation *statement e* which is, or purports to be, about some observed thing or event, and an associated observation *report e'* that reports the making of the observation statement *e*. Employing this distinction, we might restate Agassi's thesis as the claim that what scientific hypotheses should explain (and, presumably, be testable against) is not observation statements but observation reports. It seems to me that this proposal would complicate the task of science without disposing of the epistemological problem of an empirical basis. True, we would no longer be concerned with the truth or otherwise of an observation statement e; but we would instead be concerned with that of the associated observation report e': we would need to speak of *its* acceptance and, presumably, of the justification of its acceptance. So no philosophical simplification would be achieved. On the other hand, the task of science would be complicated and distorted by being given a sociopsychological twist. Popper had suggested that in some cases, such as that of flying saucers, 'the explanation needed may not be of flying saucers, but of reports of flying saucers' (1957, p. 191); and Agassi's proposal in effect generalised this to all cases; for instance, what Kepler should have explained is not the observed positions of the planets, as reported by Tycho Brahé, but why Tycho Brahé made those reports.

The problem of the empirical basis that faces us, given what is said elsewhere in this book, can be set out as follows.

1. We want our theories to be under the (negative) control of experience.

2. In line with Hume we are assuming that level-0 reports are the only nonanalytic statements that may rightly be regarded by their authors when they make them as infallibly true.

3. Later in this chapter we will find that it is not possible so to augment the premises of a physical theory that it becomes testable against level-0 reports only.

4. So level-1 statements are indispensable for the testing of such theories. *But can there be rational acceptance of level-1 statements?* If not, our attempt to defeat rationality-scepticism fizzles out. For a theory T to be rationally accepted it must satisfy (B*) better than its rivals, and a necessary condition for this is that it is unrefuted. But if the acceptance and rejection of level-1 statements is nonrational, the difference between being "refuted" and being "unrefuted" becomes a matter of nonrational convention.

5. Now there could be rational acceptance of level-1 statements if they were validly inferrable from suitable level-0 premises. But we saw (§3.2) that even phenomenalism fails to secure this. And if level-1 statements are construed physicalistically, then it would seem that the relation of them to level-0 premises does not even fit any standard inductive pattern. Schlick wrote: 'the enlargement of knowledge obtained by induction by its nature extends always to instances of the same kind only. . . . Inductions become . . . absolutely impossible and unmeaning if they attempt to leap over over into an entirely new sphere'

(1926, p. 106). But a progression from level-0 to level-1 would involve such a leap. And Ayer seems to have agreed that this gap could not 'be bridged by a legitimate process of inductive reasoning. . . . It is admitted that the inferences which are put in question . . . are not inductive, in the generally accepted sense' (*1956*, p. 80). So level-1 statements would need to be inferred from level-0 premises by some kind of quasi-inductive inference, quite contrary to our proposition (III), the deductivist thesis.

6. So our problem is: can we show, *without* recourse to any quasi-inductive view, that there can be rational acceptance of level-1 statements?

7.2 Popper's Account

Popper (*1934*, chap. 5) tackled the problem of the empirical basis in a bold way, and we must now consider whether he there provided a solution for our problem. He posed it in the form of a trilemma due to Fries:

> He [Fries] taught that, if the statements of science are not to be accepted *dogmatically*, we must be able to *justify* them. If we demand justification by *reasoned argument, in the logical sense*, then we are committed to the view that *statements can be justified only by statements*. The demand that *all* statements are to be logically justified . . . is therefore bound to lead to an *infinite regress*. Now, if we wish to avoid the danger of dogmatism as well as an infinite regress, then it seems as if we could only have recourse to *psychologism*, i.e. the doctrine that statements can be justified not only by statements but also by perceptual experience. (*1934*, pp. 93–94)

He overcame the Fries trilemma with an account of the acceptance of basic statements that involves harmless versions of each of its three ingredients, namely, dogmatism, infinite regress, and psychologism. There is a touch of dogmatism in that they are accepted by what is, from a logical point of view, 'a free decision' (p. 109); but such a decision, which he likened to a jury's verdict, is revocable; if there is a challenge to a basic statement at which the testing process had stopped, then the testing process will start up again. There is a potential infinite regress, in that the testing process always *could* be carried a step further. But this is harmless, since the tests are not intended to establish the truth of the basic statements in question. Finally, there is a touch of psychologism:

> And finally, as to *psychologism*: I admit, again, that the decision to accept a basic statement, and to be satisfied with it, is causally connected with our *experiences*. But we do not attempt to *justify* basic statements by these experiences. Experiences can *motivate a decision*, and hence an acceptance or a rejection of a statement, but a basic statement cannot be justified by them—no more than by thumping the table. (P. 105)

Popper also posed the problem of the empirical basis by way of a sharp criticism of Otto Neurath's account of what he called protocol sentences. Neurath's (1933) had been directed against Carnap's view (which we considered in §3.22). Neurath invoked the idea of a 'universal slang' containing both ordinary and scientific language and having a physicalist character; and he held that "protocol" sentences should be formulated, not in an 'experiential' or 'phenomenalistic' language as Carnap had proposed, but in this physicalist language; and he insisted that previously accepted protocol sentences may get discarded, or deleted.

Popper declared this 'a notable advance'; but he added that this step

> leads nowhere if it is not followed up by another step: we need a set of rules to limit the arbitrariness of 'deleting' (or else 'accepting') a protocol sentence. Neurath fails to give any such rules and thus unwittingly throws empiricism overboard. . . . Every system becomes defensible if one is allowed (as everybody is, in Neurath's view) simply to 'delete' a protocol sentence if it is inconvenient. (P. 97)

So we need to ask whether Popper filled the gap left by Neurath. Decisions on basic statements, he said, 'are reached in accordance with a procedure governed by rules' (p. 106). What rules did he provide? Well, he explicitly formulated only two rules. He laid down (p. 102) a rule as to what requirements a statement must satisfy to count as a basic statement, but this rule says nothing about acceptance and rejection of basic statements. And he spoke of 'a rule which tells us that we should not accept *stray basic statements*—*i.e.* logically disconnected ones—but that we should accept basic statements in the course of testing theories' (p. 106). I have misgivings about this rule. Consider the discovery of Uranus by William Herschel in 1781. Holton and Roller describe it as follows. One night Herschel was searching the sky with his homemade, 10-ft. telescope. For years he had patiently looked at and re-examined every corner of the heavens, and he was becoming well-known among astronomers for his discoveries of new stars, nebulae, and comets. On that particular night he observed a hitherto uncatalogued object of such "uncommon appearance" that he suspected it to be a new comet. Through the Royal Society the news spread. (*1958*, p. 196)

Now it is perfectly true that *later* observations on Uranus, as this uncatalogued object was subsequently called, were made in the course of testing a theory, namely Newtonian theory, and with famous results. But what about Herschel's original report? It seems to me entirely possible that it was a "stray report" in Popper's sense. Of course, it could be said that Herschel was testing the "theory" that there is no object at a certain place in the heavens. But an answer along those lines would tend to suggest that *no* basic statements are stray, since there will always be *some* hypothesis with which a basic statement conflicts and which its author may be said to have been testing.

But suppose that my misgivings are baseless and that this is a good rule; even so, we do not yet have any rules governing the *acceptance* of basic statements, and it is not easy to see how a noninductivist could provide such rules. However, Popper also outlined a procedure for testing disputed basic statements. Before I turn to this I will consider two possible interpretations of the passage I quoted earlier about decisions on basic statements being 'causally connected' with perceptual experiences. One interpretation would be that perceptual experiences are here being assigned a *merely* causal role and are not being regarded as providing any sort of *reason* for acceptance. The other interpretation would be that they are both causes of and (inconclusive) reasons for acceptance of basic statements. On the first interpretation, perceptual experiences are no more given an epistemological role than are brain cells that may be causally connected with a person's decision on basic statements. On the second interpretation, they do play an epistemological role. The differences between the two interpretations may be indicated in the following way. Suppose, first, that we are informed that *X* was motivated to say what he did by fear, or else that what he said was causally connected with his having taken a certain drug. This information would give us no reason to assume that what *X* said was true. But now suppose that we are further informed that he was motivated by fear of activating a lie detector, or else that the drug in question was a highly efficacious truth drug. *This* information would have some tendency to suggest that what he said was true. Or suppose that a radar device is linked to a computer that prints out statements like: 'At 02h. 13m. 21s. GMT small object bearing 030°, distance 37 miles, height 100 feet above sea level, speed 650 m.p.h., course 210°'. Given that this device is in good working order, the fact that it has been *caused* to print out such a statement would give us good *reason* to accept the statement.

I used to think that the first interpretation of Popper's (*1934*) account was correct; but some things he has said since, both in (1974b) and in personal communication, favour the second interpretation. So I will consider his (*1934*) procedure for dealing with disputed basic statements first under the first interpretation, and then under the second interpretation.

There is a good deal of evidence in (*1934*, chap. 5) in favour of the first interpretation, according to which perceptual experiences lie outside the domain of epistemology. For instance, at one place (p. 99) Popper said that the only problems that interest epistemologists concern the logical connections between scientific statements, feelings of perceptual assurance, etc. being of interest only to psychologists. At another place, after saying that the events reported by basic statements must be '*observable*' (p. 102), he hastened to add:

> No doubt it will now seem as though in demanding observability, I have, after all, allowed psychologism to slip back quietly into my theory. But this is not so. . . . I am using [the concept of an *observable event*] in such a way that it might just as well be replaced by 'an event involving position and movement of macroscopic bodies'. (P. 103)

And I have already quoted his statement that 'a basic statement cannot be justified by [perceptual experiences]—no more than by thumping the table'. It does seem to me that all this quite strongly suggests that Popper wished in (*1934*) to view the whole of science, including its empirical basis, as an objective system in which neither perceptual experiences nor any other subjective or psychological elements have any place, though we may, as a matter of psychological and extra-epistemological fact, have been motivated by perceptual experience to insert this or that component into the system.

Popper's procedure for handling disputed basic statements is essentially the following: a disputed basic statement is tested by 'using as a touchstone any of the basic statements which can be deduced from it with the help of some theory, either the one under test, or another'; and this testing process should 'stop only at a kind of statement that is especially easy to test' and over which 'the various investigators are likely to reach agreement' (p. 104). Let b_1 be a disputed basic statement, and let b_2 be one that is deduced from b_1 with the help of additional assumptions A_1. (It seems to me that, in addition to some theory, A_1 will, typically, need also to include some new initial conditions, for instance that the event reported by b_1 was photographed.) Now merely *deducing* b_2 from $b_1 \wedge A_1$ does not yet constitute any sort of test on b_1. (Einstein did not test his theory when he derived the star-shift prediction; rather he showed the way to a new kind of test, which was carried out later by Eddington and others.) So if b_1 is to be tested via b_2, b_2 itself must be put to the test. How? On the present interpretation, according to which perceptual experiences play a causal and not an epistemological role, we may not answer that b_2 should be tested against the perceptual experiences of investigators in a position to observe what it describes. And this seems to have been endorsed by Popper: 'we stop at basic statements which are easily testable. Statements about personal experiences—*i.e.* protocol sentences—are clearly *not* of this kind' (pp. 104–105). So if b_2 is to be tested, it seems that at least one more basic statment, say b_3, will have to be deduced from b_2 with the help of further additional assumptions, say A_2. Suppose that this is done and that b_3 is regarded as 'especially easy to test' and that the various investigators reach agreement and accept it. But if it was especially easy to test, ought they not, before they accept it, make one last effort and actually *test* it? If a neighbour tells you that there is a hippopotamus in his garage you might regard his statement as easily testable, but would you therefore accept it without more ado? But how are they to test the easily testable b_3? As we have just seen, the testing process is not supposed to stop with statements about perceptual experiences. The only way to test it officially recognised by Popper in (*1934*) would be by 'using as a touchstone any of the basic statements which can be deduced from it with the help of some theory'; in other words, from b_3 and a suitable A_3 we derive a b_4. And then the same difficulty would arise. All we are getting, under this first interpretation, is a lengthening chain of derivations: no *tests* are being made.

Now let us turn to the second interpretation whereby perceptual experiences are both causes of and reasons for the acceptance of basic statements. This would obviate the difficulty just considered; investigators might halt the testing process, say at b_3, on the ground that their perceptual experiences give them reason to accept this easily testable statement. This interpretation is supported by what Popper said in (1974b) about his (*1934*) position. He was there replying to Ayer who had said that an observational experience can supply 'me not only with a motive but also with a ground for accepting the interpretation which I put on it' (1974, p. 688); he added: 'there seems to be no good reason why we should not regard our experiences as directly justifying ... the sort of statements that Popper treats as basic. We cannot hold that they verify them conclusively; but this is not a bar to our holding that they give us an adequate ground for accepting them' (p. 689). So Ayer was advocating what I earlier called the quasi-inductive view. In the course of his reply, Popper wrote: 'Our experiences are not only motives for accepting or rejecting an observational statement, but they may even be described as *inconclusive reasons*. They are reasons because of the generally reliable character of our observations; they are inconclusive because of our fallibility' (1974b, p. 1114). I can discern no significant difference between this view and Ayer's quasi-inductive view. (If the claim that our observations are 'generally reliable' means, as it presumably does, that they have given, and *will continue to give*, pretty good information about the external world, it is obviously an inductivist assumption. Popper had previously invoked the fact that organisms' sense organs are well adapted to decode and interpret their environment. We saw earlier (§3.31) that W.K. Clifford was careful to use the hypothesis of natural selection only as an explanation, and not as a justification, for the widespread belief in the uniformity of nature.)

I conclude that a solution of our problem, which would show that there may be rational acceptance of level-1 statements but *without* bringing in any inductive or quasi-inductive assumption, has not been provided by Popper.

7.3   The Role of Perceptual Experience

I will now try to provide a solution of our problem; or perhaps I should say a quasi-solution, for the account I will give will show only that there may be a quasi-rational acceptance of level-1 statements. What I mean by 'quasi-rational' may be illustrated by an example. A chess master plays, and wins, a game of "speed"-chess with only a second or two per move. Afterwards he and some colleagues subject the game to a very patient and thorough analysis; *and they fail to find a better alternative to any of the moves he made.* (I don't think that this is too fanciful. A gifted young chess player once told me that only once, in the course of the many postmortems he had made on his games of "speed"-chess, had he found what now seemed to him a wrong move of his.) Then we could say that although, in the actual game, he had had almost no time for conscious calculation, it was *as if* his moves were the result of intensive deliberation.

7.31 *Preconscious Interpretative Processes*

It may be objected that if, to revert to Jeffrey's example, I am woken by the sun shining on my face, then it is absurd to claim that my judgment that the sun is shining is rational or even quasi-rational; is not the process completely involuntary and *causal?* The objector might add that Locke was essentially right when he declared that 'in bare, naked perception the mind is, for the most part, only passive' (*1690*, II, ix, 1). Well, it is generally agreed among psychologists and physiologists concerned with perception that in ordinary perceptual acts mind and brain are highly active: an enormous amount of unnoticed and very rapid activity goes into these acts. Hume said that we get an impression of the page of a book we are reading and that the other perceptions excited by it are ideas. But does a page of print obtrude an impression of itself into a reader's mind? When you have read this sentence, please fix your gaze steadily on a word lower down and discover how little you now clearly perceive. I conjecture that it will be about the area of a small coin. (I have borrowed this experiment from Grey Walter, *1953*, p. 77.) And even when you deliberately fix your gaze on something, your eyes, physiology tells us, continue to make rapid little scanning movements; if the image on the retina is artificially stabilised, after a second or two, parts of it will fade and before long the image will disappear. According to Kathleen Wilkes, it is the area of the frontal lobes that,

> by controlling the saccadic eye movements, ensures that a complex visual object is scanned efficiently and economically, and is actively searched for clues to its identity; it forms hypotheses [Footnote: The use of this apparently anthropomorphic description will be defended shortly] about what the object might be ... and in terms of these hypotheses it guides the searching movements; in short, it makes perception a goal-directed activity. (1980, pp. 121–122)

It is, I think, agreed by all investigators of perceptual processes that their extreme rapidity excludes from our conscious awareness an enormous amount of psychophysiological processing and interpretation. For instance, N.F. Dixon, who likens visual perception to the building up of an identikit picture, says: 'So rapid is the normal act of perceiving that the underlying stages remain unanalysable except by special methods' (1966, p. 47). Perhaps the best way to bring home the powerful role of interpretative assumptions on which we unnoticingly rely in our perceptions is to begin by considering cases where we succeed in seeing things as they (presumably) are, though we would not so perceive them if perception were the kind of passive recording process that Locke took it to be; and then proceed to cases where a certain exploitation of those assumptions tricks us into seeing what is not there and even into seeing what we know very well cannot possibly be there.

It is well known that one of these assumptions is an assumption of rectangularity: if you are in a normal, medium-sized room, the wall in front of you will appear rectangular; and no doubt it *is* rectangular. But now concentrate your gaze on a corner where it meets the ceiling and a side wall. You will observe three angles, each greater than a right angle, and averaging 120°. And the same will of course hold for its other three corners. So you see the wall as a rectangle each of whose angles is greater than a right angle! (See Schlick, *1918*, p. 258.) A photograph of the wall, taken from where you are standing with a wide-angled lens, would show a curvaceous expanse; but you see it as the rectangle it in fact is.

It is also well known that we are equipped with interpretative propensities that enable us to see things in their true colours, offsetting the distorting effects of sunglasses, dark shadows, etc. Dixon reports the following result:

> A person is asked to match a pair of black and white discs, which are rotating at such a speed as to make them appear uniformly grey. One disc is standing in shadow, the other in bright illumination. By adjusting the ratio of black to white in one of the discs the subject tries to make it look the same as the other. The results show him to be remarkably accurate, . . . he has made the proportion of black to white in the brightly illuminated disc almost identical with that in the disc which stood in shadow. But . . . when the matched discs, still spinning, are photographed . . . [t]he disc in shadow is obviously very much darker than the other one. (1966, pp. 45–46)

A more striking example of people seeing things that are there but which, if I may so put it, they should not see, is provided by the following finding of Trevarthen (reported by Wilkes, 1980, p. 118): if commissurotomised patients are shown, say, a photograph of a whole face, but in such a way that an image of only one-half of the face can be transmitted to the primary visual cortex, they unhesitatingly report seeing a whole face.

But our interpretative tendencies can cause us to see what is not there. For instance, our tendency to link up separate perceptions can do so: a light flashes on a dark screen and a moment later another flashes a short distance from the first; and what we see is *one* light moving from the first position to the second. (According to Vernon, *1952*, chap. 8, this effect was discovered by Wertheimer in 1912.) More striking examples are provided by optical illusions. A rather eerie feature of most of these is that they do not relax their hold over us when we know very well that what we are seeing is not there. I have heard Richard Gregory say that although he has had an inverted face mask, with the nose receding at the back, in his office for many years, he still cannot help seeing it as uninverted. Still more striking testimony to the power of the interpretative categories that govern our perceptual processes is provided by, for example, the Penroses' "impossible" figures, for in their case we cannot help seeing them as objects that we know cannot exist. Taken together these examples suggest that Kant was importantly right and importantly wrong in holding that perceptual experience is structured by certain interpretative categories that are (i) inflexible and (ii) infallible.

Gregory has put forward the following explanation for our alternating perceptions of such ambiguous figures as the Necker cube and the charming girl/old hag:

> we do not perceive the world merely from the sensory information available at any given time, but rather we use this information to test hypotheses of what lies before us. . . . We see this process of hypothesis-testing most clearly in the ambiguous figures, such as the Necker cube. . . . Here the sensory information is constant (the figure may even be stabilised on the retina) and yet the perception changes . . . as each possible hypothesis comes up for testing. Each is entertained in turn, but none is allowed to stay when no one is better than its rival hypothesis. (*1972*, p. 222)

257

As to nonvisual perception: I will mention only one phenomenon, to do with hearing, the well-known "cocktail party" effect. You, let us suppose, are absorbed in conversation with a charming companion at a cocktail party; all other voices in the crowded room merge into a collective buzz. However, if someone in your vicinity should, without raising his voice, happen to mention your name you will notice it with a slight shock. Dixon comments on this: 'even the non-perceived flow of information must have been continuously scanned below the level of consciousness' (1966, p. 65).

7.32 *Perceptions as Explananda*

My aim in what follows is to argue for the rationality, or rather the quasi-rationality, of ordinary perceptual judgments in a way that is not vulnerable to Humean scepticism. This obliges me not to rely on any particular psychological or physiological hypothesis about whatever interpretative categories may be involved in perceptual processes. However, as I said at the outset of this book, Humean scepticism allows that I can have a certain amount of egocentric knowledge about my own beliefs and experiences; and in what follows I shall rely on the following bits of egocentric knowledge:

1. I have perceptual experiences that grow spontaneously into perceptual judgments. For instance, I have the visual image of a white, pencil-like shape slowly lengthening across a blue expanse and I unhesitatingly say to myself, 'A jet aircraft leaving a vapour trail'.

2. In such cases my perceptual judgment could not possibly be generated by nothing but my perceptual experience.

3. Therefore my perceptual experiences must be rather strongly interpreted by me. However, I am not usually conscious of any interpretative process. Hence this must be a largely subconscious process.

Carnap (in *1928*) said something very similar to this: our formation of 'representations of things . . . from the given, does not, for the most part, take place according to a conscious procedure' (p. 158). And he added that his constructional system 'is a rational reconstruction' of a process 'which, in cognition, is carried out for the most part intuitively' (ibid.). I too am going to attempt a rational reconstruction of the largely intuitive process of perceptual judgment. (It will, however, be very different from Carnap's.) For this purpose I introduce an imaginary character called John Wideawake. He is the person I would be if I could somehow carry out consciously, but with no loss of time or interference with my other activities, the various mental operations involved in my perceptual judgments. (I should mention in passing that I consider it in principle impossible that J.W., as I will call him for short, could achieve *complete* self-awareness. See Ryle *1949*, pp. 195–198 on 'The Systematic Elusiveness of "I" ', and F. A. Hayek 1963, pp. 60–63 on the impossibility of articulating *all* the rules governing conscious action.) Like me, J.W. accepts proposition (III), the deductivist thesis. He prefers assured truth to dubious conjecture where he can get it, but he would hate to pretend to himself that some proposition that he accepts, and that is really only a conjecture, is an assured truth. In areas where unverified conjectures are all he can get, he accepts our proposition (IV*), which says that, given evidence *e*, it is rational for a person to adopt a hypothesis *h* if *h* is possibly true and the best explanation he can find for *e* (and to retain *h* unless he discovers new evidence adverse to it or hits upon a still better explanation for *e*). Indeed, he accepts our aim (B*) and the desirability of proceeding to ever deeper explanations.

But what, it will be asked, is J.W. entitled to regard as given evidence? Well, he does suppose himself to possess incorrigible *knowledge* concerning his own current perceptual experiences. He is in agreement here with both Hume and Descartes. Thus Hume wrote: 'For since all actions and sensations of the mind are known to us by consciousness, they must necessarily appear in every particular what they are, and be what they appear' (*1739–40*, p. 190). And Descartes had likewise insisted that no demon could deceive him into believing that he was having perceptual experiences that he was not really having. In the second Meditation, after supposing 'that all the things that I see are false', he went on: 'Let it be so; still it is at least quite certain that it seems to me that I see light, that I hear noise and that I feel heat. That cannot be false' (*1642*, p. 153). J.W. takes a similar view. If he makes a truthful level-0 report about his own current sensory experience, then he *knows* it to be true: not even a Cartesian demon could have deceived him over *this*.

On the other hand, J.W. accepts that the principles he relies on in interpreting his perceptual experiences are by no means infallible; for he has found that they sometimes trick him into perceptual judgments that turn out to be false. For instance, he is aware that he interprets shadows on the assumption that the illumination is from above; so that if he looks at the rounded rivet heads on some boiler plating that is illuminated from below, they look like rounded cavities. He is also aware of a strong expectation of rectangularity in buildings and other artifacts; and he has found that this can lead him to judge a short lady to be taller than a tall man if he sees them in the far corners of what is in fact an Ames Distorted Room. Far from regarding his interpretative principles as synthetic a priori categories in Kant's sense, he regards them as a system of hypotheses with the help of which he tries to account for his perceptual experiences.

Let us ask what J.W. would make of this well-known passage by the late J. L. Austin:

> The situation in which I would properly be said to have *evidence* for the statement that some animal is a pig is that, for example, in which the beast itself is not actually on view, but I can see plenty of pig-like marks on the ground outside its retreat. If I find a few buckets of pig-food, that's a bit more evidence, and the noises and the smell may provide better evidence still. But if the animal then emerges and stands there plainly in view, there is no longer any question of collecting evidence; its coming into view doesn't provide me with more *evidence* that it's a pig, I can now just *see* that it is. (*1962*, p. 115)

J.W. would agree that his perceptual experience of seeing a pig does not provide him with evidence for the conclusion that there is a pig out there; he does not engage in any kind of quasi-inductive inference from inner experience to outer reality. He would also agree that Austin has nicely caught the plain man's naive realism and his habit of using '*I see* . . . ' as a success-word that effortlessly closes the gap between inner experience and outer reality. But J.W. himself is too conscious of the variety of hypotheses he uses in his perceptual judgments to be a naive realist. He uses 'I see . . . ' rather as someone pressing one eyeball with his finger might say 'Now I see two cups' or as someone having his eyes tested might say, after the oculist has inserted another lense, 'Now I see a second red line parallel to the first.'

However J.W., if we suppose him to be in the situation depicted by Austin, *knows* that he is having the perceptual experience of seeing (in the above noncommittal sense) a pig. Instead of regarding a level-0 report of this perceptual experience as a premiss from which he may quasi-inductively infer the level-1 conclusion that there is a pig a few yards in front of him, he regards it as an explanandum calling for an explanans. He responds to this perceptual experience in rather the same sort of way that I respond to *perplexing* experiences. For instance, I am in conversation with someone when I get a fright: his teeth seemed momentarily to *flicker*. What's happening? Am I hallucinating? And then, with relief, I hit upon the conjecture that he is wearing dentures that are a little loose. Or take the very first occasion on which I saw a pencil-shaped cloud slowly lengthening across the sky. I cannot actually remember, but I like to think that I said to myself: 'That's a strange phenomenon. What can it be? Perhaps it's one of those new jet aircraft leaving a vapour trail.' However, whereas my "explanations" of perplexing experiences consist essentially just of a level-1 statement ('He's wearing dentures', 'It's a jet aircraft'), J.W.'s explanations of his perceptual experiences are far more elaborate and systematic. Level-1 statements will typically (though not invariably, as we shall see) play an essential role in them, but these will now be embedded in a considerable theoretical network. Remember that J.W. subscribes to (B*), which says, among other things, that for an explanans h to be a good explanation of an explanandum e, h should be deeper, or ontologically richer, than e; that is, h should introduce things of kinds not mentioned in e. In the present context, e is a level-0 perceptual report: its content is purely "autopsychological", to use Carnap's term. If h is to provide a deductive explanation for e, it must yield this autopsychological content; but if h is to comply with (B*) and be ontologically richer than e, it must also introduce things of kinds not mentioned in e.

Thus if e reports one of J.W.'s visual perceptions, he will at least make a beginning in complying with his aim (B*) if by way of a conjectural explanation of e he advances a hypothesis h consisting of: (i) a level-1 statement about the position of a physical object (or combination of objects) relative to himself; (ii) a hypothesis, which may be rather rudimentary, about the source of the light illuminating the object and the way the object reflects the light; (iii) a hypothesis, which may again be rather rudimentary, about the optical effects on his eyes and optical nerves of the light reflected from the object(s); (iv) some neurological-psychological bridging assumption that correlates neurological changes with sensory experiences; plus (v) any other hypotheses that I would have relied on unnoticingly but which J.W. explicitly formulates. What such a conjectural explanation essentially says is that it is because of the object (or combination of objects) in (i), together with the circumstances and regularities described in (ii) to (v), that J.W. had the perceptual experience reported by e.

Having arrived at such an explanatory hypothesis, it will be rational for J.W. to retain it unless he afterwards finds a better alternative to it or later evidence tells against it. As a matter of fact, this does not often happen. But is does occasionally. For instance, if what had appeared to him to be a very small man in a far corner of a room should appear to grow larger as he walks towards the lady in the other far corner, J.W. would revise his previous perceptual judgment; for one of his hypotheses in category (v) above is an assumption about size constancy, to the effect that people, buses, etc. do not rapidly expand or contract.

I said that J.W.'s conjectural explanations of his experiences do not invariably include a level-1 statement. One obvious example is where the explanation is that the experience was a dream experience. (I may perhaps mention another kind of perplexity that sometimes obliges me to proceed in something like a wideawake manner. It results from my sometimes having dreams of a very humdrum and realistic kind. Afterwards, I have a clear memory image, say of a colleague telling me that he has now marked those scripts; and I hesitate between the level-1 hypothesis that he did tell me this and the hypothesis that it was a dream experience; I have sometimes had to ask him.) Again, in the case of certain auditory experiences, J.W. may be undecided whether a level-1 statement is or is not called for. He wakes in the night and hears a persistent, low, humming noise. Is it the water cistern? Or is it only a humming in his ears?

An important question that remains is under what conditions, if any, it is rational for J.W. to accept level-1 statements, put forward by other people, that he is not in a position to check against his own perceptual experience. But it will help to prepare the way for that if we first ask how theory-laden a rationally acceptable level-1 statement may be.

7.4 "Scientific Facts" and the Duhem-Quine Problem

Consider the following pairs of statements: 'The pointer is pointing at 10' and 'A current of 10 amps is passing in this circuit'; 'He is looking poorly' and 'He has a virus infection'; 'This device made a click' and 'This Geiger counter registered the presence of an α particle'. We might say that each pair consists of a "thin" and a "fat", or of a "naive" and a "sophisticated", observation statement. In Duhem's terminology, a naive observation statement, if true, records a *practical* fact whereas a sophisticated one records a *theoretical* fact (*1906*, p. 151). Poincaré drew a rather similar distinction between a *crude* fact and a *scientific* fact (*1905*, pp. 115f.; and see Giedymin, *1982*, pp. 123f.).

I will argue later for a thesis already touched on, namely that the account presented in the previous section has the consequence, and I will claim that it is a desirable consequence, that sophisticated level-1 statements are no less candidates for rational acceptance into the empirical basis than naive ones. In the meanwhile, let us examine some of the consequences of the view that only naive level-1 statements are admissible into the empirical basis.

7.41 *Poincaré on Crude Facts*

Poincaré was understandably horrified by the thesis that 'the facts of science and, *a fortiori*, its laws are the artificial work of the scientist' (*1905*, p. 112), a thesis he had found in Le Roy (1899–1900). But it was not so easy for him to counter this thesis; for he agreed that scientific facts, as he called them, are indispensable for science, and he conceded to Le Roy that much that is conventional enters into a scientific fact. But he insisted that what, at bottom, makes something a *fact* is its being open to sensory verification:

> A statement of fact is always verifiable, and for the verification we have recourse either to the witness of our senses, or to the memory of this witness. This is properly what characterizes a fact. (P. 118)

But does not a statement of scientific fact elude the possibility of sensory verification? Consider this example, given by Poincaré:

> I observe the deviation of a galvanometer by the aid of a movable mirror which projects a luminous image or spot on a divided scale. The crude fact is this: I see the spot displace itself on the scale, and the scientific fact is this: a current passes in the circuit. (Pp. 116–117)

But we surely cannot *see* the current passing in the circuit. If science is to rest on a basis of hard and verified facts, must not so-called scientific "facts" be excluded from it? Poincaré tried to evade this unwelcome conclusion by what has seemed to many, myself included, a transparent piece of cheating.

> What difference is there then between the statement of a fact in the rough and the statement of a scientific fact? The same difference as between the statement of the same crude fact in French and in German. (P. 119)
>
> *The scientific fact is only the crude fact translated into a convenient language.* (P. 120)

One might liken a crude fact to a man in overalls and a scientific fact, as here construed, to *the very same man* dressed in his Sunday best.

7.42 *Duhem on Practical Facts*

Duhem impatiently dismissed Poincaré's attempt to reduce scientific facts to crude facts (*1906*, pp. 149f.). A statement of "scientific" or "theoretical" fact is essentially richer than one of "crude" or "practical" fact. No one has insisted more strongly than Duhem on the theory-laden, even theory-drenched, character of experimental reports in physics (ch. 4); and he accepted the conclusion that such a report, unlike a naive observational report, can never be certainly true:

> An experimental result in physics does not have the same order of certainty as a fact ascertained by non-scientific methods, through mere seeing or touching by a man of sound body and mind. . . .
>
> . . . the account of an experiment in physics does not have the immediate certainty, relatively easy to check, that ordinary, non-scientific testimony has.
>
> Ordinary testimony, which reports a fact established by the procedures of common sense and not by scientific methods, *can be certain*. . . . (*1906*, p. 163, my italics)

I think it is right to say that, before Popper and Neurath, virtually all philosophers concerned with the nature and status of statements at the empirical basis of science required that they be certain or that the basis be rocklike. Both Poincaré and Duhem would, I imagine, have been startled and shocked by Popper's arresting image of science rising, 'as it were, above a swamp . . . like a building erected on piles. The piles are driven down from above into the swamp but not down to any natural or "given" base' (*1934*, p. 111). In our account, the requirement that they be certain is replaced by the requirement that they be rationally, or quasi-rationally, accepted in the sense explained in the previous section. On this view there is no reason why rather sophisticated statements should not be accepted into the empirical basis. Suppose that I attach two galvanometers to an electric circuit and that both their pointers come to rest at 10; then it may well be that the best explanation I can think of for my perceptual experiences will include the premiss 'A current of 10 amps is passing in this circuit'.

We may pick out the following three views concerning the empirical basis: (1) it must consist only of level-0 reports since only these can be quite certain; (2) it may contain naive (or "thin") level-1 reports on whose truth or falsity a layman could decide, since these can be virtually certain; (3) it may contain sophisticated (or "fat") level-1 reports on whose truth or falsity a layman could not normally decide, although these cannot be certain. Now a scientific theory typically yields predictions that are testable against sophisticated level-1 statements. Duhem would surely have agreed that *if* these are admitted into the empirical basis then, while the various hypotheses that compose a scientific theory may not be separately testable, the theory as a whole will be testable. Let T be an axiom set for such a theory and assume that T does have PFs (potential falsifiers) under option (3). Drawing on Duhem's (*1906*) I will now argue that if, having renounced option (3) in favour of option (2), we seek to augment T in such a way that wherever it had previously been testable against sophisticated, it now becomes testable against naive, level-1 statements, this will lead to what I call unmanageable largism: not only will the axiom system become hopelessly unwieldy, but the task could never be completed (I owe the term 'largism' to Worrall). In what follows I will use e's for sophisticated and o's for naive level-1 statements. We assume that T yields an array of SPIs (singular predictive implications) of the form $e_i \rightarrow e_j$ but, as yet, none of the form $o_i \rightarrow o_j$; and our task is so to enrich it that wherever it yields an $e_i \rightarrow e_j$, it now also goes on to yield an $o_i \rightarrow o_j$. Let us begin by tackling one of its SPIs. An example from Duhem (*1906*, p. 146) will serve. Let $e_1 \rightarrow e_2$ say that if the pressure on this battery in our laboratory increases by so many atmospheres, its electromotive force increases by so many parts of a volt. In our laboratory we have an instrument for measuring pressure and another for electromotive force. Let α_1 describe the design of the first instrument and the theories relied on in its design; and let α_2 do the same for the second instrument. Then we may assume that α_1 entails that *if* the instrument is in good working order and connected up, *then* the pointer moves through so many intervals if and only if the pressure increases by so many atmospheres. Letting o_3 stand for the antecedent clause in the above, we may represent this consequence of α_1 by $o_3 \rightarrow (o_1 \leftrightarrow e_1)$; and α_2 may likewise be assumed to entail $o_4 \rightarrow (o_2 \leftrightarrow e_2)$, where o_4 says that the second instrument is in good working order and connected up. Then T augmented by α_1, α_2, o_3 and o_4 entails $o_1 \rightarrow o_2$. Or to put it another way, $o_3 \wedge o_4 \wedge o_1 \wedge \sim o_2$ is a PF of $T \wedge \alpha_1 \wedge \alpha_2$.

So far, so good: we have made a beginning in the task of augmenting T so that it becomes everywhere testable against o's rather than e's. But remember that we have so far dealt with only one of the many kinds of singular predictive implication yielded by our theory. And there is a more serious difficulty, which was highlighted by Duhem: our treatment of this one is as yet far too restricted. We have in our laboratory, let us say, an open-arm manometer for measuring pressure and a galvanometer for electromotive force: it is to these two instruments that our α_1 and α_2 apply. But there are all sorts of other ways of measuring pressure and electromotive force (p. 149). This would not matter too much if we were in a position to specify all of these alternatives in advance. That would mean that instead of augmenting T with α_1 which, in conjunction with o_3, yields $o_1 \leftrightarrow e_1$, we should augment it with α_1, α_1', α_1'' . . . which, in conjunction with, respectively, o_3, o_3', o_3'' . . . , yield respectively $o_1 \leftrightarrow e_1$, $o_1' \leftrightarrow e_1$, $o_1'' \leftrightarrow e_1$ . . . But according to Duhem we could never enumerate all these alternatives: 'A single theoretical fact may then be translated into an infinity of disparate practical facts' (p. 152). I would express this a little differently by saying that there is no way of knowing what new methods of measuring a certain kind of magnitude are going to be invented. (How amazed the Babylonian geometers, or earth measurers, would have been by most of our modern ways of measuring distance.)

Thus to demand that our scientific theories be testable always against naive rather than sophisticated level-1 statements pushes us towards unmanageable largism. It does this at the behest of a demand for observational certainty. But does it achieve observational certainty? Not if the argument in §3.1 was right. Duhem pictured a lay onlooker watching a physicist perform a measurement. The onlooker is supposed to observe an oscillating iron bar, with a mirror attached, sending a beam of light to a celluloid ruler. The physicist observes the electrical resistance of a coil (*1906*, p. 145). But did the layman *know* that the bar is made of iron or that the spot of light on the ruler has been sent there by the mirror? Surely not. If we want certainty we must turn to option (1) and an empirical basis consisting only of level-0 reports.

7.43 *Quine and the Tribunal of Sense Experience*

This is what Quine did. He regarded what I am calling naive level-1 statments as essentially *theoretical*: pigs, ink bottles, and pointers are on a par with Homer's gods or the physicist's molecules.

Physical objects are conceptually imported into the situation as convenient intermediaries . . . as irreducible posits comparable, epistemologically, to the gods of Homer. (1951, p. 44)

Considered relative to our surface irritations, which exhaust our clues to an external world, the molecules and their extraordinary ilk are thus much on a par with the most ordinary physical objects. . . . So much the better, of course, for the molecules. (*1960*, p. 22)

Man has no evidence for the existence of bodies beyond the fact that their assumption helps him organize experience. (*1966*, p. 238)

Now consider the task of completing a further augmentation of the original T so that it becomes testable everywhere against level-0 reports. I begin with a concession: let us forget the obstacle encountered on the previous round and assume that T has already been augmented so as to be testable everywhere against naive level-1 reports. This time I will use o's and e's to denote respectively level-0 and naive level-1 statements; or to put it in Quine's (*1960*) terminology, an o_i describes a glimpse and an e_i the thing allegedly glimpsed. Suppose that T, as previously augmented, has yielded $e_1 \rightarrow e_2$, this being a conditional statement about pointer readings. Can we augment T further so that it yields $o_1 \rightarrow o_2$, this being a conditional statement about sense-data? On the previous round we introduced an α_1 and an o_3 such that α_1 entails $o_3 \rightarrow (o_1 \leftrightarrow e_1)$. Can we proceed analogously here? The previous α_1 described the inner working of, and theories behind, a measuring instrument. On the present round the role of measuring instrument is being filled by a human observer. Thus the present α_1 would need to be a psychophysical theory that yields conclusions of the form: whenever a human organism suffers such-and-such surface irritations, to use Quine's phrase, it experiences such-and-such perceptions. It seems to me rather unlikely that an adequate α_1 would be forthcoming. For one thing, human observers, even when they are in good working order, are rather variable in what they attend to. They do not click with the reliability of a Geiger counter whenever, say, there is a misprint in what they are reading. But suppose for argument's sake that an adequate α_1 has been formulated. We now need an analogue of the previous o_3, which said of a measuring instrument that it is in good working order and connected up. We clearly need something analogous here; for just as a statement about change of pressure cannot be correlated with one about a pointer moving in a certain way unless we make a suitable assumption about the presence of a measuring instrument, neither can a statement about a pointer moving be correlated with one about perceptual experience unless we make a suitable assumption about the presence of an observer. But what should o_3 now say? We might try replacing 'is in good working order' by 'X is "a man of sound body and mind" ', to use Duhem's phrase, and 'is connected up' by 'X is gazing at a certain pointer'. But this would not do; these statements relate a psychophysical organism to a physical object; and physicalistic components are inadmissible in a level-0 statement, which is just a first-person, present-tense report about its author's current experiences. If we seek level-0 versions of the above, we shall find that they cannot do what is required of them. We might try replacing 'in good working order' by 'I feel fine and am having lively perceptions'; but the latter is consistent with its author having taken an LSD tablet. As for 'connected up': there could be no level-0 counterpart for this because a level-0 statement made by me speaks only of *me* and cannot relate me to something external to me. Thus even if we could find an adequate α_1 we could *not* find a suitable o_3 such that α_1 entails $o_3 \rightarrow (o_1 \leftrightarrow e_1)$. On the previous round we did succeed in augmenting T, with α_1 and α_2, so that $o_3 \wedge o_4 \wedge o_1 \wedge \sim o_2$ became a PF of $T \wedge \alpha_1 \wedge \alpha_2$. But on the present round we cannot carry out the analogue of this because we have no equivalent for o_3 and o_4.

In a famous passage Quine wrote: 'our statements about the external world face the tribunal of sense experience not individually but only as a corporate body' (1951, p. 41). I say that this statement is too *optimistic*. Theories about the external world cannot be brought into relation with the perceptual experiences of an ego without the help of a singular premiss, or conjunction thereof, that locates this ego in a physical organism and relates this organism to the external world. This premiss will not be supplied by the theory: it is not the business of Newtonian mechanics to keep us informed of the whereabouts of John Flamsteed, Edmond Halley, and co., and of the positionings of their telescopes. Nor can this premiss be supplied by perceptual experience.

We must agree with Quine that our statements about 'ordinary physical objects' are conjectural hypotheses. But this faces us with a choice. We can cling to the idea that the empirical basis should contain nothing conjectural and that such hypotheses must be expelled from it. But as we have just seen, to take this line is to make the best the enemy of the good: to insist that theories about the external world should be tested only against the *best* kind of observational knowledge, namely infallible perceptual experience, has the unintended consequence that even the corporate body of our statements about the external world becomes quite untestable.

The other alternative is to say that, since it is self-defeating to demand that our theories be tested against test statements that are certain, we will demand only that they be tested against observational knowledge which, though fallible, is *good* in the sense that the level-1 statements composing it have been rationally, or quasi-rationally, accepted.

7.5 Test Statements as Mini-theories

I will now present an argument, which I would have liked to address to Duhem, for option (3) as opposed to option (2). It is that option (2), in comparison with option (3), would deprive the empirical basis of valuable *observational* content. A "theoretical fact", for Duhem, consists of a populist component and an elitist component. More specifically, it consists of phenomena that all normal observers could observe and a theoretical interpretation imposed on them by the scientific experimenter:

> An experiment in physics is the precise observation of phenomena accompanied by an *interpretation* of those phenomena. (*1906*, p. 147)

This suggests that we could strip away the interpretation leaving the phenomena unchanged; such a detheorising would give us the basic "practical fact"; interpretation is, as it were, above and external to the plain matter of fact being interpreted. But the truth is, surely, that interpretation penetrates and colours and shapes observations. This was insisted on at length, and without any claim to originality, in §7.3. Other things being equal, the richer is our interpretation of what we are observing, the sharper and more perceptive is our observation of it. This will come as a truism to students of perception, but it needs to be emphasised in the present context. Let me illustrate it with a personal experience. A good many years ago I was in an audience looking at a screen on which was projected, though I did not know this at the time, the "Hidden man" picture designed by Dorothy Archbold (Porter, 1954). We were asked what we saw. All we could see was a lot of black blobs on a white background; it looked like an elaborate inkblot. We were then told that it contains a picture of a man's head. This was followed in due course by one or two gasps and exclamations of 'I see him'. But most of us could still see only black and white patches. We were then given further assistance; a new slide was projected showing a drawing of a man's head, and we were told that this corresponded to the concealed head. When the original slide was restored, the gasps and exclamations increased. And suddenly I saw him: a sombre figure whose coal-black eyes seemed to be burning into my soul. (The figure has been described as Christ-like but I took him for a seventeenth-century Spanish nobleman.) It would do a big injustice to this experience to say that what I was now seeing was what I had seen before but with an interpretation superimposed. The interpretation dramatically enhanced and changed what I now saw. Nor was there any possibility of stripping the interpretation away and reverting to the naive phenomenon; try as I might, those eyes refused to subside into black patches.

If I wake up, around dawn, in a room I have not slept in before, I am sometimes uncertain where I am. I see a grey shape, a dull glint . . . but I do not understand what these are. And then, a few moments later, I remember where I am, and take a firm interpretative hold; I now see a dressing gown hanging on the door, the mirror above the washbasin . . . The transformation is a bit like that of an inkblot into a picture of a man's head. And there is again no possibility of setting aside the interpretation and reverting to seeing no more than a grey shape, a dull glint . . .

Let us transpose Duhem's watchful layman, who looks on while an experimenter makes a scientific measurement, to some other contexts; he looks on while, say, a pilot checks his instrument panel, or a musician reads a score, or a forensic expert examines some stained clothing, or an intelligence officer examines aerial reconnaissance photographs. Would anyone really claim that the layman *observes* all that the expert observes though without superimposing an interpretation? On one of the photographs, say, there is a darkened area, about 1 by 3 millimetres. The intelligence officer interprets it as the shadow cast by a camouflaged radar tower. It is most unlikely that the layman even noticed this area since he had no reason to attend specifically to it.

It is sometimes possible to compare, under controlled conditions, how much has been observed by lay and expert observers. Take a group whose members show no significant differences of perceptual or intellectual competence outside the field in which some of them are recognised experts. In particular, ensure that the experts do not have a superior short-term memory. Then expose them briefly to observable situations within the experts' field and ask them to reconstruct what they observed. Tests of this kind have been made with chess masters and weaker players (see de Groot, *1965*, and Chase and Simon, 1973). When exposed for five seconds to a chessboard on which the pieces had been placed at random, the chess masters performed no better than the others in reconstructing the positions of the pieces; they did not reveal a superior short-term memory. (Nor, rather surprisingly, did they reveal a superior calculating power.) But in one respect they were decisively superior to the weaker players: if the position on the chessboard came from an actual game (though not one in which they had participated), after examining it for five seconds they could reconstruct it almost faultlessly, far better than the weaker players. In those five seconds, obviously, they had observed more than the others had.

When we examine something, what we observe depends on what we attend to, and this depends very much on what intellectual preparation we bring to it. This is a commonplace but it needs repeating here. I consider it highly unlikely that Duhem's imagined layman would have observed an oscillating mirror sending a beam of light to a celluloid ruler (unless of course he had been instructed to attend to them). Why should he, confronted by a 'table crowded with so much apparatus: an electric battery, copper wire wrapped in silk, vessels filled with mercury, coils, a small iron bar carrying a mirror' (p. 145), focus on just the things Duhem imagined him to focus on? If we ask why Duhem himself picked on just these things, the answer is obvious: he brought to the situation a theoretical understanding that indicated which bits, out of all that apparatus, were involved in measuring the electrical resistance of the coil. The layman, presumably, did not bring to it assumptions that would direct his attention to these bits. He might have observed, say, a used coffee cup, some volumes on a shelf, and a white-coated figure bending over some vaguely discerned apparatus.

I would have liked to put the following argument to Duhem: 'There is only a difference of degree between your "practical facts" and your "theoretical facts" or between naive and sophisticated level-1 judgments. The former do not achieve genuine certainty; they are not infallible. True, a layman may *feel* quite certain that there is an instrument with a pointer in front of him; but then an experimenter may feel quite certain that there is an ammeter in front of him and that it is measuring the current in a circuit. But philosophically considered, the layman's naive judgment is a conjecture generated by the unnoticed and spontaneous application of various interpretative assumptions to perceptual data, and so is the experimenter's sophisticated judgment. And the latter is richer than the former, not just in theoretical but also in *observational* content; an empirical basis containing only naive level-1 statements will be *empirically* poorer than one that also contains sophisticated experimental reports. Some philosophers, including the one with whom your name is often linked nowadays, call for an empirical basis containing only level-0 reports of the "Blue here now" variety, reports of whose truth their authors can be quite certain. But that has the self-defeating result of rendering our theories about the external world altogether untestable. If we want testability we must forswear certainty at the observational level. But if we forswear certainty and renounce option (1), then why should we not, on the principle that one may as well be hanged for a sheep as a lamb, swing right over to option (3)? You would be the first to agree that it is option (3) rather than option (2) that accords with scientific practice: when an experimenter sends in a report, he does not clutter it up with details about iron bars, celluloid rulers, etc. Why should not our philosophical account of the empirical basis for science reflect its real nature?'

Duhem would presumably have replied that, although naive level-1 statements may not measure up to some philosophical standard of certainty, they are in fact more nearly certain than sophisticated level-1 statements, and that to allow the latter into the empirical basis seriously increases the risk of error entering into it. I agree that it does increase this risk: the theoretical assumptions involved in an experimenter's report may be false and may distort it into an unfaithful rendering of what was observed. A nice (or should I say nasty?) example of this has been given by Glymour:

> John Mayow, the seventeenth-century chemist, performed the familiar experiment that consists of burning a candle over water under an inverted beaker; Mayow noted at first that the volume of water in the beaker increased, and that the volume of air decreased. But almost immediately he redescribed what had happened as a case of the air having lost its "spring", and thereafter it was the decreased spring of the air and not the decreased volume that Mayow gave as the result of the experiment and the phenomenon to be explained. (*1980*, pp. 121–22)

The excess content of a sophisticated over a naive level-1 statement does indeed mean an increased risk of error. But when we turn from the chances of error getting in to the chances of detecting errors that have got in, the boot is on the other foot. The main way of checking on an experimental report is of course to *repeat the experiment*. There can of course be no such thing as an *exact* repetition of an experiment; and it would be rather futile to try to reproduce the physical detail of the first experiment, so that if a grey galvanometer with a dial was used then, we must not use a green voltmeter with a digital display. No, to repeat an experiment one needs to understand, and reproduce, its *un-*

derlying structure, preferably with a more refined apparatus; and this can be done only if one is equipped with a theoretically sophisticated description of the original experiment. If a physicist restricted his experimental reports to statements about the movement of a spot of light on a celluloid ruler, or whatever the appropriate naive level-1 statements might be, he would deny to others the theoretical plan of his experiment and they would not know how to repeat it.

It seems to me that option (2) is an unhappy compromise that gives no real satisfaction to anyone. It pushes us towards unmanageable largism without giving us certainty; and although it reduces, without eliminating, the risk of error, it also reduces our hope of detecting errors by repeating experiments. Option (1) does at least satisfy those who want observational certainty, though at a disastrous cost. Option (3) satisfies those who want to avoid unmanageable largism: it rehabilitates the possibility of rational scientific choice between definite theoretical units, namely theories that are sufficiently fleshed out with auxiliary assumptions to yield predictions about, say, the bending of light rays grazing the sun or the small precession of the perihelion of Mercury, but not sufficiently to yield predictions about dots on photographic plates, pointer movements, etc.

7.6 The Responsibility of the Experimenter

We might say (with apologies to Pericles) that a scientific community is a democracy in which, although only a few may originate new theories, all are free to criticise and test them. A theorist and an experimenter may of course coexist in one person, but suppose them separate, and consider their respective responsibilities first under option (2) and then under option (3). (I will not consider option (1) since it is unworkable.) I assume that it is *the task of the theorist* to put forward a testable theory and the task of the experimenter to test it. But under option (2), according to which it is to be tested against naive level-1 statements, it will be for the *theorist* to try to think of all the sophisticated instruments that might be employed in testing it and to incorporate theories about them (theories of the kind denoted by α_1, α_1', $\alpha_1'' \ldots \alpha_2$, α_2', $\alpha_2'' \ldots$ in §7.42) into the axioms of his theory. Under option (3) it is enough if the theorist puts forward an axiom system T that yields sophisticated SPIs of the form $e_i \rightarrow e_j$ about, say, the bending of light rays grazing the sun; it is for the experimenter to use his ingenuity in devising ways to test such predictions. Indeed, far from being confined to observing pointer readings etc., it may be the experimenter who hits upon a novel way of bringing a theory to test. Consider the fundamental assumption of Descartes's theory of light, namely that light is transmitted instantaneously. The problem of how to test this was unsolved for a long time. Descartes had a shot at it and so did Galileo, but without success. The problem was rather like finding a way of forcing the Loch Ness monster, if indeed this monster exists, to give a clearcut indication of its presence. This problem was solved by Ole Roemer. (Roemer may have been anticipated by Cassini; see Sabra *1967*, p. 205.) If light is transmitted instantaneously, then it takes no longer to reach the earth from Jupiter when these two planets are furthest apart than when they are nearest each other. But if its speed is finite, then of course it does take longer when they are furthest apart. Now Jupiter's innermost moon is eclipsed when Jupiter is between it and the earth. (I am relying here on Jaffe's excellent *1960*.) Roemer introduced the subsidiary assumption that this moon orbits Jupiter in a uniform way: at any rate, it does not speed up, or slow down, as the earth approaches or recedes from Jupiter. On this assumption, if the duration of an eclipse of this moon were found to be shorter when the earth is between the sun and Jupiter than it is six months later when the earth is furthest from Jupiter, then Descartes's fundamental assumption is false; and by 1676 Roemer had found its duration to be about twenty-two minutes shorter. Adding a further subsidiary assumption about the magnitude of the diameter of the earth's orbit round the sun, Roemer was able to calculate the approximate speed of light; and his calculation led to new corroborations:

> On the basis of his calculated speed of light, Roemer succeeded in forecasting the times of certain eclipses months in advance. In September 1676, for example, he predicted that in the following November a moon of Jupiter would reappear about ten minutes late. The tiny satellite co-operated, and Roemer's prediction came true to the second. (Jaffe, *1960*, p. 28)

Another famous experiment that exploited the difference in the earth's situation after an interval of six months was the Michelson-Morley experiment. We touched earlier (§5.25) on Maxwell's idea of an electromagnetic medium which is at the same time the medium in which light waves undulate. This developed into the idea that absolute space is filled by a stationary ether that is not dragged round by the earth as its atmosphere is. How could this elusive monster, if it exists, be forced to give some indication of its presence? Maxwell himself despaired of the possibility of testing the ether hypothesis. But theoreticians do not have the last word in such matters. Albert Michelson hit upon the following idea. On a stable platform (a massive stone slab floating in mercury) there is a light source that sends a beam of light to a half-silvered mirror A at 45° to it. At A half the beam is deflected 90° towards a mirror B and half the beam passes through A to a mirror C. At B and C the beams are reflected back to A where they recombine and travel together to an interferometer. After positioning B and C as nearly equidistant as possible from A, one of them is finely adjusted so that there is, at the present time, no interference. Now if this apparatus is moving relative to a stationary luminiferous ether both because of the earth's daily rotation and because of its annual revolution around the sun, then if this experiment is repeated at regular intervals during each day and daily for a year, there should be times when an interference pattern appears. More specifically, as Jaffe put it,

> in the course of a year there would be two days when the maximum effect, if indeed there was any effect, could be observed. On one day the earth would be traveling in exactly the opposite direction from its path on the other day. Hence, on one of these days the earth would be moving against an ether wind, and exactly six months later it would be rushing along with the ether wind. (P. 87)

In experiments of this kind, as Popper said, 'Theory dominates the experimental work from its initial planning up to the finishing touches in the laboratory' (1934, p. 107). And it is essential that the experimenter be allowed to operate with sophisticated, theory-laden level-1 statements. To know the result of the Michelson-Morley experiment we need to be told whether *interference patterns* were observed in the course of the year. To demand that the ether hypothesis should have been tested against only naive level-1 statements would be rather like demanding that the hypothesis that the figure in the "Hidden man" picture stares straight at the beholder should be tested against statements about black-and-white patches.

Someone who takes what I called a quasi-inductive view of the relation of level-1 statements to perceptual experiences is likely to want the former to "go beyond" the latter no more than is necessary; for the more they transcend their basis in experience the less safe they become. In other words, he will want statements that are accepted into the empirical basis to be as thin and naive as they can be without ceasing to be physicalistic statements. But the view presented in §7.3, according to which quasi-rationally accepted level-1 statements play a quasi-explanatory role, puts us under no pressure to keep them as thin and naive as possible. This view defeats rationality-scepticism concerning the empirical basis without reneging on proposition (III), the deductivist thesis, and without requiring an impossibly large augmentation of the premises of a scientific theory if it is to be rendered testable.

7.7   The Acceptability of Other People's Reports

The imaginary John Wideawake who was introduced in §7.31 was there depicted as a rather lonely figure, intent on seeking explanations, typically involving some level-1 statement in an essential way, just for his own perceptual experiences. Nothing was said there about his acceptance or otherwise of level-1 statements of which he is not the original author. It is to this question that I now turn. It is obviously an important question in view of the enormous reliance we all in fact put on the observational experience of other people.

Now if, as I have argued, it is permissible for J.W. to introduce, in an attempt to explain certain perceptual experiences, not just the "thin" level-1 statement that there is in front of him a white dial with a black pointer, but the "fat" statement that this is the outward face of a sophisticated measuring device, an ammeter say, then it is likewise permissible for him to introduce, in appropriate circumstance, not just the "thin" statement that there is a pinkish dial with certain protuberances and indentations before him, nor the somewhat less "thin" one that there is a face before him, but the "fat" statement that this is the face of an intelligent human being. Again, he has no compunction about enriching the statement that there is a page of print before him with the assumption that it is an expression of human thought.

Suppose that J.W. accepts the observation report (in Hesse's sense) e' that a person X has put forward the observation statement e: under what conditions is it rational for J.W. to accept e? If J.W. is in a position to check e against his own perceptual experience, no special difficulty arises; he can proceed as in the one-person case, with the difference that instead of inventing the hypothesis e himself he has taken it from another source. So suppose that e reports something outside the range of his own perceptual experience. Now among the various interpretative principles that J.W. consciously employs (and which I usually rely on unnoticingly) is one that says that, in the absence of evidence to the contrary, human behaviour should be assumed to be purposive. So J.W., having accepted that X asserted e, will assume that this was a purposive act. The question now is what sort of purpose X may have had. Now there are lots of hypotheses concerning X's purpose which, if adopted by J.W., would incline him *not* to proceed from e' to e. But suppose that J.W.'s conjecture is that X is a scientific observer whose aim was to publish the truth about the result of an experiment he has conducted. In that case, J.W. conjectures that X had quasi-rationally accepted e, or that it is as if e had been a key premiss in the best explanation that X could find for some of his perceptual experiences. J.W. will now ask himself whether, if he had been in X's shoes and had had those experiences, he might have come up with a different explanation. Suppose that the e in question had been John Mayow's statement (referred to near the end of §7.5) that when a candle was burnt over water under a beaker the air *lost its spring*; J.W. might have preferred to explain what Mayow saw in terms of a reduction in the volume, rather than the springiness, of the air. In that case J.W. would proceed from his acceptance of the observation report e' to the acceptance of an observation statement somewhat different from e itself. Alternately, J.W. may conclude that if he had been in X's shoes, he could not have come up with any better explanation for the perceptual experiences he would have had than one that includes e as a main component; and in that case he proceeds from his acceptance of e' to the vicarious acceptance of e itself. Having once accepted a level-1 statement in this way, J.W. retains it unless some new consideration obliges him to reconsider it.

One can imagine an ideal scientific community in which the report of an experimental result is never sent in by an individual but always by a team, each of whose members has found it satisfactory after critical examination and testing against perceptual experience, so that it has, to borrow from Popper (*1945*, ii, p. 217), the objectivity that comes from intersubjectivity. Once agreed by the team, it is flashed to the other members of the community, who may challenge it. If no one challenges it, it goes into a central databank as a new item in the community's empirical basis. If it is challenged, reasons must be given. A conference between the challenger, the team, and any other interested parties then takes place. All hypotheses that the various individuals have brought to the matter in question are pooled, discrepancies between them are identified, and a collective attempt is made to resolve the dispute. If this succeeds, a new report is sent in by this enlarged team.

A statement that has been accepted into the empirical basis may be challenged subsequently but the onus is on the challenger to make a good case for reopening the question (for instance, that the experiment has been repeated, with a different result; or that in the original report, a hypothesis was relied on that has since been superseded). If this happens, a conference is again convened to try to resolve the dispute.

Although this is an idealised picture, I think that it has some resemblance to what actually happens in science at least in those cases where an experimental report has important theoretical implications.

8

....

Corroboration

8.1 Corroboration and Verisimilitude

We now have both an aim for science, claimed to be the optimum aim, and the idea of an empirical basis for science, consisting of level-1 statements which, though conjectural, are well tested against perceptual experience and quasi-rationally accepted. (Henceforth, when I speak of *evidence* I will mean something reported by a statement incorporated into the empirical basis.) The question now is whether, equipped with these, we can answer Hume by defeating rationality-scepticism while retaining probability-scepticism. We could defeat rationality-scepticism if we could *know*, at least in a good many cases, which one of a set of competing hypotheses, given the present evidence, best satisfies (B*); for we surely have the best possible reason to adopt the one that best satisfies the optimum aim for science. (I say 'in a good many cases' because we must allow for the possibility that two are tied for first place, or that the situation is messy in a way that does not, at this stage, permit a verdict to be reached. Cricket matches occasionally end in a tie and often in a draw, but that does not make it a futile game; if no side *ever* won, the game would lose its point.) But it is one thing to have an aim for science and another to have a method by which we can, in favourable circumstances, actually pick out the hypothesis that best satisfies that aim. Is there such a method? I will argue in §8.3 that the one that is best corroborated, in essentially Popper's sense, is the one that best satisfies (B*).

The question 'Why do corroborations matter?' has often been addressed to Popper by his critics (for example: Salmon, *1966*, pp. 26–27, 1968, p. 28, 1981, pp. 119f; Lakatos, 1974, pp. 154f; Ayer, 1974, p. 686; Putnam, 1974; Grünbaum, 1976, pp. 234–236, 246–247; O'Hear, *1980*, p. 42). Since (*1963*) Popper has held that the aim of science is to progress towards the truth with theories that are ever closer to the truth, or have an ever higher degree of verisimilitude; and he claimed that his theory of corroboration 'is the proper methodological counterpart' of the aim of increasing verisimilitude (*1963*, p. 235). A good many philosophers who incline more to inductivism than to Popper's epistemology have accepted from him that, provided an adequate explication of it is found, increasing verisimilitude should be taken as the aim of science. (I am thinking particularly of the Finnish philosophers Ilkka Niiniluoto, Risto Hilpinen, and Raimo Tuomela; and perhaps the English philosophers L.J. Cohen and W.H. Newton-Smith should be included here.)

This raises an issue that I cannot avoid. For I have claimed that (B*) is the optimum aim for science; but (B*) says nothing about verisimilitude. Its component (A*) says that what science aspires after is *truth* rather than approximations thereto; and on the question of the truth of the theories you accept, it says only that they must be possibly true for you at the present time, in the sense that you have not, despite your best endeavours, found any inconsistencies in them or between them and evidence available to you. It does not say that a theory T_j should be preferred to T_i only if T_j seems closer to the truth than T_i. So there are two possibilities. Either (B*) omits something that could and should have been included, and is therefore not the optimum aim for science, or the aim of increasing verisimilitude fails to satisfy at least one of the (I trust uncontroversial) adequacy requirements laid down in §4.1 for any proposed aim for science. The present section will have the negative purpose of showing that Popper's theory of corroboration should *not* be geared to the aim of increasing verisimilitude, and that this aim fails to satisfy our third adequacy requirement, namely that a proposed aim for science should *serve as a guide* in the making of choices between competing hypotheses. In the section after this, §8.2, I will show that component (A*) of (B*) generates a *demand for tests* on theoretically promising theories. Then I will ask: when does evidence *corroborate* and, more especially, *strongly* corroborate a theory, and when is one theory *better corroborated* than a rival theory? In §8.3 I will argue that (B*) calls for the acceptance of a theory T if T is, at the time, the *best corroborated* theory in its field. Thus corroborations do matter, from the standpoint of (B*); indeed, they are of decisive importance. After that I will test this theory of corroboration against various more or less well known difficulties and "paradoxes" that have been found to beset many theories of confirmation.

I turn now to the bearing of corroboration appraisals on verisimilitude appraisals. Let us write $\mathrm{Vs}(T_j) > \mathrm{Vs}(T_i)$ to mean that T_j is a better approximation to the truth, or has greater verisimilitude, than T_i; and assume, for the sake of the present argument, that this notion is well defined. And let us write $Co(T_j) > Co(T_i)$ to mean that T_j is better corroborated, at the present time, than T_i. Quite what this should be taken to mean we will consider in due course. For the present we may take it as saying that T_j has, so far, performed better under tests than T_i, perhaps because it has been exposed to, and has stood up under, a wider variety of tests. The question now is whether, given that $Co(T_j) > Co(T_i)$, we have any justification for concluding that $Vs(T_j) > Vs(T_i)$. There would seem to be three main alternatives: (a) we have *every* justification: we would *know* that $Vs(T_j) > Vs(T_i)$; (b) we have *no* justification: we could only *guess* that $Vs(T_j) > Vs(T_i)$; (c) we have *some* justification: although we could not *know* we would have *good reason to suspect* that $Vs(T_j) > Vs(T_i)$. Popper never countenanced answer (a); and he sometimes seemed to reject it in favour of answer (b). For instance, he said that if asked how he knows that one theory has more verisimilitude than another, his answer is: 'I do *not* know—I only guess' (*1963*, p. 234). But I think that he really preferred answer (c). For instance, after pointing out that we may have $Vs(T_j) > Vs(T_i)$ in cases where both T_i and T_j are refuted, he added that the fact that the refuted T_j has withstood tests that T_i did not pass 'may be a *good indication*' (p. 235, my italics) that $Vs(T_j) > Vs(T_i)$. His preference for answer (c) became more pronounced in other works. In his (*1982a*), the text of which was virtually complete in 1962, Popper wrote:

> If two competing theories have been criticized and tested as thoroughly as we could manage, with the result that the degree of corroboration of one of them is greater than that of the other, we will, in general, have *reason to believe* that the first is a better approximation to the truth than the second. (P. 58)

(It is puzzling that in his 1982 introduction to this work, Popper now says that the idea of verisimilitude, or of one theory being a better approximation to the truth than another, 'is not an essential part of my theory' (p. xxxvii). In that work the idea of one theory being better than or superior to another had been *equated* with its appearing 'to come *nearer to the truth*' (p. 25). At another place in this work, he had said, in connection with the problem of induction, that while there is *no* reason to believe in the *truth* of a physical theory, 'there may be reasons for preferring one theory as a better *approximation to the truth*', adding: 'This makes all the difference' (p. 67). And in his *1977* (with

281

John Eccles) he had *equated* scientific progress with increasing verisimilitude (p. 149). Earlier (in *1963*, p. 232) he had said: 'I believe that we simply cannot do without something like this idea of a better or worse approximation to truth.')

His preference for answer (c) was again made very clear in his (*1972*). For instance, he there wrote:

> I intend to show that while we can never have sufficiently good arguments in the empirical sciences for claiming that we have actually reached the truth, we can have strong and reasonably good arguments for claiming that we may have made progress towards the truth. (Pp. 57–58)

And later in the same book he went further still:

> But the critical discussion can, if we are lucky, establish sufficient reasons for the following claim:
> 'This theory seems at present, in the light of a thorough critical discussion and of severe and ingenious testing, by far the *best* (the strongest, the best tested); *and so* [my italics] it seems the one nearest to truth among the competing theories.'
> To put it in a nutshell: . . . we can, if we are lucky, rationally justify a preference for one theory. . . . And our justification . . . can be the claim that there is *every indication* [my italics] at this stage of the discussion that the theory is *a better approximation to the truth* than any competing theory so far proposed. (P. 82)

8.11 *Lakatos's Plea*

A section of Lakatos's (1974) is entitled *A plea to Popper for a whiff of 'inductivism'* (p. 159). His view was that "the game of science" for which Popper's methodology provides rules will remain a game to be played for its own sake (*la science pour la science*, so to say), unless those rules are governed by an epistemological aim lying beyond them. With this much I of course agree, though I would add that the proposed aim should satisfy the adequacy requirements set out in §4.1. (And I would further add that the aim should, ideally, be the highest of those that satisfy these adequacy requirements.) However, the aim Lakatos envisaged was *increasing verisimilitude*. His idea was that while the move from, say, NM to the better corroborated GTR must be esteemed an excellent move in the game of science judged by Popper's methodological rules, it can be esteemed an advance in *knowledge* only if corroboration appraisals are linked to verisimilitude appraisals by some inductive principle:

> the link has to be re-established between the game of science on the one hand and the growth of knowledge on the other. (P. 157)

> This can be done easily by an inductive principle which connects . . . verisimilitude with corroboration. (P. 156)

> Only some such conjectural metaphysics connecting corroboration and verisimilitude would separate Popper from the sceptics and establish his point of view, in Feigl's words, 'as a *tertium quid* between Hume's and Kant's epistemologies'. (Pp. 163–164)

Whether it would be so easy to formulate such an inductive principle may be doubted. Popper's idea was that in a progressive sequence of scientific theories, T_1, T_2, T_3 . . . , we may have $Vs(T_1) < Vs(T_2) < Vs(T_3)$ . . . although T_1, T_2, T_3 . . . are all refuted. But if they are all refuted, each has a degree of corroboration of -1. But suppose that any such difficulties are overcome and that a happy formulation of this *IP* (inductive principle) is arrived at. Then this *IP* would be exposed to the familiar objections, rehearsed in §3.3, to which *any IP* is exposed. So far as answering Hume is concerned, we would be back to square one. Hume said, in effect, that some inductive assumption is (i) indispensable and (ii) unjustifiable. Popper (in *1934*) rejected (i) and retained (ii). Carnap and others retained (i) and rejected (ii), arguing that induction is a *logical* process involving probability logic. Lakatos, who had no illusions about the probabilist programme's ability to eliminate the problem of induction (see his 1968), was *in effect urging the retention of both* (i) and (ii).

If Popper's (*1972*) had appeared before Lakatos wrote his (1974) he would have realised that there was no need to beg Popper to *introduce* a whiff of inductivism into his philosophy; it was already there. Or so I will now argue.

8.12 *An Inductive Progression*

It is an incidental merit of the concept of incongruent counterparthood, defined in §5.1, that it enables us to give a perfectly clear meaning to verisimilitude comparisons in one special case, namely where c_i and c_j are incongruent counterparts. For in that case every consequence of c_i has a counterpart among the consequences of c_j and vice versa: their consequences are in one:one correspondence. And $Vs(c_i) > Vs(c_j)$

must then mean that a true consequence of c_i is paired with a false consequence of c_j more frequently than the converse. In what follows I will confine myself to cases where c_i and c_j are incongruent counterparts. My argument will be very simple. It boils down to this: (i) a comparative corroboration appraisal of the form 'c_i is better corroborated than its incongruent counterpart c_j' is a historical report about the past performance of these two hypotheses, and does not, by itself, carry any predictive implications about their future performance; (ii) a comparative verisimilitude appraisal of the form $Vs(c_i) > Vs(c_j)$ *does* carry predictive implications about future performance, at least in the case of incongruent counterparts; hence (iii) a progression from the premiss that c_i is better corroborated than its counterpart c_j to the conclusion that $Vs(c_i) > Vs(c_j)$ is an inductive progression.

Popper has always insisted on (i) above: 'As to degree of corroboration, it is nothing but a measure of the degree to which a hypothesis h has been tested, and of the degree to which it has stood up to tests' (*1959*, p. 415). And he reiterated this very emphatically in (*1972*): 'Corroboration (or degree of corroboration) is thus an evaluating *report of past performance*. . . . Being a report of past performance only, . . . *it says nothing whatever about future performance*' (p. 18). However, Popper seems to have overlooked (ii), perhaps because he has usually been concerned with competing theories of unequal strength. But if c_i and c_j are incongruent counterparts, we may think of their respective consequence classes on the analogy of two urns containing equal numbers of balls, some white (true) and the others black (false), the ratio of white to black being higher in the case of c_i. And we may think of tomorrow's crucial experiment on the analogy of a selection of an equal number of balls from each urn. If the Angel of Truth were to inform me that $Vs(c_i) > Vs(c_j)$, and if a prize were offered for guessing correctly which will pass the crucial experiment (no prize being awarded if both fail), then I would be mad not to bet on c_i. I would realise that it remains possible that c_j will pass and c_i fail; the experiment may happen to be directed at a point where c_i is wrong and c_j is right. But if I have no additional information (such as that tomorrow's experiment is only a repetition of an earlier one that c_j passed and c_i failed), then I must conclude from what the Angel of Truth has told me that it is more probable that c_i will pass tomorrow's test than that c_j will.

Thus to say that a corroboration report can, if we are lucky, give us *every indication* that one theory is *a better approximation to the truth* than another theory is to say that a report that *says nothing whatever about future performance* may give us every indication that one theory *is likely to perform better than another in the future.* As I said, from $Vs(c_i) > Vs(c_j)$ we can conclude only that the prediction that c_i will pass and c_j fail tomorrow's test is more *probable* than the converse. But Popper has rightly insisted that Hume's argument against the possibility of valid inferences from premises about observed instances to conclusions about unobserved instances remains no less cogent if 'the word *"probable"* is inserted before "conclusions" ' (*1972*, p. 4).

In case I am sounding holier-than-thou I should add that I too claimed (in 1968b) that in some cases a corroboration appraisal may support a verisimilitude appraisal. I should also add that Popper had a logical argument for his claim. He proved (in 1966) a theorem to the effect that an increase in content, as would occur in moving from a hypothesis h_i to a stronger hypothesis h_j that strictly entails h_i, is always accompanied by an increase in truth content but not always by an increase in falsity content. From this he concluded that

> the stronger theory, the theory with the greater content, will also be the one with the greater verisimilitude *unless its falsity content is also greater.* (*1972*, p. 53)

He added:

> We try to find its weak points, to refute it. If we fail to refute it, . . . then we have reason to suspect, or to conjecture, that the stronger theory has no greater falsity content than its weaker predecessor, and, therefore, that it has the greater degree of verisimilitude.

But we know now that the only case where an increase in content is not accompanied by an increase in falsity content as well as of truth content is the case where the stronger theory is *true.* If h_i and h_j are false, then there is no such asymmetry as Popper had supposed between truth content and falsity content; both increase with content. In 1974 Tichy, Harris, and Miller each showed that if h_j is stronger, and seems intuitively closer to the truth, than h_i, which it does not entail but largely corrects, but is nevertheless false, then h_j will have false consequences that are outside the falsity content of h_i, and h_i will have true consequences that are outside the truth content of h_j; thus h_i and h_j will be incomparable for verisimilitude by Popper's original definition, according to which $Vs(h_j) > Vs(h_i)$ if the falsity content of h_j is contained in that of h_i and the truth content of h_i is contained in that of h_j, at least one of these containments being strict. (Informal proof, based on the above three men's 1974 papers: since h_j corrects h_i, there will be at least one true consequence, call it t_j, of h_j that is not a consequence of h_i, and at least one false consequence, call it f_i, of h_i that is not a consequence of h_j. Since h_j is false, it will have at least one false consequence, call it f_j. Then the falsity content of h_j will include $t_j \wedge f_j$, which is not a consequence of h_i; and the truth content of h_i will include $f_j \rightarrow f_i$, or $\sim f_j \vee f_i$, which is not a consequence of h_j.)

My erroneous idea in (1968b) was that, if h_i and h_j are very unequal with respect both to content and to past performance, then we may have a priori reasons for doubting $Vs(h_i) = Vs(h_j)$ and a posteriori reasons for doubting $Vs(h_i) > Vs(h_j)$, leaving $Vs(h_j) > Vs(h_i)$ as the most plausible alternative. My mistake, as Miller duly pointed out (1975, p. 191), was to overlook the all too likely possibility that h_i and h_j are incomparable for verisimilitude.

My present objection to any progression from corroboration appraisals to verisimilitude appraisals can be summarised, very simply, as follows: if we *cannot* legitimately proceed from a premiss p to a conclusion r; and if we *can* legitimately proceed to r from some other premiss q; then we *cannot* legitimately proceed from p to q. Here, p is a comparative corroboration report that says of two hypotheses (of equal strength) that one of them has performed better in the past than the other; r is a prediction that says that the former hypothesis will, on the average and other things being equal, perform better than the latter in the future; and q is a verisimilitude appraisal that says that the former hypothesis is closer to the truth than the latter.

If we exclude as illegitimate any sort of progression from corroboration appraisals to verisimilitude appraisals, we are forced back to a position that Popper took at one point in (*1963*) and which I quoted earlier, namely that we can only *guess* that the better corroborated of two competing theories is the more verisimilar. The question now arises whether such guesses, though never positively justified, are at least under some sort of negative critical control. Popper seemed to suggest that they are. For instance, he wrote:

> But I can examine my guess [that the theory T_j has a higher degree of verisimilitude than the theory T_i] *critically*, and *if* it withstands severe criticism, then this fact may be taken as a good critical reason in favour of it. (*1963*, p. 234, my italics)

Now if someone puts forward as a conjecture that $Vs(T_j) > Vs(T_i)$, there is indeed one criticism to which it might be exposed, namely that

T_i and T_j are not comparable for verisimilitude. But *that* criticism, if it applied at all frequently, would undermine the claim that increasing verisimilitude is the proper aim of science. So let us suppose, for the sake of the argument, that T_i and T_j are comparable for verisimilitude; on that supposition, a criticism of $Vs(T_j) > Vs(T_i)$ would be a consideration that has some tendency to justify $Vs(T_j) \leq Vs(T_i)$. But if all verisimilitude appraisals are guesses that are never positively justified, there is no more possibility of justifying the latter appraisal than the former; a "criticism" of the former conjecture would be no more than a rival and equally unjustified guess, one metaphysical claim pitted against another. I conclude that if verisimilitude appraisals can only be conjectural, and if two theories are comparable for verisimilitude, then a comparative verisimilitude appraisal of them cannot be under any genuine critical control.

I now revert to the question whether (B*) is suboptimal in failing to include increasing verisimilitude as part of the aim of science or whether the latter fails to satisfy our adequacy requirements, set out in §4.1, for any proposed aim. The aim of increasing verisimilitude certainly satisfies requirement 4, that it should be impartial, and requirement 5, that it should involve the idea of truth. And if a formally and materially adequate explication of 'closer to the truth' is arrived at, this aim might satisfy requirements 1 and 2, for coherence and feasibility. But if verisimilitude appraisals can only be a matter of uncontrolled guesswork, then this aim cannot satisfy requirement 3, that it should *serve as a guide* when we try to make rational choices among competing hypotheses. After quoting a passage in which Popper had written that 'we hope to learn from our mistakes' and that we can have 'the serious purpose of eliminating as many of these mistakes as we can, in order to get nearer to the truth' (*1963*, p. 229), Lakatos commented that this passage 'amounts to no more than the assertion that we must play the scientific game *seriously*, in the hope of getting nearer to the Truth. But did Pyrrho or Hume have anything against being "serious" or entertaining "hopes"?' (1974, p. 161). A proposed aim for science that fails requirement 3 *provides no answer to rationality-scepticism.* Suppose that the aim is to progress to ever ϕ-er theories (for example, to theories that are ever closer to the truth), and that the methodological rule is to adopt (accept, prefer) whichever theory is, at the present time, the ψ-est (for example, the best corroborated) in its field (and to suspend judgment where there is no ψ-est theory). Assume that this methodological rule is a workable one in the sense that we can, in a good many cases, *identify* the ψ-est of a set of competing theories. However, which of them is the ϕ-est is a matter of uncontrolled guesswork. Then while we could indeed make clear-cut decisions in accordance with our rule, we would have no reason to suppose that our decisions conform with our aim. Asked why we take ψ-ness so seriously, we could answer only that we *hope* that it is an indication of ϕ-ness. But a sceptic would *fear* that it is not. To erect increasing verisimilitude into that by which we ultimately try to discriminate between competing theories would be like trying to steer a course through the ocean of uncertainty by a star that is permanently behind cloud.

Let us now consider why corroborations are important from the standpoint of (B*).

8.2 Why Corroborations Matter

Popper created a stir in (*1963*) by introducing a 'third requirement' to be satisfied by a new scientific theory for it to be a major advance on its predecessor(s). His first two requirements were:

1. 'The new theory should proceed from some *simple, new, and powerful, unifying idea*' (p. 241).
2. The 'new theory should be *independently testable*' (ibid).

And the third was:

3. The new 'theory should pass some new, and severe, tests' (p. 242).

His first requirement is reflected in component (B1-2), the demand for greater depth-cum-unity, of our (B*), and his second requirement is taken up in component (B3-4), the demand for greater predictive power.

Some inductivists hailed this third requirement as a major concession to their point of view; and some of Popper's followers deprecated it for the same reason (see Agassi, *1975*, pp. 26–27). But it is hard to see what the fuss was about. The second requirement says, in effect, that the new theory should be more corrobor*able* than its predecessor(s) and the third adds that it should, moreover, go on to become better corrobor*ated*. Popper conceded too much when he said that he had never previously 'explained clearly the distinction between what are here called the second and third requirements' (p. 248n.). He had drawn the distinction between degree of corroborability (testability, falsifiability) and degree of corroboration quite clearly in (*1934*). For instance, he there wrote: 'Of course, the degree of corroboration actually attained does not depend *only* on the degree of falsifiability: a statement may be falsifiable to a high degree yet it may be only slightly corroborated, or it may in fact be falsified' (p. 268). One argument that Popper gave for his third requirement was, in effect, an appeal to what, in §5.11, I called the antitrivialisation principle; it would be trivially easy to make theoretical "progress" in accordance with just the second requirement:

> The mere fact that the theory is independently testable cannot as such ensure that it is not *ad hoc*. This becomes clear if we consider that it is always possible, by a trivial stratagem, to make an *ad hoc* theory independently testable, *if we do not also require that it should pass the independent tests in question:* we merely have to connect it (conjunctively) ... with any testable but not yet tested fantastic *ad hoc* prediction. (*1963*, p. 244)

However, it is not open to an adherent of (B*) to argue that corroborations are needed in order to block such trivial increases in testability, since these are already blocked: (B1-2) requires the greater predictive power of a deeper theory to be generated with the help of its richer theoretical core, and the theory has to satisfy the organic fertility requirement, which it would not do if one of its axioms were merely a tacked on prediction. Then how can an adherent of (B*) explain the need for corroborations?

Rather as Popper called his first and second requirements 'formal' and his third a 'material requirement' (p. 242), I will call components (B1-2) and (B3-4) of (B*) prior demands, and component (A*), which requires the hypotheses accepted by X at any one time to be possibly true for X at that time, a posterior demand. I will say that T_j is (B*)-superior to T_i if T_j is (B)-superior and, moreover, T_j satisfies (A*).

As a way of showing that corroborations matter for anyone who accepts (B*), I will first look at the matter from the point of view of someone who accepts only (B), or (B*) minus (A*). Let T_j be a new theory that is (B)-superior to an existing theory T_i; but no independent tests have so far been made on T_j. Assume that person Y, as well as accepting (B), is a Humean who accepts probability-scepticism. Then Y may deprecate proposals to test T_j. He may reason as follows:– As things now stand, T_j is clearly a theoretical advance on T_i. If tests on T_j go ahead, either (i) they will all corroborate T_j, or (ii) some of them will falsify T_j. If (i), then probability-scepticism implies that these corroborations will not raise the probability that T_j is true or enhance its epistemological status in some other way, such as by justifying the favourable verisimilitude comparison $Vs(T_j) > Vs(T_i)$. Its epistemological status will be what it is now: T_j will still be just an unrefuted conjecture. If (ii), then its epistemological status will be badly damaged. As Miller put it: 'The passing of tests therefore makes not a jot of difference to the status of any hypothesis, though the failing of just one test may make a great deal of difference' (1980, p. 113). So it looks as though, while refutations matter a lot, corroborations do not matter at all. Y may conclude that to put this splendid new theory to the test will be a case of 'Heads I win nothing, tails I lose much'.

But the matter looks quite different when (B) is remarried to (A*). To make my point I need to assume that scientists have thought up at least one independent test on T_j, which has not so far been independently tested. After all, it would be rather surprising if, despite its (B)-superiority to T_i, no one had been able to think of a way of testing its excess empirical content. To begin with I will suppose that a test on T_j has not only been conceived but is physically set up and will go ahead shortly unless something intervenes. (T_j might be Einstein's GTR and the test might be Eddington's "star-shift" experiment at Principe on May 29, 1919. We may imagine that Eddington and his team have got everything ready and are only waiting for the eclipse of the sun to become total.) Now (A*), it will be remembered, says that X may adopt T_j only if, despite his best endeavours, he has not succeeded in detecting any inconsistencies within it or between it and evidence available to him. Assume that T_j is, so far as X knows, both internally consistent and consistent with the evidence now in his possession: it is, for him at this time, a possibly true hypothesis. However, a new piece of evidence is now available to him, though it is not yet in his possession: namely the outcome of the test that is about to be carried out. If he somehow prevented it being carried out, he would wilfully deprive himself and others of evidence that may conflict with T_j; he would have violated (A*), which requires him to endeavour as best he can to detect any such conflicts. Clearly, the situation would be essentially the same if the test had been designed but not yet physically prepared: a new piece of evidence would now be available to X in rather the same sense that if I run out of tobacco, but I know that a tobacconist is open not too far away, then tobacco is now available to me even though it will be a while before I can fill my pipe.

So X, who accepts (B*), will agree with Y, who accepts only (B), that T_j is a theoretical advance over T_i; and he will further agree that they may, at this present time, accept T_j in preference to T_i since T_j is, for them at this time, a possibly true hypothesis; but unlike Y, X will want this test to go ahead. Suppose now that, instead of just one test, lots of tests on T_j have been envisaged. In a scientifically ideal world, X would want all these tests to go ahead forthwith. It would be as if there were many stones before him, each of which may be concealing evidence adverse to T_j, and (A*) enjoins him to leave none of these stones unturned in the search for refuting evidence. But in the actual scientific world human and technical resources are not unlimited, and it is rather likely that only a relatively small proportion of all these tests could be carried out, at least in the near future. Is there any principle by which X could determine which of all these many stones should be turned? If there were an infallible way of grading the tests on T_j that have so far been envisaged according to their likelihood of turning up refuting evidence, then (A*) would clearly behove X to give priority to those with the highest likelihood. In the absence of such an infallible method is there a second-best principle of selection?

8.21 *Background Knowledge*

If we could judge one proposed test on a theory to be harder or more severe than another one, then (A*) would oblige us to give it priority over the other. But how could such a judgment be justified? There are some easy cases. If there are two ways of testing the same quantitative consequence of a theory, one of them being more stringent than the other in the sense that it gives numerical values within a smaller interval, then it is a more severe test. Alternately, suppose that we can apply a very stringent measuring method to two different quantitative consequences of a theory, one of these being more precise than the other. Then other things being equal, a test on the former would be the more severe test. But this does not get us far. Suppose that the more precise consequence coincided with a well-tested consequence of an earlier theory whereas the less precise one was a novel prediction; then other things would be far from equal and a test on the latter might very well be judged the more severe.

Popper introduced a measure of the severity of tests with the help of his concept of 'background knowledge' (*1963*, pp. 238f.). If e is the predicted outcome of a test, h is the theory under test, and b is background knowledge, the severity of the test is given by $p(e, h.b) - p(e, b)$. But all this poses some serious difficulties. One concerns the evaluation of $p(e, b)$. (The value of $p(e, h.b)$ will usually be 1.) If $p(e, b)$ is to have a determinate value other than 0 or 1, we presumably need to know just what is in b. Popper spoke of 'the vast amount of background knowledge which we constantly use' (*1963*, p. 168). We surely could never claim to have articulated all this; how could we ever know, even of some vast set of statements, that it contains *all* of our background knowledge? Recall the difficulties (discussed in §2.42) that attend the requirement of total evidence. As Popper truly remarked, 'I cannot know what my total knowledge is' (*1968*, p. 137). Nor can I know what my total background knowledge is; and this means that it is always possible that I have overlooked some bit of it that bears on e, so that $p(e, b)$ is indeterminate.

Another difficulty concerns the entry conditions for admission into "background knowledge". Popper said that it consists of all that is being taken as unproblematic when a theory is being tested. But this can only be a sociopsychological matter; for according to Popperian fallibilism, nothing in background knowledge really *is* unproblematic in any objective sense. We have no ground for rejecting Miller's assertion that background knowledge 'is all too likely to include all variety of unexamined prejudice and presumption' (1982, p. 37). If that is so, it would clearly be preferable if we could make the theory of corroboration independent of this dubious and largely unknown quantity.

8.22 *Diminishing Returns*

A further difficulty connected with Popper's use of background knowledge was pointed out independently by Musgrave (1975) and O'Hear (1975). If a theory passes a severe test, and the test is subsequently repeated several times with similar results, then presumably these later tests are of diminishing severity. Popper had written:

> A serious empirical test always consists in the attempt to find a refutation, a counter example. In the search for a counter example, we have to use our background knowledge; for we always try to refute first the *most risky* predictions. . . . (*1963*, p. 240)

At this point Popper quoted from a passage from Peirce which is worth quoting more fully:

> We all know that as soon as a hypothesis has been settled upon as preferable to others, the next business in order is to commence de-

ducing from it whatever experiential predictions are extremest and most unlikely among those deducible from it, in order to subject them to the test of experiment . . . ; and the hypothesis must ultimately stand or fall by the result of such experiments. (1901, 7.182)

The passage from Popper continues:

. . . which means that we always look in the *most probable kinds* of places for the *most probable* kinds of counter examples—most probable in the sense that we should expect to find them in the light of our background knowledge. Now if a theory stands up to many such tests, then, owing to the incorporation of the results of our tests into background knowledge, there may be, after a time, no places left where (in the light of our new background knowledge) counter examples can with a high probability be expected to occur. But this means that the severity of our tests declines. This is also the reason why an often repeated test will no longer be considered as significant or as severe: there is something like a law of diminishing returns from repeated tests. (Ibid.)

Musgrave supposed the same kind of test on a hypothesis h to be carried out ten times. Let $e_1, e_2, \ldots e_{10}$ be the results predicted for these ten tests by h together with background knowledge, and assume that in the event all ten predictions were borne out. Let $b_1, b_2 \ldots b_{10}$ be the state of background knowlegde on respectively the first, second . . . tenth test. Assume that the first test was a severe one. Now whether or not we want to say that there was steadily diminishing severity, we surely do want to say that the tenth test was markedly less severe than the first; and this requires $p(e_1, b_1) \ll p(e_{10}, b_{10})$. But what has b_{10} got that b_1 did not have and that enables it to give e_{10} a markedly higher probability than b_1 gave e_1? The obvious answer is that b_{10} incorporates the favourable results of the previous nine tests. But, as Musgrave said, to hold that those nine observed instances raise the probability of the next, as yet unobserved, instance 'clearly involves a straightforward inductive argument' (1975, p. 250). What are the alternatives? One that both Musgrave and O'Hear considered is to repudiate any inductivism here and deny that $p(e_{10}, b_{10}) > p(e_1, b_1)$. (As O'Hear mentioned, we might insist that no test is ever an exact repetition of a previous one.) This would mean, as O'Hear said, that 'every test of a given type should be considered as severe as every other test of that type' (1975, p. 275). Another alternative, which Musgrave was inclined to favour, is that after 'sufficiently many' repetitions, some low-level generalisation g is incorporated into background knowledge so that the latter, thus strengthened, henceforth predicts the experimental results without assistance from h; whereupon the severity of this kind of test abruptly falls to zero. Grünbaum (1976, p. 237) called this the "saltation" version of the law of diminishing returns. If, as I suggested earlier, there is no objective criterion for whether to include a statement in background knowledge, it will be a matter for arbitrary decision if and when such a g is included.

So we seem to face a trilemma: (i) diminishing severity inductively sanctioned (perhaps in a Bayesian way: see Grünbaum 1976, Urbach 1981); (ii) undiminished severity in perpetuity; (iii) undiminished severity up to an arbitrarily chosen point and then an abrupt drop to zero severity. How this trilemma should be resolved I will consider shortly.

8.23 *Strong, Moderate, and Weak Corroborations*

We are seeking a way of discriminating among possible tests on a theory in the hope of giving priority, as (A*) demands, to those with a better chance of turning up refuting evidence. In place of Popper's idea of background knowledge I propose to use, in this connection, the idea of a *historical record of tests* carried out in the domain of the theory in question. Before turning to this I may mention that we do not need anything like background knowledge to play the other role it plays in Popper's system. As well as providing a measure of the severity of a test on a hypothesis h, it was also to provide the subsidiary assumptions that in conjunction with h will, typically, entail the prediction e that is to be tested (see Worrall, 1978, p. 66 n. 6). Thus in the case of a severe test, b must, in isolation, bear very unfavourably on e, so that $p(e, b)$ is close to zero; but b must also be consistent with h and must, moreover, in conjunction with h, bear very favourably on e, so that $p(e, h.b)$ is one. Where others might say that a theory T, reinforced by auxiliary assumptions A and initial conditions e_1, both supplied by background knowledge, has yielded the prediction e_2, I say that a fleshed out theory T, consisting of fundamental assumptions T_H and auxiliary assumptions A, entails without extralogical assistance the SPI (singular predictive implication) $e_1 \rightarrow e_2$.

The historical record of tests, as I envisage it, will have the following character. For every experimental test that has been carried out in the domain of the theory or theories now under consideration, it will describe the set up of the experiment in a quite specific way. It will say: (i) what initial conditions were intentionally realised by it; (ii) what kind(s) of outcome or effect the apparatus was capable of registering; (iii) with what degree of precision it could measure such effects; and (iv) what the actual outcome was.

Imagine such a historical record to be stored in a computer's memory. We have before us a new theory T from which we have derived the experimental generalisation g. This latter, let us suppose, has the form $\alpha = f(\beta, \gamma)$ where α, β, and γ are variable magnitudes of a measurable kind. We now ask the computer whether g would have been at risk from any recorded test if g had been formulated before the test was carried out. If there has in fact been an experiment in which specific values of β and γ were realised, and values of α were measured, the answer will of course be yes (and the computer will say how g would have fared under that test). If the answer is no, I say that g is an *empirically novel* consequence of T.

I turn now from empirical novelty to theoretical novelty. Let T_j be a new theory and for ease of exposition let all its predecessors be represented by T_i. I say that a low-level experimental generalisation g_j that is entailed by T_j is theoretically novel if g_j has no counterpart among the consequences of T_i; in other words, T_j breaks new ground here by making a predictive assertion in an area where its predecessor is silent. And I say that g_j is theoretically challenging if it has an incongruent counterpart g_i among the consequences of T_i and, moreover, g_i and g_j are discernibly different, in that they diverge sufficiently for a crucial experiment between them to be possible.

We can assume that if g_j is theoretically novel it will also be empirically novel; for tests are not normally made where there is as yet nothing to test. If g_j is theoretically challenging, it may also be empirically novel but it need not be. If the counterpart consequence of T_i that it challenges, namely g_i, has not been tested, then they are both empirically novel. But the more interesting cases are ones where g_i has been tested. Suppose that g_i has failed tests that g_j would have passed. In that case T_j, in challenging T_i in an area where the latter is already known to be in trouble and in a way that avoids this trouble, is not risking much and g_j is not empirically novel. The most interesting case is where g_i has been tested and has *passed* all tests, but these tests did not refute g_j because they were not stringent enough to have been able to discriminate between these only very slightly diverging experimental generalisations. In this case T_j is challenging T_i at a place where T_i has been performing well. And I say, in this case, that g_i becomes empirically novel once measuring methods have been refined to the point where empirical discrimination between g_i and g_j is possible.

In what follows I will assume that the measurements made in the course of testing theories have a pretty narrow interval of imprecision. Galileo's water clock may not have been a perfect time measurer but it helped to make the inclined plane experiment a pretty stringent test. I will say that a test on a theory is *hard* if (as well as involving pretty stringent measuring methods) it is aimed at an empirically novel experimental generalisation entailed by the theory, and that it is *soft* if it can be regarded as a mere repetition of a test that the theory has already passed or would have passed if it had been formulated when the original test was made. A test is neither hard nor soft but *medium* if aimed at a g that has already passed tests but the present test is not a mere repetition of any of them, being a new kind of test or at least more stringent than its predecessors. I admit that the borderline between *soft* and *medium* tests is a bit vague just because the idea of a "mere repetition" of a test is a bit vague. There is of course no such thing as an exact replica of an earlier test: some of the conditions are bound to have changed, and even a small change may be significant. However, this vagueness will not matter too much since the more important distinction is that betwen *soft-to-medium* and *hard*.

It will be said that some inductivist assumption is lurking behind these distinctions. Is not a soft test on a theory one we have inductive grounds for expecting it to pass, since it has passed similar tests before? And is not a hard test one we have no inductive grounds for expecting it to pass, since no test had ever been aimed at this part of its predictive content? But I do not think that this objection is fair. A hard test on T always increases the *variety* of tests to which T has been subjected. A medium test either increases the variety or it involves increased *stringency*. A soft test involves no (significant) increase in variety or stringency. Now no inductivist assumption is needed to establish that a theory is more severely tested by being subjected to a variety of tests than to repetitions of the same test. Presented with ten different low-level consequences of T, all empirically novel, and a choice between either (a) carrying out one test on each of these or (b) repeating ten tests on one of them, we know a priori that T will be at greater risk if we choose (a). Again, we know a priori that T will be more at risk the more stringent is a test on a precise g. Our classification of tests reflects the idea that if, as (A*) requires, we are concerned to detect any conflicts between T and evidence available to us, then we should try to subject T to tests that are as variegated and as stringent as we can make them.

I will say that a theory gains a strong, moderate, or weak corroboration from passing, respectively, a hard, medium, or soft test. (Of course, if the "test" is such that the theory cannot help but "pass" it, no matter what its outcome may be, then it is a pseudo-test and the theory gains no corroboration from it.)

Let us now consider the bearing of all this on the Musgrave and O'Hear trilemma concerning diminishing returns. (I pass over the scholastic adhockery with which I tried to wriggle out of this trilemma in 1978c.) The present account endorses a "saltation" version with the difference that, after *one* hard test on a theory, all "mere repetitions" of it are soft. To put it another way: given that a test of a certain kind on an empirically novel consequence of a theory has been thought up, as Einstein thought up the star-shift experiment, it is very important to know whether the theory has stood up to this kind of test. If it has, it is credited with a strong corroboration. For it to have done so it must, of course, have passed the test on at least one occasion. Passing it on subsequent occasions does not count for nothing; the theory gains at least a weak corroboration and will gain a moderate corroboration if it is repeated using more accurate and discriminating methods of measurement. But what really matters is simply whether it has passed this *kind* of test.

8.24 *The Heritability of Corroborations*

I now turn to the following question. Let T_j be a new theory in a certain field, and again let T_i represent all its predecessors. Assume that T_j is (B)-superior to T_i. However, T_j revises the empirical content of T_i only very slightly and goes beyond it only to a small extent. The historical record shows that T_i passed all but one or two of the many and varied tests on it. I will suppose that T_j was not born refuted (the claim that all theories are born refuted will be dealt with in §8.6). In other words, if T_j had been available, it would have passed all the tests T_i passed *and* the one or two that T_i failed; the latter would have constituted crucial experiments between them and would have told in favour of T_j. From what has already been said it is clear, first, that T_j may gain a strong corroboration from a test on one of its few testable consequences that go beyond those of T_i and, second, that T_j could gain only a weak corroboration from a "mere repetition" today of any of the many tests that T_i passed or of the one or two that it failed. The question is: can we regard T_j as having been tested, as it were by proxy, when T_i was tested? Is T_j entitled to inherit, as strong corroborations for itself, those earlier test results?

The relation as I have here depicted it between T_j and T_i is exemplified by the relation of GTR (General Theory of Relativity) to NM (Newtonian mechanics). As Einstein put it: 'This agreement [with Newtonian mechanics] goes so far, that up to the present we have been able to find only a few deductions from the general theory of relativity which are capable of [experimental] investigation, and to which the physics of pre-relativity days does not also lead, and this despite the profound difference in the fundamental assumptions of the two theories' (*1920*, p. 124). In our terminology, Einstein was saying that despite the radical conflict between the theoretical cores of the two theories, only a few places have been found where the empirical content of GTR diverges sufficiently from that of NM to be independently testable (in the sense that the test might go against GTR without hitting NM). Einstein went on to give the three famous independently testable consequences, concerning: (a) the precession of the perihelion of Mercury; (b) the deflection of light rays from a star by the gravitational field of the sun ("star-shift"); and (c) the displacement of the spectral lines of light from dense stars towards the red ("red-shift"). Evidence for (a) had been known since Leverrier. As to (b), the deflection predicted by Einstein's theory was 1.74″, twice the 0.87″ predicted by Newton's theory. Earman and Glymour summarised the outcome of the two eclipse expeditions, to Principe and Sobral, in 1919 as follows:

> The natural conclusion from these results is that gravity definitely affects light, and that the gravitational deflection at the limb of the sun is somewhere between a little below 0.87″ and a little above 2.0″. If one kept the data from all three instruments, the best estimate of the deflection would have to be somewhere between the Newtonian value and the Einstein value. If one kept only the results of the Sobral 4-inch instrument [whose plates were unequivocally the best], the best estimate of the deflection would be 1.98″, significantly above even Einstein's value. The conclusion that the Astronomer Royal announced . . . on November 6, 1919, was stronger: Einstein's prediction . . . had been confirmed. (1980a, p. 76)

As to (c), the situation seems obscure. In 1915 Einstein wrote privately that the "red-shift" prediction 'has already been brilliantly confirmed' (quoted in Earman and Glymour, 1980b, p. 197). Yet in (*1920*) he wrote: 'It is an open question whether or not this effect exists' (p. 131), adding that if it 'does not exist, then the general theory of relativity will be untenable' (p. 132). His English translator added a note saying that "red-shift" was 'definitely established by [Walter S.] Adams in 1924, by observations on the dense companion of Sirius, for which the effect is about thirty times greater than for the sun.'

This paucity of independently testable consequences of GTR has led one commentator to pronounce that marvellous theory a 'largely fruitless' achievement. Kuhn has said: 'The equations embodying [GTR] have proved so difficult to apply that ... they have so far yielded only three predictions that can be compared with observation. Men of undoubted genius have totally failed to develop others. ... Einstein's general theory remains a largely fruitless, because unexploitable, achievement' (1961, pp. 188–189). What Kuhn was overlooking here is that, besides yielding those three predictions, GTR *also* reproduced, virtually unchanged, the entire empirical content of NM. And what is at present lacking from our account of corroboration is a way of recognising this latter achievement.

It might be said that, on a Popperian view of corroboration, GTR is better corroborated than NM because a few discorroborations outweigh any number of corroborations, and evidence concerning (a) and (b) discorroborated NM. But I do not think that this answer suffices. There is a sense in which it was something of a lucky accident for GTR that NM got discorroborated. Consider first the situation with respect to (a). Just as it was fortunate for Kepler that the solar system contains Mars, so it was fortunate for Einstein that it contains Mercury. GTR predicts a precession of the perihelion of *all* planets; but the amount is too small to be detected by present methods except in the case of Mercury. As to (b): as we saw, it is in any case not absolutely clear that the 1919 results did discorroborate NM. If the photographic plates had been a bit more blurred, or the sky had been cloudier, and if Mercury had not existed, there might have been no discorroborations of NM, which would have passed many hard tests, and GTR would have passed very few independent tests. I stand by my position that merely repeating, *as a test on GTR, a test that NM had previously passed and that GTR would also have passed if it had been formulated then*, would be a soft test on it. But the question arises as to whether we might regard GTR as having undergone, as it were by proxy, a hard test when this test was originally made on NM.

8.25   *The Zahar-Worrall View*

The question of the heritability by T_j of strong corroborations gained by T_i is not quite on all fours with the question: Can evidence that T_j explains, and that *dis*corroborated T_i, strongly corroborate T_j? I will begin with this latter question. For instance, was GTR strongly corroborated by already known evidence concerning the precession of the perihelion of Mercury? This problem has been keenly examined by Elie Zahar (1973, pp. 101–104). Zahar of course agreed with Popper that a theory is corroborated if it makes a successful prediction (such as GTR's concerning "star-shift", or "red-shift") that is temporally novel, in the sense that nothing of the kind predicted is recorded in background knowledge. But he insisted that novelty should not be equated just with temporal novelty. Then when should a prediction (retrodiction) concerning an event of a kind already known be considered novel? Zahar answered that there are two possibilities: the theory may have been designed to fit events of that kind, for instance by suitably adjusting a free parameter; or else the prediction may have fallen out, without contrivance, as an unintended consequence. Zahar concluded that in order to assess whether evidence that was already known supports a theory that predicts it *'one has to take into account the way in which [the] theory is built and the problems it was designed to solve'* (p. 103).

Worrall succinctly summarised this idea thus:

This methodology [of scientific research programmes] embodies the simple rule that one can't use the same fact twice: once in the construction of a theory and then again in its support. But any fact which the theory explains but which it was not in this way pre-arranged to explain supports the theory *whether or not the fact was known prior to the theory's proposal.* (1978, pp. 48–49)

In the methodology of scientific research programmes, as set forth by Lakatos and developed by Zahar and Worrall, great importance is attached to the positive heuristic of a research programme and to the way in which, under its guidance, theories are successively constructed. They hold that, in making the question of empirical support for a theory by previously known evidence depend on whether that evidence was used in the construction of the theory, they do not make it a 'person-relative' question, as Musgrave at one time claimed (1974, pp. 13–14). I will not go into that issue. But it would clearly be more satisfactory, from the standpoint of this book, if we could determine whether a theory is corroborated by evidence that has discorroborated its predecessor without investigating the process by which T_j was constructed. For our (B*) only lays down desiderata that a completed theory should satisfy; it says nothing about the process of construction. It may be difficult to discover much about this process. Zahar insisted that his new criterion

> for novelty of facts ... implies that the traditional methods of historical research are even more vital for *evaluating experimental support* [my italics] than Lakatos had already suggested. The historian has to read the private correspondence of the scientist whose ideas he is studying; his purpose will not be to delve into the psyche of the scientist, but to disentangle the heuristic reasoning which the latter used in order to arrive at a new theory. (1973, pp. 103–104)

But what if any private correspondence is lost and we cannot disentangle his heuristic reasoning?

8.26 *Were the Theory's Fundamental Assumptions Involved?*

Suppose that, in a case where T_j explains the already known evidence e that had refuted its predecessor T_i, we were in a position to say, not merely that T_j was not in fact adjusted to fit e, but that there can be no question of its having been so adjusted. In that case we could surely say that e provides a strong corroboration for T_j. But when would we be in a position to say this? In §5.23 a distinction was drawn between the fundamental and the subsidiary assumptions of a theory; and part of our answer must surely be that T_j's fundamental assumptions play a key role in its explanation of e. But this needs reinforcing since, as it stands, it leaves open the possibility that T_j's fundamental assumptions are merely those of T_i after being tinkered with a little to get them to yield, in conjunction with the same subsidiary assumptions, results in conformity with e. Something like this nearly happened to NM (Newtonian mechanics) when it was still in trouble over the "misbehaviour" of Uranus and shortly before it was released from these difficulties and triumphantly vindicated by the discovery of Neptune. Some astronomers mooted the suggestion that the inverse square law was not quite correct, the small error in it becoming detectible only at very great distances (see Musgrave, 1976, p. 460). If Neptune had not been discovered, a revised version NM' of NM might have been developed whose fundamental assumptions were the same except that $1/r^2$ was replaced by, say, $1/r^{1.99997}$ or whatever was needed to secure a reasonable fit with the evidence e concerning the behaviour of Uranus. In that case the fundamental assumptions of NM' would have played a key role in its explanation of e, but we would not have wanted to say that NM' was corroborated by e since it had been adjusted in a pretty ad hoc way to e.

I propose to exclude cases of this kind by requiring, not only that T_j's fundamental assumptions be involved in its explanation of e, but also that its corroborable content be greater than that of T_i; or to put it in our earlier notation, $CT(T_j) - CT(A_j) > CT(T_i) - CT(A_i)$. For if these two conditions are satisfied, then T_j's ability to explain e is part and parcel of its superior unity and explanatory power.

Let us now turn to the question of the corroboration of T_j by evidence that had previously *corroborated* T_i. I start with the following important passage from Popper:

> Newton's theory unifies Galileo's and Kepler's. But far from being a mere conjunction of these two theories ... *it corrects them while explaining them.* The original explanatory task was the deduction of the earlier results. Yet this task is discharged, not by deducing these earlier results but by deducing something better in their place: new results which, under the special conditions of the older results, come numerically very close to these older results, and at the same time correct them. Thus *the empirical success of the old theory may be said to corroborate the new theory* [my italics]; and in addition, the corrections may be tested in their turn—and perhaps refuted, or else corroborated. What is brought out strongly, by the logical situation which I have sketched, is the fact that the new theory cannot possibly be *ad hoc* or circular. Far from repeating its *explicandum*, the new theory contradicts it, and corrects it. In this way, even the evidence of the *explicandum* itself becomes independent evidence for the new theory. (1957, p. 202)

I agree with this, though I would have put the main emphasis on the *unity* of NM; it is corroborated by those two, seemingly disjoint, sets of earlier results because it explains close approximations of both from one and the same set of fundamental assumptions; and its revisions of them, far from being mere tinkerings, are induced *systematically* by those fundamental assumptions. More generally, we may say that in cases where T_i gained a strong corroboration from a test result e that is explained by T_j, the latter may inherit this corroboration if it is deeper

302

and more unified than T_i and its fundamental assumptions were involved in its explanation of e.

8.27 *Corroborable Content*

The requirement that the theory's fundamental assumptions must be involved can be extended from the inheriting of earlier corroborations to the gaining of fresh ones. We suppose an axiom set T for a unified theory to be equivalent to $T_H \wedge A$ where T_H and A denote respectively its fundamental and its auxiliary assumptions. Suppose that g is a low-level generalisation that is entailed just by A. Then it is rather unlikely that g is empirically novel (namely, the historical record contains no test that would have been a test on g if g had been formulated when the test was carried out). But suppose that it is and that it now passes a novel test with flying colours. (I have not investigated the matter myself but I have heard it said that some of the "corroborations" claimed for Velikovsky's theory resulted from tests on predictions that were not logically related to the central ideas of his theory.) However, since its fundamental assumptions were not involved, the *theory* does not thereby gain a corroboration: the corroborable content of T is represented not by $CT(T)$ but by $CT(T) - CT(A)$.

The foregoing account of corroboration applies only to scientific *theories*, and not to experimental laws and empirical generalisations, which do not permit of a partitioning into fundamental and subsidiary assumptions. Its special features stem from the fact that we want a later theory to be corroborated by evidence, which it explains, gained from tests on earlier theories *provided* there is no question of its having been adjusted in a more or less ad hoc way to that evidence. But the matter of adhockery is much less urgent when we turn from theories to laws and generalisations. While we want an explanatory theory to be non-ad hoc vis-a-vis the laws and generalisations explained by it, we are not too worried if the latter are rather ad hoc vis-a-vis experimental measurements and singular observations. If a proposed curve does not fit a set of measurements too well, we have very little compunction about trying to secure a better fit, provided we can achieve this by a reasonably simple general formula. If black swans turn up in Australia we unashamedly modify 'All swans are white' accordingly.

Freed from our previous concern with adhockery we can afford a much simpler concept of corroboration for laws and generalisations. If h is a law or generalisation I will say that h is well corroborated if it has been well tested and is as yet unfalsified, and if $h_1, \ldots h_n$ constitute a set of competing laws or generalisations I will say that h_1 is better corroborated than the others if (1) it is well corroborated, and (2) either (a) $h_2, \ldots h_n$ are all refuted or (b) those of them that are unrefuted have been less well tested than h_1 (for instance, because they are less general or less precise).

8.3 Is the Best Corroborated Theory Always the Best Theory?

Henceforth I will speak of a theory having gained a corroboration irrespective of whether it won it directly or inherited it. Assume that T_i and T_j are unified theories and well tested. I say that T_j is at the present time better corroborated than T_i, or for short $\mathrm{Co}(T_j) > \mathrm{Co}(T_i)$, if these two conditions hold: (i) T_j is unrefuted, and (ii) T_j's corroborations dominate T_i's in the sense that no test result is less favourable, and at least one test result is more favourable, to T_j than to T_i. (Something like (ii) was mooted by Musgrave, 1978, p. 184.) A test result is more favourable to T_j if either it refutes T_i but not T_j, or it corroborates T_j but not T_i. (It might seem that there is a third possibility, namely that it corroborates T_i but less strongly than it corroborates T_j, say by giving the former a weak or moderate corroboration and the latter a strong corroboration. But in fact this cannot happen: a weak or moderate corroboration for a theory presupposes the earlier occurrence of a strong corroboration for that theory. Let e_2 be a test result from which T_i gains only a weak or moderate corroboration. This means that T_i entails a low-level generalisation g_i which had been tested before and has now been tested again. Let e_1 be the result of the original test on g_i when g_i was empirically novel. If e_1 and e_2 were both favourable to g_i, and if T_i gained a weak or moderate corroboration from e_2, then T_i gained a strong corroboration from e_1. If T_j is corroborable by tests on a consequence g_j that is empirically indistinguishable from g_i, then T_j likewise gains a strong corroboration from e_1 and a weak or moderate one from e_2. It cannot happen that either e_1 or e_2 corroborates one of these two theories strongly and the other only weakly or moderately.)

Some people, especially if they subscribe to the thesis that all theories are born refuted, may say that we should dispense with condition (i) above and rely on (ii) alone, thus allowing that a refuted theory may be the one in its field that is best corroborated. A debating point that could be made in reply is that (ii) by itself would mean that $\mathrm{Co}(T_j) > \mathrm{Co}(T_i)$ in cases where T_j is what Tichy would call a lousy theory, having been falsified by ever so many test results, and T_i is even lousier, having been falsified by all of these and more besides. But my main reason for retaining condition (i) is the following. The big question before us is: if there is one theory that is best corroborated in its field, can we safely conclude that it is the one that best fulfils (B*)? Now (B*) contains (A*), which says, in effect, that a refuted theory is not, as it stands, even a candidate for being the one that best fulfils (B*).

Let us now address the big question. What corroborations a theory has gained depends very much, of course, on what tests have been carried out in its field, and we need to make some reasonable assumption about this. As we saw, (A*) calls for priority to be given to hard tests on a promising theory, a hard test being a test on an empirically novel, low-level generalisation g entailed by the theory but not by its auxiliary assumptions alone. And it was suggested that if g is theoretically novel it will also be empirically novel (since tests are not made where there is as yet nothing to test). I will assume that if a theory has theoretically novel g's among its consequences, then at least one test has been made on at least one of them. (If T_i in this way exceeds T_j in one direction while T_j exceeds T_i in another direction, then our assumption is that each of them will have had a test on its excess corroborable content.) I will refer to this as our *test assumption*.

In a set of competing theories, the number of unrefuted ones may be zero, one, or more than one. If it is zero we have the disappointing case where no candidate of sufficient merit for the prize of (B*)-superiority has so far presented itself. If it is one, and the sole survivor is a respectable theory and well corroborated, we have the unproblematic case where the prize goes automatically to it. So let us attend to the interesting case where there are two (or more) unrefuted competing theories, say T_i and T_j. Now it may be that neither of these is better corroborated than the other; in which case the question of which deserves the prize will have to be postponed until the situation becomes clearer. So let us assume (i) that T_i and T_j are both unrefuted and (ii) that T_j is better corroborated than T_i, or $Co(T_j) > Co(T_i)$. Assumption (ii) means that no test result has been more favourable to T_i than to T_j while at least one has been more favourable to T_j. As we recently saw, there are only two ways in which a test result can be more favourable to T_j than to T_i. One is that it refutes T_i but not T_j; but that is here excluded by assumption (i) (that both T_i and T_j are unrefuted). The other is that it corroborates T_j but not T_i. For a test result to have corroborated T_j without either refuting or corroborating T_i there must be *at least one place* where the corroborable content of T_j exceeds that of T_i. On the other hand our assumptions imply that there is *no place* where the corroborable content of T_i exceeds that of T_j. For suppose there were such a place; then according to our test assumption, there would have been at least one test on it and this test would have led either to a refutation of T_i, contrary to assumption (i) above, or to a corroboration of T_i but not T_j, contrary to assumption (ii). Thus assumptions (i) and (ii), together with our test assumption, imply that T_j's corroborable content exceeds T_i's but not vice versa, or that

$$\mathrm{CT}(T_j) - \mathrm{CT}(A_j) > \mathrm{CT}(T_i) - \mathrm{CT}(A_i)$$

where A_j and A_i are the auxiliary assumptions of the two theories. But we found in §5.35 that this is a sufficient condition for T_j to be (B1-2)-superior to T_i; and if it is (B1-2)-superior it is automatically (B3-4)-superior as well.

Thus given (i) that T_i and T_j are the only unrefuted theories in their field, and (ii) that T_j is better corroborated than T_i, and (iii) that they have been tested in the way that our test assumption requires, it follows that T_j is (B*)-superior to T_i and that T_j is the theory in its field, at the present time, that best fulfils the optimum aim for science. So we have very good reason to accept T_j in preference to its unrefuted and, of course, its refuted competitors.

However, our answer to rationality-scepticism relies on the assumption that, at least in a good many cases, we can speak of *the* best corroborated theory in its field. But arguments have been advanced to the effect that the uniqueness condition is never satisfied here. It would never be satisfied if there were always *indefinitely many* equally well corroborated alternatives to any theory, and also if the number of well-corroborated theories were always *zero*. We must now look into arguments for each of these alternatives.

8.4   Are There Paradoxes of Corroboration?

The field of confirmation theory has been discovered, during the last forty years or so, to be something of a minefield, and we have to consider whether our theory of corroboration avoids its well-known dangers. One of these has often been touched on earlier in this book: whenever some evidence *e* confirms a hypothesis *h*, will it not simultaneously confirm an indefinite series of alternative hypotheses *h'*, *h"* ... all standing in a similar relation to *e*? Or rather, will not its confirming force be dissipated among all these alternatives so that it does not actually *confirm* any of them?

A noninductive theory of corroboration enjoys one major advantage, in this connection, over probabilist theories of confirmation. Many examination systems use an "absolute" standard of marking: if an examiner gives a candidate a high mark he does not subsequently revise it downwards on finding that other candidates have performed equally well. But one can imagine a system in which there is a fixed total of marks to be apportioned among all the candidates, so that a given candidate's mark depends partly on his own performance but also very much on how many other candidates there are and on how they perform.

Corroborations are like the former: if a theory passes a hard test, aimed at a novel consequence g, it gains a strong corroboration; and if another theory also entails g or something empirically indistinguishable from it, it too gains a strong corroboration. Corroborations are not rationed out.

But probabilist theories of confirmation are like the latter system. There is a fixed sum, namely 1, to be distributed over an exhaustive set of mutually exclusive hypotheses; and unless special measures are taken to obviate this, a share will have to go to all unformulated members of the set. Suppose that h_1 and h_2 are seriously proposed hypotheses and that the evidence E strongly favours h_1. This does not mean that h_1 gets a high mark. If h_1 and h_2 are at all interesting, the language in which they are formulated will undoubtedly allow the formulation of many further alternatives, say $h_3, \ldots h_n$. And in a logical system of probability, such as Carnap's, both the initial probabilities of all these hypotheses and their posterior probabilities on E are objectively determined by, eventually, the initial weighting of the system's basic units (in Carnap's system, by the equal weighting of all structure-descriptions). The unformulated members of the set are *there*, making their presence felt by demanding their proper share of the probability. Perhaps there are, among the unformulated $h_3, \ldots h_n$, a large number that would have a posterior probability as high as that of h_1; in which case the latter will be very low. As Shimony put it: 'undiscriminating impartiality towards hypotheses which no one has seriously suggested is a veiled kind of skepticism, since it would leave very little . . . probability to be assigned to any specific proposal' (1970, p. 89).

But if such an 'undiscriminating impartiality' is dictated by a non-arbitrary distribution of initial probabilities over basic units, how can a probabilist keep out those unwanted and unformulated hypotheses?

One solution is to scrap the idea of probability logic as a system as hard and steely as classical logic, of which it was intended to be a generalisation, and to replace it with a soft and rubberlike system that is easy to manipulate. The personalist theory of probability allows people to assign whatever initial probabilities they like to hypotheses, provided only that these are "coherent", that is, collectively satisfy the axioms of the probability calculus. This of course allows them to assign zero probability to all unformulated hypotheses. It also allows them to assign zero probability to a promising rival hypothesis that threatens ones they personally favour.

Shimony (1970) developed a Bayesian position, which he called 'tempered personalism', that was intended both to avoid the sceptical implications of logical probability and to be less libertarian than an untempered personalism. The tempering consists in superimposing the requirement that people should adopt an open-minded attitude, so far as their initial probabilities are concerned, towards *all seriously proposed hypotheses* in a field of investigation, but not towards unformulated or frivolous hypotheses (though they should allow for the possibility that all the seriously proposed ones will turn out to be false). In line with this, they should assign a positive initial probability to each of these hypotheses, these probabilities summing to less than 1, the remainder accruing to what Shimony called the catch-all hypothesis, namely the negation of the disjunction of all the former hypotheses. Within these limits people are free to express their personal preferences among the hypotheses in differing initial probabilities; but Shimony claimed 'that persons who disagree sharply regarding prior probability evaluations and yet conform to the prescription of open-mindedness ... are able to achieve rough consensus regarding posterior probabilities relative to a moderate amount of experimental data' (pp. 96–97). As more evidence comes in, so disagreements over initial probabilities will tend to be "swamped". In §3.4 we considered an earlier attempt to avoid an 'undiscriminating impartiality' towards hypotheses, namely the proposal of Jeffreys to order their initial probabilities according to their simplicity. Shimony claimed that the concept of simplicity had been given a burden too great to bear, adding that 'one of the advantages of the tempered personalist formulation of scientific inference is that it uses a different primary criterion for comparing hypotheses, namely, that of being or not being seriously proposed' (p. 155).

What is to count as a "seriously proposed" hypothesis? If this is left to the subjective judgment of individuals, then, as Shimony pointed out

(p. 110), we would, in effect, be back with an untempered personalism. Shimony's answer is complicated by the fact that for him it is neither a sufficient nor a necessary condition that someone should actually have seriously proposed it. A cranky hypothesis put forward by a fanatical scientologist might be seriously proposed by him and yet not "seriously proposed"; and the catch-all hypothesis is "seriously proposed" (p. 133) (otherwise it would not get a positive initial probability) although it is unlikely that any scientist is seriously proposing it.

Rather than investigate his complex and somewhat inconclusive answer to this question, I prefer to point out that what Shimony was struggling to achieve within a probabilist framework is achieved straight off, without perplexity or complexity, by our theory of corroboration. We do not have to labour to keep out unformulated rivals to a seriously proposed theory, because corroborations, unlike probabilities, are not rationed out among competitors. From a corroborationist standpoint, unformulated alternatives are *not* already there, making their presence felt. Nor do we have to vet formulated theories for crankiness, frivolity, etc. Any theory can enter the corroboration stakes. How it performs will be determined objectively, by the interplay between its corroborable content and the test record. Nothing personal, tempered or otherwise, influences the result. Finally, we are under no pressure to repudiate logical probability because of its sceptical implications. We simply accept those implications as correct and assess scientific theories according to an aim that they can fulfill, instead of clinging to some sort of probability as our aim and trying to render this feasible by allowing initial probabilities to be manipulated so that they may, perhaps, give us something not too unlike what we want.

8.41 *Upwards and Sideways Proliferation Arguments*

However, the fact that a theory's corroborations are not disturbed by the existence of other theories that may be corroborated by the same test results does not completely defuse proliferation arguments, as I will call arguments to the effect that, for any given theory T that is well corroborated by a body of test results E, there are always ever so many alternative theories T', T'' ... likewise corroborated by E. In particular, it does not dispel the claim that there never is one best corroborated theory.

Glymour has claimed that hypothetico-deductive theories of confirmation are exposed to what I will call an *upwards* proliferation argument (*1980*, pp. 30f.). His claim is based on the assumption that a hypothetico-deductive theory of confirmation will comply with what Hempel (1945, p. 32) called "the converse consequence condition". This condition, which Hempel rejected, says that if e confirms h then e also confirms any stronger hypothesis that is consistent with e and of which h is a consequence. (By contrast, the consequence condition, which Hempel accepted, says that if e confirms h then e confirms any weaker hypothesis that is a consequence of h. He pointed out that if *both* conditions were accepted, any e would confirm any h consistent with e.) Slightly adjusted to bring it to bear on our theory, Glymour's point can be put like this:– Let T denote the premises of a theory that is claimed to be the best corroborated theory in its field and let E be the evidence that is supposed to corroborate it; let $h_1, h_2 \ldots$ be various hypotheses that are logically independent of T, and E, and each other. Then $T \wedge h_1$ and $T \wedge h_1 \wedge h_2 \ldots$ are sets of premises that likewise entailed the predictive consequences, tests on which are recorded by E, and these expanded theories are likewise corroborated by E. Whether we regard the series $T, T \wedge h_1, T \wedge h_1 \wedge h_2 \ldots$ as lengthening indefinitely, or regard it as eventually reaching saturation point when it has become something like "the whole of science", there is no one member of it that we can identify as the one best corroborated by E.

Suppose that one tries to ward off this objection by pointing out that h_1 and $h_2 \ldots$ were all redundant for the derivation of the predictive consequences of T, tests on which are recorded in E. This answer is exposed to an argument, which I will call the *contracting target* argument, whose effect is roughly the opposite of the upwards proliferation argument. For any given testable consequence of T, say g, there will be, as Glymour points out, a subtheory of T that suffices to entail it; we can always split T into T' and T'', where $T' \wedge T''$ is equivalent to T, and T' suffices for the derivation of g, T'' being no more needed for its derivation than were $h_1, h_2 \ldots$ Indeed, we could let T' be g itself and let T'' be $g \rightarrow T$. The upwards proliferation argument together with the contracting target argument jointly suggest that what E corroborates may be anything from "the whole of science" down to some little subtheory of T.

My answer to both of the above arguments is that corroborations are gained, not by arbitrary conjunctions of premises, but by unified scientific theories that satisfy the organic fertility requirement. And this means that our theory of corroboration will not, in general, comply with the converse consequence condition. Suppose that T is widely regarded as the best theory in its field; and then someone claims that he has a bigger and more powerful theory that is at least as well corroborated and that yields T as one of its consequences. But it turns out that there is a permissible partition (see §5.34) of his "theory" into T and h such that $CT(T \wedge h) = CT(T) \cup CT(h)$. Then it would be as if he had submitted, for the prize at a garden fete for the biggest marrow, two marrows stuck together.

Of course, someone may succeed in augmenting a unified axiom set T with an additional assumption h so that $T \wedge h$ satisfies the organic fertility requirement. (An example is the augmentation of Newtonian mechanics with the assumption that light rays consist of fast-moving particles with a positive mass. The marriage of this T with this h generates consequences about the bending of light rays close to the sun that are not consequences of either T or h alone.) If he does so, and if his $T \wedge h$ stands up to tests on its novel consequences, he will have made, not an objection to our theory of corroboration, but a contribution to science.

As to the contracting target argument: suppose that a unified theory T has gained a corroboration from a test on a low-level consequence g. Now there are bound to be theorems of T that are weaker than it and stronger than g and which suffice to entail g. But, in the first place, such a theorem may very well fail the organic fertility requirement (as it presumably would do if it were $g \wedge c$ where c is some other consequence of T rather remote from g); in which case it is not a candidate for gaining corroborations. But suppose that a theory T, for instance Maxwell's theory, has a theorem T', for instance Fresnel's theory, that satisfies OFR. In that case we can allow that the theorem T' gains a corroboration. There is nothing disconcerting about this. As we have recently seen, corroborations are not rationed out: that T' gains a corroboration from a test result does not mean that T is thereby deprived of a corroboration from that test result. And when we examine the overall pattern of the corroborations gained by T and T' we will find, given that neither has so far been discorroborated and that there have been tests on the excess corroborable content of T over T', that the former dominate the latter so that $Co(T) > Co(T')$.

It is interesting in this connection to consider certain doubts that have been raised as to whether GTR really was corroborated by the detection of "red-shift". Earman and Glymour (1980b) have investigated this question. One source of doubt was a certain obscurity as to how the "red-shift" prediction was derived within GTR. Obviously, if this prediction was not, after all, a consequence of the theory's premises, then the theory was not, after all, corroborated when this prediction survived tests. On this point the authors say that 'the leading theoretical physicists of the period, while grasping the essential ideas, still did not manage to produce a clean and unambiguous formal derivation of the spectral shift' (p. 189); but they add that this was put right in 1935 by J.L. Synge. But there were doubts of another kind. The authors write:

> It is frequently claimed that the solar red shift provides no evidence at all for the field equations of general relativity. The grounds most commonly given are that the red shift can be deduced from principles other than the field equations. . . . This very common opinion is also very curious. For . . . the claim seems to be that a phenomenon does not confirm an hypothesis if that phenomenon can be deduced from some weaker hypothesis entailed by the first. (P. 204)

They rightly add that such a principle is untenable. Let g, T_H and A be respectively the "red-shift" prediction, the fundamental, and the auxiliary assumptions of GTR. Given (i) that $T_H \wedge A$ does indeed entail g, the only way, according to our theory, to show that GTR was not corroborated by the positive results of tests on g would be to show (ii) that g is entailed by A alone; and one does not show this by calling attention to a theorem T' of $T_H \wedge A$ that is weaker than it and stronger than g and that suffices to entail g. As we saw, it is a logical triviality that there are bound to be such theorems.

Suppose, however, that while g is not entailed by A alone, it is entailed by $T_H' \wedge A$ where T_H' is a proper subset of T_H; in other words, only some of the fundamental assumptions of T were needed for the derivation of g. It may be asked why in that case the theory as a whole should gain a corroboration from a successful test on g: should not the corroboration accrue just to that proper subset of its premises that were required for the derivation of g?

If a haphazard conjunction of premises has yielded a novel prediction that passed a test, it would be right to separate off the premises responsible for the prediction and concentrate the credit on them. It would be rather as if a few individuals in a small crowd of bystanders had grabbed a bank robber; just those individuals, and not the crowd, would deserve the praise. But I see the fundamental assumptions T_H of a theory T more on the analogy of the top management of an enterprising business in a competitive economy. One might want to investigate just which members of the top management were involved in a particular branch of its activities. But the main question, at least for an investor, is whether the company as a whole is making a good profit and, more particularly, whether it is competing successfully with its rivals. Rather similarly, one might want to investigate just which of a theory's fundamental assumptions were involved in a particular prediction (perhaps they all were); but the main question is whether the theory as a whole is well corroborated and, more particularly, whether it is better corroborated than its rivals.

8.42 *Goodman's Paradox*

Between hypotheses that are as yet quite undreamt of and ones that are being seriously canvassed, there is an intermediate category of ones that are trivially easy to manufacture from an existing theory or hypothesis that has been seriously proposed, though no one but a philosopher would want to. These provide what I called sideways proliferation arguments that would, if successful, have the embarrassing consequence that, to any theory h that is regarded as the best corroborated, there will correspond a host of alternatives h', h'' . . . each having an equal amount of testable content, and standing in a similar relation to the evidence that corroborated h, but differing significantly from h. (We have frequently met this pattern before, for instance in §3.3.)

The most famous of these sideways proliferation arguments is, of course, due to Goodman: evidence that confirms 'All emeralds are green' must equally confirm 'All emeralds are grue', an emerald being grue if it is either (i) examined before t_o and green or (ii) not examined before t_o and blue (*1954*, p. 74), where t_o is some arbitrarily chosen point of time, usually in the near future.

One point worth noticing is that the trick in this form cannot be played on any scientific theory that relates the behaviour of observed things to that of unobserved things; for tampering with what it says about unobserved things would have repercussions on what is said about observed things. According to NM, for example, the movement of an observed mass such as the moon is a function of the net force acting on it, and this in turn depends on the gravitational pulls exerted on it by all masses, observed and unobserved. Newton himself sought to block any such breakdown in uniformity with this Third Rule:

> The qualities of bodies, which admit neither intensification nor remission of degrees, and which are found to belong to all bodies within the reach of our experiments, are to be esteemed the universal qualities of all bodies whatsoever. (*1729*, p. 398, italicised in the original)

However, something like "grue" variants of such theories can be constructed. For ease of exposition let such a theory be represented by $\forall x(Fx \rightarrow Gx)$. Let t_o be a point of time in the future, say in the year 2000. Then we could change our theory to

$$\forall x \forall t((t \leq t_o) \rightarrow (F(x,t) \rightarrow G(x,t))) \wedge$$
$$\forall x \forall t((t > t_o) \rightarrow (F(x,t) \rightarrow G'(y,t)))$$

where G and G' are incompatible. This says that up to t_o anything that is F is G and that thereafter anything that is F is not G but G'.

Bartley suggested (1968, pp. 54f.) that hypotheses like 'All emeralds are grue' need not be seriously entertained because a hypothesis should be seriously entertained only if it is directed at the solution of some problem; and there is no problem to which "grue-ish" hypotheses are directed. But Goodman could retort that if T is a scientific theory that is directed at a serious problem, and if T' is a "grue-ish" variant of T, then T' is an alternative (and no doubt perverse) solution of that same problem. And I would add that a Popperian concern for problematicality should extend to statements that *pose* problems as well as to ones that offer solutions of them; and Goodman's hypotheses certainly pose a problem.

If a theory is well corroborated, is a "grue-ish" variant of it equally well corroborated? Musgrave (1978, p. 182) pointed out that on Popper's theory of corroboration it will not have been corroborated by the results of tests on the original theory that were carried out before this perversion of it was constructed, since they will have been incorporated into background knowledge; to which Hübner retorted that this defence assumes that "grue-ish" perversions always come afterwards. But suppose a quirky genius got in first with a "grue-ish" yet brilliant theory, which won some spectacular corroborations before it was straightened out; would not the boot be on the other foot with the straightened out version failing to win corroborations won by its crooked predecessor?

Our answer to the present proliferation argument is in line with our previous answers to other proliferation arguments: the above "theory" is no *theory* in our sense. Let T' and T'' denote respectively the part that contains $(t \leq t_o)$, and the part that contains $(t > t_o)$, as its antecedent clause. T' is silent about anything after t_o as is T'' about anything up to t_o. There is no temporal overlap; consequently, conjoining T' and T'' can yield no testable consequences not already entailed by T' or by T''; we have $CT(T' \wedge T'') = CT(T') \cup CT(T'')$. It would be as if someone had stuck half a marrow to half a cucumber.

But wait: could not an analogous trick be played on a respectable scientific theory (Worrall)? Suppose we split a theory T, which I will again represent by $\forall x(Fx \to Gx)$, into

$$\forall x \forall t((t \le t_o) \to (F(x,t) \to G(x,t))) \land$$
$$\forall x \forall t((t > t_o) \to (F(x,t) \to G(x,t))).$$

Call the first conjunct T' and the second T''. Clearly, $T' \land T''$ is equivalent to T. Will we not again have $CT(T) = CT(T') \cup CT(T'')$? If we did, the condition I am relying on to deny corroborations to "grue-ish" variants would equally hit the original theories. Moreover, our organic fertility requirement would break down, since every theory would fail it.

A genuine scientific theory makes possible predictions across a span of time: it will have singular predictive implications of the form $e_1 \to e_2$, where e_1 describes initial conditions obtaining at one time, and e_2 describes an occurrence at a later time. Thus NM made possible predictions of returns of Halley's comet, and Kepler's laws made it possible to predict, after Uranus had been discovered and its mean distance from the sun had been estimated, when it would return to its present position after completing one orbit. And to establish that $CT(T) > CT(T') \cup CT(F'')$ we have only to find a consequence $e_1 \to e_2$ of T where e_1 describes initial conditions obtaining prior to t_o and e_2 an occurrence after t_o; for such an $e_1 \to e_2$ will not be a consequence of either T' or T''.

8.43 *"Deoccamisation"*

Glymour produced a sideways proliferation argument that involves what he calls, with acknowledgments to David Kaplan, *deoccamisation*: it multiplies theoretical predicates unnecessarily. For instance, we might replace the concept 'force' whenever it occurs in NM by the concepts 'gorce' and 'morce', of which we are told that 'the sum of the gorce and morce acting on a body is equal to the mass of the body times its acceleration' (*1980*, p. 356). We could in this way proceed from NM to NM', NM'' . . . where each of these variants has the same testable content as NM but a different theoretical ontology. If E is evidence that corroborates NM will it not equally corroborate NM', NM'' . . . ?

This causes no difficulty to an empiricist who holds that its Ramsey-sentence captures the entire content of a scientific theory containing theoretical predicates; for if the only difference between NM, NM',

NM'' ... is that theoretical predicates in one are replaced by different ones in the others, they can all have the same Ramsey-sentence. But we cannot take this line; nor can we employ here the argument we used against "grue"-type variants of a theory; if NM is a unified theory, so are NM', NM''. ...

When we considered the conditions under which a theory should inherit, as corroborations, the result e of a test that had corroborated, or perhaps discorroborated, its predecessor, we found that it was not enough to require that its fundamental assumptions be involved in its explanation of e; for these might have been those of its predecessor but tinkered with a little to secure a better fit with e. (There was, it will be remembered, a suggestion that Newton's inverse square law might be slightly adjusted to secure a better fit between NM and observations on Uranus.) To exclude this we required it in addition to have more corroborable content, and hence to be more unified, than its predecessor(s).

Now NM itself was indeed more unified than its predecessors, Galileo's and Kepler's laws, and so was entitled to inherit, as corroborations, the results of tests on them. NM was born corroborated. But a 'gorce-and-morce' variant NM' of NM is not more unified than NM: it has been got from NM by a peculiar kind of tinkering with the latter's fundamental assumptions that leaves the corroborable content unchanged. So even if NM' had been put forward before NM had gained any new corroborations from independent tests, NM' would *not* have been entitled to inherit the corroborations of NM that the latter was entitled to inherit from tests on Galileo's and Kepler's laws, and we would have Co(NM) > Co(NM').

This concludes my review of those well known "paradoxes" and difficulties in the field of confirmation theory that suggest that there never is just one best corroborated theory because there are always ever so many equally well corroborated alternatives. Before I turn to arguments which suggest that the number of well corroborated theories, far from being large, is always zero, I will look at a famous "paradox of confirmation" of a different kind.

8.44 Hempel's Paradox

Hempel's Raven Paradox has been very extensively discussed. I will here ask only whether our theory of corroboration resolves it satisfactorily.

Suppose one says that for a hypothesis of the form $\forall x(Fx \rightarrow Gx)$:

(1a)   anything that is $F \wedge G$ is a confirming instance;

(2a)   anything that is $F \wedge \sim G$ is a refuting instance;

(3a)   anything that is $\sim F$ is a *neutral* instance.

Substituting 'raven' for F and 'black' for G, this says that (1) black ravens confirm, (2) nonblack ravens refute, and (3) nonravens are neutral to, 'All ravens are black'. But suppose that we instead substitute 'nonblack' for F and 'nonraven' for G. The above now say (1') nonblack nonravens confirm, (2') nonblack ravens refute, and (3') black things are neutral to, 'All nonblack things are nonravens'. But the latter is logically equivalent to 'All ravens are black'. (For an even more peculiar consequence of (1a) see Horwich, 1978.)

Hempel resolved these contradictions by saying that it is a mistake to suppose that 'All ravens are black' tells us something about ravens only: it tells us something about *every* thing, namely that every thing is no-raven-or-black; and anything that *is* no-raven-or-black is a confirming instance, or:

(1b)   anything that is $F \wedge G$ is a confirming instance;

(2b)   anything that is $F \wedge \sim G$ is a refuting instance;

(3b)   anything that is $\sim F$ is a *confirming* instance.

Our theory of corroboration has the following implications. If a theory T entails the low-level generalization $\forall x(Fx \rightarrow Gx)$ then:

(1c)   something that is $F \wedge G$ *may* provide corroborating evidence;

(2b)   anything that is $F \wedge \sim G$ provides refuting evidence;

(3c)   something that is $\sim F$ *may* provide *corroborating* evidence.

Between this (1c) and (3c) and Hempel's (1b) and (3b) there is one important parallel: so far as their confirming (corroborating) capacity is concerned, things that are $F \wedge G$ are placed on a par with things that are $\sim F$, the difference being that (1b) and (3b) require everything that is $F \wedge G$ or $\sim F$ to provide confirming evidence whereas (1c) and (3c) allow but do not require them to provide corroborating evidence.

When I wrote something along these lines a good many years ago, Scheffler retorted that this reminded him of the unmarried girl's claim to be only a little pregnant (1961, p. 19); and on a later occasion he added that while this may *limit the scope* of the paradox it does nothing to *eliminate* the paradox itself (*1963*, p. 275). And I daresay that some readers will be inclined to agree that while my (3c) may be less paradoxical than Hempel's (3b), since it only allows things that are $\sim F$ to provide corroborating evidence, this is still rather paradoxical; and they may add that my (1c), which allows things that are $F \wedge G$ to *fail* to provide corroborating evidence, is definitely more paradoxical than Hempel's commonsensical (1b). I will now explain why our theory of corroboration leads to these results and, at the same time, try to show, with the help of examples, that these are the results we want. I begin with (1c).

According to our Popperian view, a scientific theory T can gain corroborations only from *tests*. (It may inherit corroborations from tests on earlier theories that would have tested it if it had already been published when the tests were made.) If the test is a hard one, and the result is favourable, T gains a strong corroboration; if medium or soft, a moderate or weak corroboration. A soft test is, as near as maybe, a repetition of a test that T had passed on at least one previous occasion; but of course T may fail it on the present occasion. Now suppose that an empirical investigation is to be carried out that, by its terms of reference, *cannot* result in a refutation of T, though it may produce evidence that appears to provide inductive support for T. We might call such an investigation a pseudo-test for T. According to our theory, T can gain no corroboration from whatever result this pseudo-test may have.

Is this too restrictive? Consider an example based on a famous aphorism of Bacon's (*1620*, i, xlvi). Let T be a theory about the role of Providence in human affairs that has the following consequence g: 'All sailors who have paid their vows escape shipwreck'. Represent this consequence by $\forall x(Fx \rightarrow Gx)$. Now suppose that evidence is presented of the form $Fa_1 \wedge Ga_1 \wedge Fa_2 \wedge Ga_2 \ldots \wedge Fa_n \wedge Ga_n$, where n is impressively large. This may look like good support for g and, indirectly, for T; and according to our theory, T would indeed gain a strong corroboration from it if this evidence was the result of a hard test. But suppose that it turns out that this evidence resulted from an investigation *confined to sailors who had escaped shipwreck*; it was already known, in the case of each a_i investigated, that a_i is G; the only question was whether a_i is F. In other words, this evidence resulted from a pseudo-test: by its terms of reference it could not have led to the discovery of an a_i that is $F \wedge \sim G$. It seems right to me that this evidence should not be regarded as confirming either g or T, contrary to Hempel's (1b) but in line with our (1c).

I turn now to our (3c), which allows things that are $\sim F$ to provide corroborating evidence for a theory T that entails $\forall x(Fx \rightarrow Gx)$. Let T be Newtonian mechanics and let $\forall x(Fx \rightarrow Gx)$ describe how the outermost planet in our solar system should behave, according to NM. From its discovery in 1781 until the 1840s, Uranus was believed by astronomers to be the outermost planet; and during this period evidence was accumulating that Uranus was not behaving as the outermost planet should; if we denote Uranus by a, it was looking as though astronomers were being obliged to accept, as a PF of NM, $Fa \wedge \sim Ga$. And then, in 1846, came the discovery of a planet beyond Uranus, namely Neptune: $Fa \wedge \sim Ga$ could now be discarded in favour of $\sim Fa \wedge \sim Ga$. Moreover, the supposed "misbehaviour" of Uranus could now be seen to provide corroborating evidence for NM. So our (3c) is right in allowing that something that is $\sim F$ may provide corroborating evidence for a theory T that entails $\forall x(Fx \rightarrow Gx)$. (For another example of this pattern of corroboration, see Agassi 1959, pp. 313–314.)

As I see it, Hempel rightly held that there is *one* main line to be drawn, namely between refuting and nonrefuting instances, and that it is a mistake to divide the latter into positive instances (things that are $F \wedge G$) and neutral instances (things that are $\sim F$). But whereas he left it at that, regarding all nonrefuting instances as equally confirmatory, our theory enables us to superimpose a fine structure on these nonrefuting instances, allowing them to provide strong, moderate, or weak corroborations, or noncorroborations, for a theory T, according to whether the evidence concerning them resulted from a hard, medium, or soft test on T or from an investigation that constituted no test on T.

8.5 The Duhem-Quine Problem Again

Arguments to the effect that 'the best corroborated theory in its field' never denotes anything because there are always ever *so* many equally well corroborated alternatives, were dealt with earlier. In the remainder of this section we will deal with arguments to the effect that it never denotes anything because there never is even one corroborated theory.

The term 'unmanageable largism' was introduced in § 7.42. By 'manageable largism' I mean the following thesis, which I accept. Let T be an axiom system for a testable and unified theory that complies with rules 1–5 in § 5.33. Then (i) the number of separate axioms in T is likely to be large; (ii) however, the task of writing them all out can be completed; (iii) those axioms that constitute the fundamental assumptions of the theory are not falsifiable either individually or collectively;

(iv) it may well be that a test on the theory is a test of the axioms T as a whole, so that if its result is negative, we cannot say that only some proper subset of them were hit; if the result is positive, then the theory as a whole gains a corroboration.

Unmanageable largism rejects (ii) above; to render a given axiom system falsifiable would require a process of augmentation that could never be completed. Holism is unmanageable largism carried to the limit: to render a given axiom system falsifiable, according to it, one would have to continue augmenting it until it had expanded into 'the whole of science' (Quine, 1951, p. 42). Unmanageable largism would, of course, nullify our theory of corroboration and hence our answer to rationality-scepticism; but manageable largism is entirely in line with (B*). I argued in § 7.4 that if Duhem had allowed what I call sophisticated level-1 statements into the empirical basis, as he should have done since they are as rationally acceptable as naive ones, then his position would have been a manageable rather than an unmanageable largism. Before I turn to positions that threaten our theory, I will first ask whether Duhem's position, when liberalised by the admission of sophisticated level-1 statements, is open to any sort of riposte. My answer will be that it is not.

8.51 *Duhem and Manageable Largism*

An experimentalist wishes to test a certain theoretical proposition, call it B_1. (In Duhem's example it is the proposition 'that in a ray of polarized light the vibration is parallel to the plane of polarization', *1906*, p. 184.) The target of his test is an experimental generalisation, call it g. ('If we cause a light beam reflected at 45° from a plate of glass to interfere with the incident beam polarized perpendicularly to the plane of incidence, there ought to appear alternately dark and light interference bands parallel to the reflecting surface.') But g is not deducible from B_1 alone; many additional assumptions are required ('that light consists in simple periodic vibrations, that these vibrations are normal to the light ray, that at each point the mean kinetic energy of the vibratory motion is a measure of the intensity of light, that the more or less complete attack of the gelatine coating on a photographic plate indicates the various degrees of this intensity. . . . ' pp. 185–186). Call these additional assumptions $B_2, \ldots B_n$. The result of the experiment, call it e, refutes g. Then, as Duhem rightly insisted, what e establishes is not $\sim B_1$ but $\sim (B_1 \wedge B_2 \ldots \wedge B_n)$. As he put it: 'the physicist can never subject an isolated hypothesis to experimental test, but only a whole group of hypotheses; when the experiment is in disagreement with his predictions, what he learns is that at least one of the hypotheses constituting this group is unacceptable and ought to be modified; but the experiment does not designate which one should be changed' (p. 187). It seems clear that Duhem, in differentiating between 'an isolated hypothesis' or a single axiom, and a 'whole group of hypotheses' or an axiom set, was relying on the idea of a "natural" axiomatisation, an idea that was elucidated in § 5.33; and in what follows I will assume that the theories in question are axiomatised in conformity with the rules given there.

Is Duhem's thesis, as so far presented, open to any sort of riposte? Given that $B_1 \wedge B_2 \ldots \wedge B_n$ just suffices to entail g, which is refuted by e, is there any hope that we might be able to narrow responsibility for the refutation or, in Lakatos's phrase (1962, p. 15), identify the guilty lemma(s)? Popper once suggested that there is one happy circumstance in which it might be 'reasonably safe' to pin the blame on a single axiom:

> Admittedly, Duhem is right when he says that we can test only huge and complex theoretical systems rather than isolated hypotheses; but if we test two such systems which differ in one hypothesis only, and if we can design experiments which refute the first system while leaving the second very well corroborated, then we may be on reasonably safe ground even if we attribute the failure of the first system to that one hypothesis in which it differs from the other. (*1957*, p. 132n.)

For ease of exposition, denote that one hypothesis by B_1 and the large number of hypotheses common to both systems by B_2; and let B_1' be the hypothesis which, in the second system, replaces B_1. What does it mean to 'attribute the failure of the first system' to B_1? If it means only that $B_1 \wedge B_2$ is falsified whereas $B_1' \wedge B_2$ is very well corroborated, then we are on *perfectly* safe ground in saying this. So presumably Popper had something more in mind. But what? Presumably that it is B_1 rather than B_2 that is responsible for the falsity of the refuted g. Given that both B_1 and B_2 were required for the derivation of g, to blame just B_1 is, presumably, to exonerate B_2. But does not blaming B_1 and exonerating B_2 imply that B_1 is false and B_2 is *true*? One could hardly say: 'Although B_1 may be true and B_2 may be false, it is nevertheless B_1 and not B_2 that is responsible for the falsity of g.' This way of countering Duhem seems to involve a kind of verificationism.

The fact that B_1' makes a better partner for B_2 than B_1 makes does not mean that B_1' is closer to the truth than B_1. Theories are like puddings: both obey what Moore called the principle of organic unity, which says that 'the value of such a whole bears no regular proportion to the sum of the values of its parts' (*1903*, p. 27, italicised in the original). Suppose that a chef makes two puddings, P and P', according to recipes that are identical except that, where cinnamon was used in P, nutmeg was used in P'; and P proves disappointing while P' proves delicious. Then nutmeg provides a better partner than cinnamon for the other ingredients common to P and P'. But this does not mean that nutmeg is, in itself, gastronomically superior to cinnamon. Perhaps by retaining cinnamon and varying some of the other ingredients, this chef could have made a pudding even better than P'. And something similar holds for theories. Perhaps B_1 is true and B_1' is false, but B_1' is the better partner for B_2 because there is an error in B_2 that is cancelled out by a compensating error in B_1'. (For a variant on this theme see Grünbaum, 1971, p. 84.)

Another possible riposte to Duhem is suggested by Glymour (*1980*). Glymour has a horror of holism, which I share, and also of what I call manageable largism, or the thesis that evidence confirms large theories rather than local parts of such theories, which I do not share. He is out to counter 'the puzzling opinion, common among philosophers, that evidence can only be evidence for *a theory as a whole*, or even for science as a whole' (p. 4, my italics). His strategy is to replace the idea of evidence confirming a hypothesis by that of evidence confirming a hypothesis relative to a theory: hypotheses 'are not generally tested or supported or confirmed absolutely, but only *relative to the theory*' (p. 122).

Suppose that a scientific system, containing both fundamental and subsidiary assumptions, is axiomatised as $B_1 \wedge B_2 \ldots \wedge B_n$. These axioms just suffice to entail the low-level generalisation g, which is to be subjected to a hard test. According to the kind of manageable largism involved in our theory of corroboration, a favourable outcome will provide a strong corroboration for this system as a whole rather than for some local part of it. Suppose, next, that we somehow *know* (perhaps the Angel of Truth tells us) that B_1, say, is true. We might differentiate it from the others by calling it a *theory* and $B_2 \wedge B_3 \ldots \wedge B_n$ a *hypothesis.* Then this theory, though needed for the derivation of g, is not at risk from the test on g: it is above the battle. According to whether the result of the test is unfavourable or favourable, it falsifies or corroborates just the hypothesis. Suppose, next, that although we cannot be absolutely certain of the truth of B_1, its 'inductive confirmation . . . is so enormously high that $[B_1]$ can be regarded as *well-nigh established*' (Grünbaum, 1971, p. 118). Then we might proceed very much as before and say that, this theory being well-nigh above the battle, the test can safely be regarded as bearing exclusively on the hypothesis.

Suppose, finally, that no B_i in the axiom set B_1, B_2, . . . B_n enjoys any epistemological superiority over the others. Nevertheless, we again split them into a "theory" and a "hypothesis" and say that the test result refutes or corroborates this "hypothesis" relative to that "theory". What would this mean? Only that we can *decide* to treat some axioms as pinned down, to use Morton White's term (*1956*, p. 279), and to concentrate the test result on the remainder. But on the present assumption of epistemological equality, such a decision would be quite arbitrary. We could just as well reverse their roles and regard the test result as refuting or corroborating the "theory" relative to the "hypothesis". Or we could localise the test result to some B_i by pinning down all the other axioms. But since we could select *any* B_i this would not achieve any genuine localisation.

This means that making confirmation a three-place relation between evidence, theory, and hypothesis provides no escape from manageable largism *unless* the cluster of statements designated by the word 'theory' enjoys a genuinely superior epistemological status. So far as I know, Glymour nowhere claims such a superior status for it. Nor, of course, is such a claim admissible from the standpoint of the present book. Probability-scepticism does not permit us to assign even a modest, let alone an 'enormously high', inductive confirmation to a universal hypothesis; nor can we separate off part of the theory and give it a protected status within unproblematic background knowledge, since we have replaced the latter by the idea of a historical record of experimental tests.

As I have said, Duhem's manageable largism has no adverse implications for our theory of corroboration. In a case like that envisaged by Popper, it will say that the *theory* constituted by $B_1' \wedge B_2$ is far superior to the one constituted by $B_1 \wedge B_2$, but it will not try to delve into the comparative merits of B_1 and B_1' considered in isolation. (I am here taking it for granted that these two hypotheses are not testable in isolation. If they were, and if tests on them had been thought up, then (A*) would call for these tests to be carried out.) Our theory of corroboration is not, as such, concerned to evaluate the individual ingredients that have gone into the making of a scientific theory. Like a gourmet inspecting the sweet trolley, it is content to appraise completed wholes.

8.52 *A Storm in an Ink Pot*

Before leaving manageable largism and turning to holism, I wish to draw attention to a certain equivocation in some current uses of the term 'scientific theory'. I do not say that my use of it, to denote a conjunction, $T_H \wedge A$, of fundamental and subsidiary assumptions satisfying the organic fertility requirement, is correct. Anyone is free to use it differently; and a good case could be made for using it to denote just a distinctive set of fundamental assumptions, or for a T_H by itself. My use has the drawback that if Newton revised some rather low-level subsidiary assumption of his system, say one concerning the relative masses of the sun, earth, and moon, I am obliged pedantically to insist that he replaced one theory, say $T_H \wedge A$, by a (very slightly) different theory, $T_H \wedge A'$. Is it not handier to let 'Newtonian theory' stand for just the main principles of his system, say his laws of motion and his law of gravitation, uncluttered by subsidiary detail?

Rather than swap horses at this late date I will introduce the new term 'core theory' to denote some T_H by itself; and I will sometimes use the term 'fleshed out theory', instead of just 'theory', for some $T_H \wedge A$. Failure to heed the important distinction between a core theory and a fleshed-out version of it can have alarming consequences, as we shall see.

In a famous passage, Popper contrasted the genuine confirmation gained by a falsifiable theory, such as Einstein's, when it survives a severe test, with the mass of spurious confirmations gained by an unfalsifiable theory such as Freud's or Adler's from 'clinical observations' (*1963*, pp. 37–38). And in a footnote he added:

> But real support can be obtained only from observations undertaken as tests (by 'attempted refutations'); and for this purpose *criteria of refutation* have to be laid down beforehand: it must be agreed which observable situations, if actually observed, mean that the theory is refuted. But what kind of clinical responses would refute to the satisfaction of the analyst not merely a particular analytic diagnosis but psycho-analysis itself?

To this Lakatos riposted:

> In the case of psychoanalysis Popper was right: no answer has been forthcoming. Freudians have been nonplussed by Popper's basic challenge. . . . Indeed, they have refused to specify experimental conditions under which they would give up their basic assumptions. . . . But what if we put Popper's questions to the Newtonian scientist: 'What kind of observation would refute to the satisfaction of the Newtonian not merely a particular Newtonian explanation but Newtonian dynamics and gravitational theory itself?'. (1974, pp. 146–147)

Lakatos had a point, here, which I would put like this. If by 'psychoanalysis itself' Popper meant the fundamental assumptions that constitute the core of Freudian theory, then his challenge was unfair. For in the case neither of Freudian theory nor of Newtonian theory can such a core be brought into direct confrontation with empirical evidence. Popper should have challenged Freudians to provide a fleshed out version of their theory that would show its fundamental assumptions meshing with subsidiary assumptions to yield conclusions that might be refuted by clinical observations. Whether Freudians would still have been nonplussed by *this* challenge I am not competent to say. I am persuaded by Fisher and Greenberg (*1977*) and by Grünbaum (1983) that there are testable hypotheses in the Freudian corpus, but whether any of these flow from what Freudians would take to be fundamental assumptions of psychoanalytic theory I do not know.

But Lakatos was not merely pointing to a certain unfairness in Popper's challenge to the Freudians; he was attacking Popper's falsifiability criterion. Later in this chapter we will look into Lakatos's thesis that all good scientific theories are born refuted. What concerns us here is his thesis that all good scientific theories are irrefutable: '*exactly the most admired scientific theories simply fail to forbid any observable state of affairs*' (1970, p. 16).

As I see it, Popper should have answered this along the following lines: 'If your thesis is about core theories, it is trivially true; if about fleshed out theories, trivially false. A core theory lacks the observational predicates needed for a possible conflict with observation reports. But the fleshed out versions of NM, as developed by Newton himself and his successors, especially Laplace, yield very precise predictions falsifiable by, for example, measurements of angle that disagree by only a few minutes of arc.' In fact it seems clear that Lakatos was here using 'scientific theory' very much in the sense of our 'core theory'; where I speak of a 'fleshed out theory' he spoke of the 'theoretical maze' of a falsifiable *system* of theories including all sorts of auxiliary theories (1974, p. 147). So Popper could have answered more shortly: 'Your thesis, which is clearly about core theories, is trivially true.'

Instead, Popper dramatically declared: 'Were [Lakatos's] thesis true, then my philosophy of science would not only be completely mistaken, but would turn out to be completely uninteresting' (1974b, p. 1005). Then did Popper take Lakatos's thesis to be about fleshed-out theories? No; although he did not use the word, he recognised that Lakatos's thesis was about the *core* of Newton's theory: it was, Popper said, the thesis that 'Newton's theory (*his laws of motion plus his law of gravitation*) is not falsifiable' (p. 987, my italics); and Popper undertook the impossible task of upholding the falsifiability of just this theoretical core. Apparently forgetting that he had once said 'Duhem is right when he says that we can test only huge and complex theoretical systems', Popper set out to devise potential falsifiers just for Newton's fundamental assumptions. His examples are absurdly overstrong considered as PFs of a properly fleshed out version of NM but, of course, not strong enough to be PFs of this irrefutable core. They mostly involve planets moving in highly erratic ways. But Newton's laws of motion plus his law of gravitation say nothing about the physical makeup of the planets; in particular they do not rule out the possibility that the planets are enormous rocketlike devices (see O'Hear, *1980*, p. 102) that can accelerate themselves in all sorts of ways. Another example has apples that have fallen from a tree rising up and dancing round the tree. But Newton's laws of motion and law of gravitation say nothing against the possibility of such a spectacle being produced with the help of invisible elastic threads. Of course, if we start fleshing out this core theory with assumptions about the physical makeup of the planets, the nonexistence of invisible elastic threads, etc., then it will become falsifiable by these examples. But if we start fleshing out the core of NM, we may as well do it properly and take a fully fleshed-out version, such as that laid out in the five large volumes of Laplace's (*1798–1825*) and his (*1824*). According to Laplace, very accurate observations made on Jupiter and Saturn fitted the predictions yielded by his formulas with an error of less than 1' of arc (*1824*, ii, p. 33). Suppose that there had been evidence of Saturn-Jupiter perturbations of the same order of magnitude as those actually observed *but in the opposite direction*: that would have been a clear-cut PF for this fleshed-out version of NM.

Lakatos was technically correct in saying that the fundamental assumptions of NM and of Freudianism are *alike* in being irrefutable. But he ought to have added that they are also *very unlike* in that only the former are *enormously fertile in testable consequences* when coupled with appropriate subsidiary assumptions. NM is a marvellous exemplar of our requirement that, in the case of a deep and powerful scientific theory, while we have $CT(T_H) = \varnothing$, we also have $CT(T_H \wedge A) \gg CT(A)$.

8.53 *Quinean Holism*

The main pressure that tends to expand a manageable into an unmanageable largism is the desire for a perfectly secure and error-free empirical basis. These issues were investigated in §7.4 and will not be reopened here. However, Quine had an argument for his holistic thesis that 'our statements about the external world face the tribunal of sense experience not individually but only as a corporate body' (1951, p. 41) that is independent of his assumption that the tribunal consists of *sense experience* rather than of rationally accepted level-1 statements. The argument is that just because (i) any statement (other than one at the very periphery) within 'the whole of science' can be held true in the face of recalcitrant evidence, therefore (ii) any previously accepted statement *p* may be declared false in the wake of the adjustments occasioned by that recalcitrant experience.

> A conflict with experience at the periphery occasions readjustments in the interior of the field. Truth values have to be redistributed. . . . Reëvaluation of some statements entails reëvaluation of others. . . . [T]here is much latitude of choice as to what statements to reëvaluate in the light of any single contrary experience. . . . (1951, pp. 42–43)

> Even a statement very close to the periphery can be held true in the face of recalcitrant experience. . . . Conversely, by the same token, no statement is immune to revision. (P. 43)

On this view the system of scientific statements is rather like a complicated nervous network: the pain caused by an injury, instead of being local, may be referred to remote areas. The difference is that on Quine's view we are free to decide down which logical pathways to transmit the repercussions of a clash with experience at the periphery. (He suggested that our choices should be governed by a taste for simplicity.)

But if $B_1, \ldots B_n$ are the axioms of a theory responsible for a falsified prediction, and if *p* is some proposition far removed from all this, why should p be at risk as a result of that falsification? Quine wrote: 'Any statement [at all remote from the experiential periphery] can be held true come what may, *if we make drastic enough adjustments elsewhere in the system*' (p. 43, my italics). The first part of this statement is true, indeed trivially true, in any case where we have a set of axioms that are collectively but not individually falsifiable or, in our notation, where $CT(B_1 \wedge B_2 \ldots \wedge B_n) > \varnothing$ but $CT(B_1) = CT(B_2) \ldots = CT(B_n) = \varnothing$. Faced by a falsification of such an axiom set we can of course always retain any given B_i and eliminate the conflict by weakening or otherwise adjusting one or more of the other axioms. Then why did Quine add the proviso italicised above? Why did he not say that any such B_i can be held true by making ad hoc adjustments elsewhere, adding that we can, if we wish, follow a conservative policy of making them in such a way that very little disturbance is caused to the rest of the system?

I think that Grünbaum (1960) provided the answer. Suppose that an experiment produces what is generally regarded as a decisive refutation of a theory. Let the theory consist of fundamental assumptions T_H and subsidiary assumptions A, and assume that $T_H \wedge A$ just suffices to entail the prediction e; but the experiment led to the acceptance of e', which entails $\sim e$. Then Quine's thesis, Grünbaum suggested, is not merely that T_H can be retained in the face of e' (for instance, by weakening A); it is that T_H can be retained and used in explanation of e' by partnering it with suitably revised auxiliary assumptions A'. However, in order to squeeze e' out of a $T_H \wedge A'$ it may be necessary to revise the previous A very drastically; and for any previously accepted statement p not entailed by T_H or by e' it is possible that $T_H \wedge A'$ will entail $\sim p$.

We might split this into two subtheses:

(i)   for any e' that has refuted $T_H \wedge A$ there will be an A' such that $T_H \wedge A'$ entails e';

(ii)   retaining T_H and partnering it with a suitable A' in an explanation of e' may require the rejection of any previously accepted statement p not entailed by $T_H \wedge e'$.

Grünbaum pointed out that thesis (i), as it stands, is trivially true: we have only to let A' be $T_H \rightarrow e'$ and $T_H \wedge A'$ will of course entail e'. But notice that in this form it provides no support for thesis (ii): such an A' would have *no* repercussions on any p logically independent of

$T_H \wedge e'$. And thesis (i) is trivially true in another way: we could let A' suffice to entail e' without the help of T_H. Indeed, we could let A' be e' itself; and again there would be no support for thesis (ii). So if thesis (i) is to support thesis (ii), it must be strengthened to say that, for any e' that has refuted $T_H \wedge A$, it is possible to combine T_H with an A' such that $T_H \wedge A'$ not only entails e' but provides a formally adequate explanation for e' in which T_H plays an essential role. But could we always know in advance that such an A' is to hand? It may be far from easy to find *any* formally adequate explanation for evidence e' that has refuted $T_H \wedge A$; and it will, presumably, be no easier to find one that satisfies the constraint that T_H figures essentially in it. Suppose that T_H is some key component of classical, deterministic physics, for instance the assumption of the immutability of atoms, and e' is evidence concerning some statistical regularity for which the currently accepted theoretical explanation is in terms of atomic decay according to the disintegration law. I do not say that an alternative explanation, essentially involving this T_H, could not be found, but I do say that the only way in which someone could oblige us to agree that it exists would be by actually producing a suitable candidate. Suppose that he does just that. We would then enquire whether, in addition to explaining the particular bit of refuting evidence e', his $T_H \wedge A'$ is (i) not refuted by other evidence and (ii) explains all that the currently accepted theory explains. If the answers to these questions are positive, then we have in this $T_H \wedge A'$ a serious rival to the currently accepted theory. And perhaps it will emerge triumphantly as the best corroborated theory in its field; in which case we should accept it; and if we come to accept it we should reject any previously accepted statement p with which it conflicts, just as we would if we came to accept, as the best corroborated theory, one that repudiates T_H in favour of some quite different core theory. As Lakatos observed, no Popperian would deny that the successor to a refuted theory may turn out to be inconsistent with a statement p 'in some distant part of knowledge' (1970, p. 98). Nor would a Popperian deny that the successor theory *may* be, though again it may not be, one that retains the core, namely T_H, of its refuted predecessor $T_H \wedge A$; in which case it will be the new A' that causes whatever disturbance is caused. The thesis that future scientific developments may oblige us to revise statements that we currently accept and Duhem's thesis that we can always retain an 'isolated hypothesis' in the face of recalcitrant evidence, are both true. But they do not add up to Quinean holism.

329

8.6 Are All Scientific Theories Born Refuted?

I know of only one remaining difficulty for our theory of corroboration and of rational scientific choice. Kuhn wrote: 'There are always some discrepancies [in the fit between theory and nature]' (*1962*, p. 81); and Lakatos declared that all theories 'are born refuted and die refuted' (*1978*, i, p. 5). Taken at its face value, this thesis, if correct, would wreck our theory, according to which it is rational to adopt that theory, where there is one, that is better corroborated than its rivals, it being a necessary condition for its being better corroborated than them that it be unrefuted. This condition, it will be remembered, derives from component (A*) of our optimum aim.

I think that Lakatos would have agreed that for a theory to be 'born refuted' it is not enough that there existed, when it was first put forward, some *as yet undiscovered* empirical fact that conflicts with it. That would render his thesis completely metaphysical, and it would no longer conflict with our (A*): a theory that is "born refuted" in *that* sense might very well be one that has remained for many years possibly true for all members of a scientific community, in the sense that, despite their best endeavours, they have not found any evidence that conflicts with it. To clash with our (A*), Lakatos's thesis must be understood to say that a theory is born refuted only if there existed, when it was first put forward, evidence conflicting with it that was already in at least one person's possession.

This objection to our theory is of a quite different character from the more or less formal difficulties considered hitherto in this section. It is a historical thesis. But despite its historical character it is strictly irrefutable; it has what I called in (1958) an "all-some" form: for all scientific theories there is, at the time they are first proposed, some refuting evidence in someone's possession. No historical investigation, however thorough, could establish that a particular scientific theory was *not* born refuted in this sense. Perhaps there was refuting evidence that was ignored or suppressed or forgotten, and no trace of it survives in the archives.

Then how should we proceed? I think that the best way is to take some well-known theory in the history of science that was under pretty constant empirical investigation, this investigation being well documented, and to ask whether the record gives us the slightest positive reason to suppose that it was born refuted. One theory that meets these desiderata is Kepler's. And it has the additional advantage that Lakatos himself referred to it in connection with his "born refuted" thesis (1971, p. 128; *1978*, ii, p. 202). A further advantage is that there is an intensive investigation covering this case in Curtis Wilson's excellent (1969).

As our starting point we may take a statement of Newton's in 'The System of the World', concerning the planets Saturn and Jupiter. After saying that the action of most planets on one another is negligible, he added that 'the action of Jupiter upon Saturn is not to be neglected'. He had calculated that when these planets are in conjunction, 'the gravity of Saturn towards Jupiter will be to the gravity of Saturn towards the sun . . . as 1 to about 211', adding: 'And hence arises a perturbation of the orbit of Saturn in every conjunction of this planet with Jupiter, so sensible, that astronomers are puzzled with it' (*1729*, p. 421). The question now is *when* astronomers first became 'puzzled with' this counter-evidence to Kepler's theory.

I begin with some dates. The Latin original of the above passage first appeared in 1687. Kepler discovered his First and Second laws in 1605 and the Third in 1618. His *Rudolphine Tables* eventually appeared in 1627, after various delays. Newton's 'The System of the World' grew out of a tract entitled *De Motu*, which went through three versions, the first written in 1680, the third in 1684. According to Curtis Wilson, it was only in 1684 that Newton became concerned about the implications of the idea of universal gravitational attraction for *interplanetary* interactions (1969, p. 161). On December 30 of that year Newton wrote to Flamsteed:

This Planet [Saturn] so oft as he is in conjunction with Jupiter ought (by reason of Jupiters action upon him) to run beyond his orbit about one or two of ye suns semidiameters. . . . Perhaps that might be ye ground of Keplers defining it too little. But I would gladly know if you ever observed Saturn to err considerably from Keplers tables about ye time of his conjunction with Jupiter. (*1959–77*, ii, p. 407)

Flamsteed was the obvious man to ask: he had recently, in 1683, reported to the Royal Society his observations of Saturn and Jupiter during three of their conjunctions. It is interesting that Flamsteed was somewhat sceptical about Newton's suggestion. In the course of his reply (5 January 1685) he wrote:

After ye next terme if not sooner I will inquire diligently. tho to confesse my thought freely to you I can scarce thinke there should be any such influence. . . . I can onely say that it seemes unlikely such small bodies as they are compared with ye sun ... should have any influence upon each other at so great a distance. (Pp. 408–409)

He promised to investigate the matter further 'as soone as our cold weather goes off & I can get a little leasure'. However, in this same letter, Flamsteed provided Newton with his own corrections to Kepler's tables, and these fitted in well with Newton's expectations. But he was not completely satisfied. Some six years later in 1691, he wrote to Flamsteed: 'I would willingly have your observations of [Jupiter] and [Saturn] for ye 4 or 5 next years at least before I think further of their Theory: but I had rather have them for the next 12 or 15 years' (*1959–77*, iii, p. 164). On May 4, 1694, David Gregory, in a memorandum compiled during a visit to Newton, wrote:

> 37. The mutual interactions of Saturn and Jupiter were made clear at their very recent conjunction. For before their conjunction Jupiter was speeded up and Saturn slowed down, while after their conjunction Jupiter was slowed down and Saturn speeded up. (In Newton, *1959–77*, iii, p. 318)

As we have so far reviewed it, then, the situation appears to be very much in line with Popper's idea of the typical way in which refutations come about. A new theory (Newton's) has been worked out, though not actually published yet, that revises a presently prevailing theory (Kepler's). Many of its revisions are quantitatively very slight, but there are areas (notably, Saturn and Jupiter in conjunction) where they are large enough to be experimentally distinguishable, or where, to put it in my terminology, the new theory has theoretically challenging consequences. At the behest of the new theory, these areas are now investigated very closely; and if all goes well (as it did in this case) the investigations result in strong corroborations for the new theory (and in discorroborations for its predecessor).

However, the historical facts do not fit this pattern quite so neatly. For one thing, Flamsteed did, it seems, have in his possession, already in 1683 before Newton inquired about it in December 1684, evidence of the perturbations, though he did not interpret it as such (and as we saw, he rather doubted whether Saturn and Jupiter could influence each other 'at so great a distance'). Thus Gregory noted, in a further memorandum during that same visit to Newton: 'Flamsteed's observations when compared with those of Tycho and Longomontanus prove the mutual attraction of Jupiter and Saturn in their most recent past conjunction in 1683' (in Newton, *1959–77*, iii, p. 337).

The next question is whether evidence adverse to Kepler's theory had come in prior to 1683. According to Wilson, although Kepler's own *Rudolphine Tables* had proved brilliantly successful with respect to Mercury and Mars, they had proved less successful with respect to Jupiter and Saturn: 'As for Jupiter and Saturn, it was recognized during the middle years of the century that KEPLER'S parameters, at least, required revision' (p. 101). In 1665–66 Riccioli and Wallis had reported in this sense, and in 1674 Flamsteed had reported that 'the places of the planet Jupiter have been, for these last two years, some 13 or 14 minutes forwarder in the heavens, than Kepler's numbers represent' (quoted by Wilson, p. 102). Is there any testimony concerning evidence obtained prior to the 1660s relating to Saturn and Jupiter and adverse to Kepler's theory? The only testimony I know of is the following. In a posthumously published paper, Lakatos wrote: 'Kepler himself realized in 1625 that Saturn and Jupiter do not move in ellipses' (*1978*, i, p. 210n.) Had he prepared this paper for publication himself, he would surely have added a reference; and I think I know what it would have been. For he had in his library Pannekoek's (*1951*); and on a heavily annotated page of it there occurs the following: 'The first practical problem was posed by the motion of Jupiter and Saturn. Kepler had already perceived, in 1625, that there was something wrong' (p. 300; John Stachel drew my attention to this passage). Exasperatingly, Pannekoek did not divulge *what* Kepler then perceived to be wrong nor did he provide any reference. For my part, I find it incredible that Kepler realised 'in 1625 that Saturn and Jupiter do not move in ellipses.' It would have been a heartbreaking discovery for him, and his candour and integrity would surely not have allowed him to suppress it. Yet so far as I know he gave no hint of any such discovery. Certainly, Max Caspar, who devoted fifty years to the study of Kepler's life and work, mentions nothing of this kind in his (*1948*) biography. (Incidentally, Kepler seems to have had no time in 1625 for serious astronomical research. He spent the first nine months of it traveling from city to city, trying to raise money for the printing of the *Rudolphine Tables*. He eventually got back to Linz only to learn that a new Counter-Reformation edict required all non-Catholics to convert or be banished. He succeeded in securing an exemption, but his library was put under seal, and soldiers were billeted where the printing press was housed.) But suppose that I am wrong and that Kepler 'realized in 1625' that his First Law was false. How does this bear on the thesis that all theories are *born* refuted? This one was born in 1605.

Followers of Lakatos may say that my labours over Kepler are in vain; for Lakatos had modified his original claim that *all* theories are born refuted to the claim that it is a 'historical fact that *most* [my italics] important theories are born refuted' (1971, p. 128), and his followers have accepted this modification. For instance, Worrall writes: 'it turns out that *nearly all* [my italics] theories during the whole history of science have had the lowest possible degree of Popperian corroboration (*minus* one)' (1978, p. 52). I am out of my depth with a claim of this kind. Even in its original form, the thesis is strictly irrefutable: the most a critic of it can do in a given case is to argue that the historical record gives us no reason to suppose that refuting evidence for a particular theory was in people's hands when the theory was first proposed. But in its modified form it becomes impervious even to this sort of counterargument. Suppose that I had carried out similar exercises, with similar results, for Archimedes's theory of the lever, Harvey's theory of the circulation of the blood, Galileo's laws, and Boyle's law. It could rightly be retorted that all this amounts to only a tiny fragment of all the theories put forward 'during the whole history of science' and is entirely compatible with the thesis that *nearly* all of them were born refuted. Lakatos and Worrall write as if they had surveyed the whole history of science, made a body count of theories refuted at birth, and can authoritatively report as a 'historical fact' that 'it turns out' that 'most' or 'nearly all' of them were born refuted. But how complete was their survey? Did they, for instance, look into Archimedes's and Harvey's theories, and if so, what were their findings? In what years did they find Galileo's and Boyle's laws to have been first refuted? I know of only one way in which a thesis of the form 'Most (or nearly all) *A*'s have as a matter of historical fact been *B*'s' can be rationally argued for. Proponents of the thesis should first list those *A*'s they have actually investigated. They should then say just which of these are claimed to be *B*'s and on what grounds. Once they have done that, a critical discussion can develop. For a start, a critic of the thesis can check their arithmetic; what is the actual percentage of the *A*'s on their list that they claim to be *B*'s? He can also ask whether there have not been serious omissions from their list and whether, in each particular case, their claim that an *A* is a *B* is well founded. In the absence of such an itemised investigation, a critic faces a merely impressionistic thesis with which he cannot get to grips.

This completes my attempt to counter arguments that tend to show that there never is *one* theory that is best corroborated in its field.

8.7   One-dimensional and Multidimensional Appraisals

We are now in a position to answer the question: Does our aim (B*) satisfy the adequacy requirements set out in §4.1? Its satisfaction of requirement 1, for coherence, was discussed in §5.38, and I will not repeat what was said there. That it satisfies requirement 2, for feasibility, has been the main preoccupation of the last three sections, in which we have combatted claims to the effect that there never is *one* best corroborated theory. That it satisfies requirement 3, namely that it can serve as a guide in choices between theories, was argued in §8.3 where it was shown that, given a certain assumption as to the kinds of tests that have been carried out, to choose the best corroborated theory is to choose the theory that best satisfies (B*). And it obviously satisfies requirement 4, for impartiality, and requirement 5, that it involves the idea of truth.

I will conclude this chapter by looking at an objection raised by Yehoshua Bar-Hillel against both Popper's and Carnap's methodologies. He wrote: 'Both of them seem to have assumed . . . that the comparison of theories is an essentially simple and one-dimensional affair' (1974, p. 334). According to Bar-Hillel, theories should be compared along several dimensions. He also complained that 'not enough attention has been paid to what seems to me definitely to be a decisive prior question, viz., *comparison for what purpose*? Neither Popper nor Carnap has ever, to my knowledge, explicitly posed this question' (ibid.). As a comment on Popper, this is unfair; as we saw, corroboration comparisons are made, according to him, in the hope of finding the theory that 'is a better approximation to the truth', as Popper duly pointed out in his reply (1974b, p. 1045). But let us consider Bar-Hillel's objection as if it had been directed against the present book.

For the sake of the argument I will go along with his claim that the appraisal of theories in terms of corroboration 'is an essentially simple and one-dimensional affair'. What is wrong with that? Bar-Hillel would presumably have answered that to assess a theory along just one dimension is necessarily to fail to take into account other important features of theories that would have to be assessed along other dimensions. A proponent of a "one-dimensional" method might reply that if we try to use a "multidimensional" method there is also the danger that theories become incomparable, one theory being better than another along one axis but worse along another. True, we might try to arrive at an overall evaluation of such theories by means of some system of trade-offs whereby a certain amount of inferiority on one axis is offset by a certain amount of superiority on another; but this would require *cardinal measures* for each of the axes as well as a weighting for each axis. And it is doubtful whether a nonarbitrary system of that kind could be constructed.

In this book we avoid both dangers. First, our (B*) is not only a *multiple* aim but also, I have argued, the most comprehensive aim for science that we can realistically adopt; this means that there is no danger that the theory that, in its field, best satisfies (B*) may nevertheless be inferior to a rival theory in some overlooked respect that ought to have been taken into account. Second, our method, which is to prefer the best corroborated hypothesis in the field, can be expected, at least in a good many cases, to pick one out quite unequivocally; and on the assumption that the hypotheses in question have been put to hard tests of appropriate kinds, this will be the one that best satisfies the optimum aim for science.

If I am right in this, we have, it seems to me, a very convincing answer to rationality-scepticism. Of course, it will not satisfy anyone who clings to something like component (A) of the Bacon-Descartes ideal. But as we saw in §4.4, clinging to *that* aim leads only to endless disappointment and frustration.

9

....

Epilogue

One daunting problem remains. In Part Two of this book Humean scepticism has been outflanked so far as theoreticians' choices between competing theories are concerned: a theoretician has a very good reason to accept a theory, inductive support for which is zero, if that theory is the one in its field that best fulfils the optimum aim for science. But being the one that best fulfils this aim by no means implies that a practical decision maker has any reason to *act* on it rather than to be guided by some alternative and conflicting hypothesis. What an agent would wish the hypotheses that guide his actions to have is not such properties as depth and unity, which the theoretician values, but *reliability*. He would like hypotheses that are, relative to his aim and problem situation, definite enough to give him clear guidance, and will not let him down. A hypothesis supplied by science would not let him down if it were *true*. But we can claim for the best corroborated theory in its field no more than that it is *possibly* true (in the sense explained in §4.5), and not that it is probably true or has more verisimilitude than rival hypotheses (see §8.1), many of which may likewise be possibly true. Then what reason, if any, is there for an agent to act on best (or well) corroborated theories and hypotheses? This is the pragmatic problem of induction and it is untouched by what has so far been said in this book.

As a way of entry into it I am going to hark back to a debate between Salmon on the one hand and Popper and myself on the other. My motive here is not at all defensive. On the contrary, I now think that Salmon was essentially right while Popper and I were essentially wrong. It is rather that I think that this debate has clarified the issues in a way that makes them amenable to a positive solution.

In (*1966*) Salmon wrote:

Popper furnishes a method for selecting hypotheses whose content exceeds that of the relevant available basic statements. Demonstrative inference cannot accomplish this task alone . . . ; instead, corroboration is introduced. Corroboration is a nondemonstrative kind of inference. It is a way of providing for the acceptance of hypotheses even though the content of these hypotheses goes beyond the content of the basic statements. *Modus tollens* without corroboration is empty; *modus tollens* with corroboration is induction. (P. 26)

I tried to defend Popper against this charge of a covert inductivism in (1968a). Although I am going to withdraw much of what I said there, I remain steadfast on one point: corroboration appraisals are noninductive in that they imply nothing about the future performance of the hypotheses in question. I expressed this then by saying that if a Popper-type corroboration appraisal of the form $Co(T_j, E) > Co(T_i, E)$ is true, then it is analytic, in much the same way that a Carnap-type confirmation appraisal is analytic. And the same verdict is delivered by the somewhat revised account of corroboration offered in §8.2 above. If T_i and T_j are two axiom sets that comply with the axiomatisation rules in §5.33 then, given an agreed partition of predicates into "observational" and "theoretical", the corroborable contents of T_i and T_j are fixed. And $Co(T_j, E) > Co(T_i, E)$ means that no test result reported in E is more favourable to T_i than to T_j while at least one is more favourable to T_j. And whether that is so is logically determined by the relation of E to the corroborable contents of T_i and T_j.

But if E is a report of past tests, and says nothing about the future, and if $Co(T_j, E) > Co(T_i, E)$ is analytic, then neither does the conjunction of this corroboration appraisal with that evidence-statement say anything about the future. To put it another way, being the best corroborated theory in its field at time t is what I earlier called (§6.22) a future-independent property; if T_j enjoys this property at t, then not even a disastrously bad subsequent performance can deprive it of the title of having been the best corroborated one at t. Conversely, its being the best corroborated one at t implies nothing as to its performance after t.

But if corroboration appraisals tell us nothing about future performance, how can an agent select hypotheses to guide his practical decisions when he tries to pursue his aim in some practical problem situation? I put forward in (1968a) two suggestions in answer to this.

1. My first suggestion was that the method of hypothesis selection in pragmatic contexts should be different from that in theoretical contexts: whereas the latter should be governed by purely cognitive considerations (such as problem-solving and explanatory power), the former

should be governed by utility considerations. I no longer agree with this but perhaps I may try to explain my former thinking here:– I was impressed (I now think overimpressed) by the fact that practical men are sometimes quite willing to base their calculations on hypotheses they consider *false*, a point which Popper has also emphasised (*1963*, p. 56; *1982a*, p. 65). For example, a ship's navigator on a foggy night may set course on the supposition that the ship is closer to a rocky coast than he believes her to be. Or consider a more abstract setup. A decision maker has a choice between, say, just two decisions, d_1 and d_2, and he is contemplating just two rival hypotheses, h_1 and h_2. Hypothesis h_1 indicates decision d_1, in the sense that if he *knew* that h_1 is true then d_1 would be his best decision in this problem situation, given his aim; and h_2 likewise indicates d_2. And he believes that h_1 is true and h_2 false. However, he notices the following assymmetry: if he chooses d_1 and h_2 happens to be true, then the result for him will be very bad; whereas if he chooses d_2 and h_1 is true, as he believes, the result will still be quite good (though less good than it would have been if he had chosen d_1). So to be on the safe side he chooses d_2. In other words he makes the decision that is indicated by a hypothesis that he believes to be *false*. But surely a theoretician, guided by purely cognitive considerations, would never prefer a hypothesis that he regards as false over a rival hypothesis that he conjectures to be true. Hence the method of hypothesis selection in pragmatic contexts must be different from that in theoretical contexts.

2. My second suggestion was that if utility considerations provide no guidance, for instance because instead of being skewed, as in the above example, the payoffs are symmetrically distributed, then the decision maker may as well fall back on the best corroborated hypothesis since he has nothing else to go on. Popper has suggested that there is nothing *better* to go on than the results of science (*1972*, p. 27).

Lakatos persuaded me to give up 1 and Salmon to give up 2.

Lakatos pointed out that whenever a theory or hypothesis that is believed to be false is nevertheless being acted upon, this is because there is in the background a superior theory or hypothesis that is not believed to be false and that endorses and controls reliance on the inferior one (1968, p. 190 n. 6). For instance, if space scientists rely on NM in calculating the trajectories of spacecraft, that is because GTR tells them that the errors will be negligible. Or consider my hypothetical navigator. No doubt he began by making his *best estimate* of the ship's position. He then tried to make a best estimate (perhaps I should say best meta-estimate) of the amount by which the former estimate might be out, given the lapse of time since his last fix. These attempts at a thoroughly realistic assessment of the situation yielded his best estimate of how close to the rocky coast the ship might in fact be; and given the absence of navigational hazards to seaward, he set course on the, probably false, supposition that she *was* as close as she might in fact be.

Again, in my abstract example, the decision maker was being as realistic as he could. It badly misdescribes him to say simply that he acted on a hypothesis that he believed false. He acted on his best estimate of the whole situation.

I now see that I was bringing in utility considerations at the wrong place. What someone in a potentially dangerous situation wants, to begin with, is the most realistic appraisal of it that he can get. *After* he has got that, he may then seek to insure against the danger, for instance by introducing safety margins; and he may want these to be large. Thus a bridge builder may try to design a bridge that could carry twenty times its maximum expected load. If he is to succeed in this he will need *good* engineering hypotheses. Utility considerations demand the safety margin but they cannot tell him whether to use this engineering hypothesis or that one.

As to 2, Salmon pointed out that there are *lots* of things, other than best corroborated hypotheses, to go on: astrology, numerology, the flip of a coin . . . (1981, p. 120). Addressing himself to Popper's claim that there is nothing *better* to go on than results obtained with the method of science, Salmon retorted:

> The question is not whether other methods—*e.g.*, astrology or numerology—provide more rational approaches to prediction than does the scientific method. The question is whether the scientific approach provides a more rational basis for prediction, for purposes of practical action, than do these other methods. The position of the Humean Skeptic would be, I should think, that none of these methods can be shown either more or less rational than any of the others. (1981, p. 121)

If we accept Humean scepticism, and if this puts all these methods on a par, then our conclusion must be that there is nothing better, *and also nothing worse*, to go on than the results of theoretical science. Salmon concluded his indictment of Popper and myself as follows:

> We begin by asking how science can possibly do without induction. We are told that the aim of science is to arrive at the best explanatory

theories we can find. When we ask how to tell whether one theory is better than another, we are told that it depends on their comparative ability to stand up to severe testing.... When we ask whether this mode of evaluation does not contain some inductive aspect, we are assured that ... since this evaluation is made entirely in terms of past performance, it escapes inductive contamination because it lacks predictive import. When we then ask how to select theories for purposes of rational prediction, we are told that we should prefer the theory which is 'best tested' ..., even though we have been explicitly assured that testing ... [has] no predictive import. (P. 122)

Game, set and match to Salmon!

Have Popperians produced any arguments that avoid such strictures? Well, Popper put forward the following one, with acknowledgements to Miller:

Let us forget momentarily about what theories we 'use' or 'choose' or 'base our practical actions on', and consider only the resulting *proposal* or *decision*. ... Such a proposal can, we hope, be rationally criticized; and if we are rational agents we will want it to survive, if possible, the most testing criticism we can muster. *But such criticism will freely make use of the best tested scientific theories in our possession.* Consequently any proposal that ignores these theories ... will collapse under criticism. Should any proposal remain, it will be rational to adopt it. (1974b, p. 1025)

But there is a concealed circularity in this argument. Suppose that h_1, h_2, and h_3 are rival hypotheses relevant to a practical decision and that, given the agent's preferences, they respectively indicate decisions d_1, d_2, and d_3. Then each of these possible decisions is under criticism from two of these three rival hypotheses. Suppose that someone now says that we should set aside h_2 and h_3 and retain h_1, adding that since d_2 and d_3 collapse under the criticism provided by h_1, it is rational to adopt the survivor d_1. We ask him why he singles out h_1 in this way. He answers that h_1, unlike h_2 and h_3, is one of 'the best tested theories in our possession'. But this answer reveals that he was relying on the principle that he was supposed to be arguing for, namely that the best corroborated hypothesis, or the one that has so far stood up best under test, provides the best guide for this future act.

Miller himself has claimed (1982) 'that the problem of induction is at last well and truly solved' by Popper's falsificationism (p. 18). His argument runs as follows:– The aim of science is simply to try to separate empirical hypotheses into those that are true and those that are false. In attempting this it relies solely on expulsion procedures; any testable but as yet unfalsified hypothesis is admissible into the body of scientific knowledge; but once admitted it is subject to rigorous attempts to expel it by subjecting it to tests which, it is hoped, will reveal it to be false if it is in fact false. 'Hypotheses that pass all the tests to which they are subjected are retained, and may be classified conjecturally as true. Hypotheses that fail tests are classified as false' (p. 25). And it is rational to act on hypotheses that have, so far, been retained rather than on ones that have been expelled, just because we have no reason to suppose that the former are false while we do have reason to suppose that the latter are false.

This view, it seems to me, is open to a simple and decisive objection: namely, that the set of as yet unfalsified but falsifiable hypotheses will contain many that mutually *conflict*. This is a point that Salmon had underlined: 'When one particular hypothesis has been falsified many alternative hypotheses remain unfalsified. Moreover, there is nothing unique about a hypothesis that survives without being falsified. Many other hypotheses remain to explain the same facts. Popper has readily admitted all this' (1968, p. 26). Thus where Miller writes: 'in the lucky event that we have an unrefuted hypothesis relevant to our practical problem we shall normally plan our actions on the assumption that this hypothesis is true' (p. 24), his sentence ought to have begun: 'in the all too likely event that we have a plethora of unrefuted and mutually conflicting hypotheses relevant to our practical problem . . . ' But how should this sentence now continue?

Later in his paper (pp. 38–40) Miller did consider the difficulty that 'according to Popper's theory many conflicting hypotheses may be equally well confirmed by exactly the same evidence' (p. 29). But he confined himself to the case where there is just one respectable unrefuted hypothesis and all its rivals are "grue-ish" variants of it; and he briskly declared that, given the former hypothesis and a "grue-ish" variant, 'one at least of the two hypotheses is redundant and may be excluded from consideration. It is perfectly clear which one that is' (p. 40). But a "grue-ish" variant diverges from the original hypothesis only in what it says about the future; and it is far from perfectly clear what *argument* a noninductivist can give for excluding it in favour of the original.

Foolhardy though it be, I am going to have another stab at this pragmatic problem of induction. Imagine laid out before us an array of alternative pragmatic principles. One of these, which I will call *PP*, says:

Actions that are ultimately guided by hypotheses that are *well corroborated* have the best chance of being successful.

(I inserted 'ultimately' to take care of cases where a decision maker bases his calculations on a hypothesis he believes to be false just because he accepts a superior hypothesis that endorses this.) Alongside *PP* are various other principles *PP'*, *PP''* ... each of which puts something else in place of *well corroborated*. Just what they put we need not go into. But I will assume that all these principles are *equally effective*: whenever *PP* singles out a hypothesis *h* to guide an action, *PP'* singles out a hypothesis *h'* to guide it, where *h'* gives as much guidance as *h*, and vice versa; and likewise for *PP''* ... I will also assume that a hypothesis endorsed by *PP'* will typically, though perhaps not invariably, differ significantly from the corresponding hypothesis endorsed by *PP*.

Let a decision maker face a choice between decisions *d, d', d''* ... Let decision *d* be indicated by hypothesis *h*; that is, if he *knew* that *h* is true then, given his aim, *d* would undoubtedly be his best decision. And let *d', d''* ... be indicated by, respectively, hypotheses *h', h''* ... And let *h, h', h''* ... be endorsed respectively by *PP, PP', PP''* ... Thus a rational ground for discriminating among the latter would enable him to make a rational decision. For ease of presentation we may simplify his problem to one where he faces a binary choice between just *d* and *d'*. A binary choice, say between pressing and not pressing a button, is one where not to decide to do one thing is equivalent to deciding to do the other thing; there is no escape into indecision. (If a motorist, who sees a hitchhiker thumbing a lift, dithers long enough, he in effect decides not to stop.) According to hypothesis *h*, which is endorsed by *PP*, the result will be very good for the agent if he chooses *d* and very bad if he chooses *d'*; and conversely, according to hypothesis *h'* endorsed by *PP'*, the result of *d* will be very bad and of *d'* very good. To make the discussion a little less abstract, we may suppose that *PP'* puts *hypotheses chosen by lot* in place of *well-corroborated hypotheses*. I am not sure that this *PP'* and our *PP* would be equally effective in the sense indicated above, but let us suppose that they are. Henceforth I will treat *PP'* as representative of all the other alternatives to *PP*.

I am going to make the historical assumption that if what they say were restricted to *past* actions, then *PP* would be true and *PP'* false. Hume would not, I think, have objected to this. After all, the pragmatic problem of induction is not about whether reliance on well-tested scientific results *has* proved relatively successful but about whether there is any *rational argument* for relying on them in the future. I will abbreviate 'actions guided by well corroborated hypotheses' and 'actions guided by hypotheses chosen by lot' to, respectively, 'corroboration-guided actions' and 'lottery-guided actions'. Let t_{-1} denote a time before the emergence of *homo sapiens*, let t_0 be a floating constant corresponding to *now*, and let t_1 denote a time by which *homo sapiens* will have become extinct. (That the human race will not endure forever is not an assumption that is essential for the coming argument, but it simplifies the exposition.) Then our historical assumption, which I will denote by E, could be expressed thus: if all actions between t_{-1} and t_0 had been corroboration-guided the success rate would have been higher than if they had all been lottery-guided. Letting S stand for 'successful', C for 'corroboration-guided action', and L for 'lottery-guided action', we could encapsulate this in the following frequency-statement:

E: During t_{-1} to t_0, $\mathrm{Fr}(S,C) > \mathrm{Fr}(S,L)$.

Since *PP* says that corroboration-guided actions have the *best* chance of succeeding, it obviously implies that they have a better chance than do lottery-guided actions of succeeding. So in the present discussion, where *PP* is being juxtaposed just with *PP'*, we could formulate it thus:

PP: $\forall t(t_{-1} < t < t_1)\ [P(S,C) > P(S,L)]$.

This says that a corroboration-guided action has at all times a better chance of success than a lottery-guided action.

How should we formulate *PP'*? It is incumbent on *PP'* to be as effective as *PP*; and this means that *PP'* must endorse the hypotheses it favours no less (and no more) forcefully than *PP* endorses those it favours. So far as I can see there are two main ways in which *PP'* could be construed so as to achieve this: as a straightforward *contrary* or as a "grue-ish" *variant* of *PP*. On the first construal, which I will denote by PP'_1, it would read:

PP'_1: $\forall t(t_{-1} < t < t_1)\ [P(S,L) > P(S,C)]$;

and on the second construal, PP'_2, it would read:

PP'_2: $\forall t(t_0 \leq t < t_1)\ [P(S,L) > P(S,C)]$.

The first says that *at all times*, while the second says that *from now on*, a lottery-guided action has a better chance of success than a corroboration-guided one. Given that our decision maker has *now* to decide between d and d', d being indicated by the well-corroborated hypothesis h and d' by the hypothesis h' that has been chosen by lot, PP'_1 and PP'_2 both endorse h' (and hence d') as forcefully as PP endorses h (and hence d). And the decision maker can make a rational choice between d and d' if he can find some grounds for rational discrimination between PP and PP' in both of its two versions. Can a Popperian philosopher who accepts Hume's point offer any ground for discriminating between them? Let us first consider PP in juxtaposition with PP'_1.

We obviously must not say that E provides inductive support for PP but not for PP'_1. However, this does not mean that we cannot make any use at all of E. If we consider PP and PP'_1 in the absence of E they are, if not perfect examples of incongruent counterparts, at any rate structurally similar and on a par contentwise. (They were, after all, required to be equally effective.) However, when we consider them in the presence of E a structural difference opens up between what they respectively imply for the future.

Understood in the light of the frequency interpretation of probability, PP is saying, concerning the "collective" consisting of all human actions, that were they all corroboration-guided then the frequency of success would tend to a limit higher than the limit to which it would tend were they all lottery-guided. And PP'_1 says exactly the opposite. And E says that, so far as past actions are concerned, the frequency of success would have been higher if they had all been corroboration-guided than it would have been if they had all been lottery-guided. Thus for PP to be true, given E, it is enough that corroboration-guided actions should continue to have the higher success rate in the future. But for PP'_1 to be true, given E, it is necessary that a time will come when the success rate of lottery-guided actions not only begins to exceed that of corroboration-guided ones but *exceeds it sufficiently* for the success rate of *all* actions, past as well as future, to tend to a limit that is higher than it would be if they had all been corroboration-guided. It is rather as if there were two competing hypotheses about a coin, one saying that it is biased towards heads and the other towards tails. The coin has already been tossed a good many times and has landed heads more frequently than tails, and it is going to be tossed many more times. The first hypothesis merely implies that the preponderance of heads will continue during these future tosses, but the second hypothesis implies that there will be a preponderance of tails during these future tosses sufficient to offset the previous preponderance of heads. I conclude that PP'_1, in the presence of E, makes *a stronger claim about the future* than PP makes.

I turn now to the second construal of PP'. Unlike PP'_1, PP'_2 in the presence of E *identifies* a time, namely *now*, when a significant change in frequencies sets in. Whereas PP'_1 in the presence of E implies that *a time will come* when they change, PP'_2 says that *the time has now come*. On the other hand, PP'_2 does *not* claim a higher success rate from now on for lottery-guided actions than PP claims for corroboration-guided actions. It is rather as if the second of the above two hypotheses about the coin, instead of saying that the coin is (permanently) biased towards tails, said that it has *now* become so biased. (If PP'_2 merely said that a time will come . . . , it would be ineffective: it would not tell our decision maker whether the time is now ripe to opt for d'.)

If we compare a hypothesis that predicts no change with one that (i) predicts that a once-and-for-all change will set in and (ii) names the individual (date, person, event, or whatever) that will inaugurate this new era, it seems intuitively obvious that the second makes the stronger claim. Take it to be a fact that few, if any, of my ancestors lived to be a hundred. I now consult two seers about my descendants. One says that few, if any, of them will live to be a hundred. The other says (i) that I will have a descendant who will live to be a hundred and that few, if any, of *his* descendants will live to *less* than a hundred; moreover (ii) this era of longevity will be inaugurated by my grandson Cazimir. Surely the second seer makes the stronger prediction.

Imagine people who believe in a Second Coming to be assembled together. A prophet rises up and announces: 'He will come on February 19, 1996.' One can imagine his cobelievers being rather incredulous; is it not wildly improbable that He will choose that precise date? And would this prophet's claim be any less improbable if, on February 19, 1996, he announces: 'He will come *today*'? Or suppose there is a philosophical sect whose members concede that corroboration-guided actions have tended to be more successful than lottery-guided ones but who insist that an era will commence when lottery-guided ones are the more successful; and then one of them announces that this era will commence on February 19, 1996. Would not his cobelievers find this precise date wildly improbable? And should they not find his claim equally improbable if, on February 19, 1996, he announces that the new era is commencing *today*? (I have heard Ernest Gellner thinking along these lines.)

We might say that of the three hypotheses, PP, PP'_1 and PP'_2, it is the first that constitutes the null hypothesis: in the presence of E it boringly predicts no change. The second predicts, in the presence of E,

(i) that a time will come when the success rate of lottery-guided actions begins to exceed that of corroboration-guided ones, and moreover (ii) that it will exceed it sufficiently to compensate for its previous poor showing. The third predicts, in the presence of E, (i) that a time is coming when the success rate of lottery-guided actions begins to exceed that of corroboration-guided ones, and moreover (ii) that that time has come *now*. It seems clear that PP' in either version is, in the presence of E, making a stronger claim about the period t_0 to t_1 than is PP.

Now a Popperian philosopher tends to be favourably disposed, in theoretical contexts, towards the stronger of two (or more) competing hypotheses, other things being equal; though he may look askance at it if it calls on us 'to accept an unexplained and inexplicable change at a certain date' (Popper, *1982a*, p. 68). But in pragmatic contexts his preference tends to be reversed; if a larger assumption would need to be satisfied for decision d' to succeed than for decision d to succeed, then he will tend, other things being equal, to prefer d as the safer decision. Of course, other things may very well be unequal. Perhaps d' offers the prospect of a glittering prize; or perhaps the assumption on which d depends, though relatively weak in some formal sense, has been falsified.

But in the situation we are envisaging, other things are not unequal in any of these ways. Our decision maker faces a binary choice between d and d'. The payoff from d will be very good if h is true and very bad if h' is true, and conversely for d'; and h and h' are respectively endorsed by PP and PP'; and he accepts evidence E. Now if, as I have claimed, PP' (in either version) in the presence of E implies a stronger claim about the period t_0 to t_1 than does PP, then this gives him a ground for rational discrimination between them: he will commit a lesser hostage to fortune by basing himself on PP. Whether this consideration tilts the balance strongly or not, it does tilt it. He can rationally choose d.

I suspect that the pragmatic problem of induction has seemed insoluble for the following reason. Hume would concede, I am assuming, that we have evidence E to the effect that reliance on well-tested results of science has paid off pretty well in the past. But surely we cannot *use* this evidence? One recalls Mill's ingenuous (and Keynes's somewhat less ingenuous) attempt to use such evidence as inductive support for some inductive principle. But if we cannot use E, or if in other words there can be no a posteriori solution of the problem, then the only remaining possibility is an a priori solution. But the choice between the null or 'no change' hypothesis PP and any alternative to it is a choice between factual, or synthetic, principles with divergent implications for the future. And if we accept proposition (I), the anti-apriorist thesis, we exclude the possibility of any synthetic a priori knowledge about the future course of nature.

The trick of the solution proposed above is that it does make use, but *not inductive* use, of *E* and that it does rely on an a priori consideration but without assuming any synthetic a priori knowledge. The evidence *E* is used only to argue that, in the presence of *E*, *PP'* makes the stronger claim about the future; and the a priori consideration is only that, other things being equal, a practical decision is safer the weaker are the assumptions on whose correctness its success depends. In other words, I use *E* to generate a formal difference between *PP* and *PP'* that tilts the balance in favour of the former.

. . . .

In the remaining pages I will first dispel (I hope) the objection that it is just not possible that this book achieves what it claims to achieve, and then assess the intellectual costs and gains of its answers to the two questions posed by Einstein that I used as a motto.

In the section of his (1974), entitled *A plea to Popper for a whiff of 'inductivism'*, Lakatos bracketed Popper and myself as 'philosophers who want the impossible: to fight pseudo-science from a sceptical position' (p. 160). In this book I have not been directly concerned to fight pseudo-science. But he would no doubt have added that it is likewise impossible to uphold genuine science from a sceptical position, which is what I have tried to do. And I think that many other philosophers would have agreed that it is impossible to construct a positive account of scientific knowledge that contains no whiff of any doctrine vulnerable to Humean scepticism; for the only raw materials the latter allows are perceptual experiences and the only inferences it allows are deductions; as well try to build a soaring cathedral from twigs tied by nylon thread as to construct science out of these. 'If science consists solely of observation statements and deductive inferences', Salmon has insisted, then science 'is barren' (*1966*, p. 24). If the view of science presented in this book avoids that barren outcome, this can only be, a critic may say, because certain metaphysical assumptions, vulnerable to Humean scepticism, have been smuggled in. Moreover, he might continue, it is rather obvious where they have been smuggled in and what they are:– That so-called optimum aim for science presupposes a strong kind of realism. *Ought* implies *Can.* If that aim is to be feasible, an external world must exist; moreover, it must be a multilayered physical reality that is amenable to ever deeper scientific exploration. But, this critic might add, the only metaphysical position compatible with Humean scepticism is phenomenalism: the former says that we cannot have *knowledge* of anything beyond perceptual experience and the latter that there is nothing beyond perceptual experience to have knowledge of.

I will take this last point first. I say that scepticism no more endorses phenomenalism than agnosticism endorses atheism; it is *neutral* between phenomenalism and realism. I have the authority of Berkeley for this. Having got Hylas to concede that someone 'who entertains no doubt concerning some particular point, with regard to that point cannot be thought a *sceptic*', Berkeley's spokesman Philonous enquires: 'How cometh it to pass then, Hylas, that you pronounce me a *sceptic*, because I deny what you affirm, to wit, the existence of matter?' (*1713*, p. 173). The scepticism that has been argued for in this book supplies no positive argument *against* the existence of an external world: it allows it to be *possibly true* that an external world, consisting partly of unobservable realities, exists.

I turn now to the first and more telling part of this imaginary critic's accusation. Let us say that a statement is a *presupposition* of an aim if it correctly identifies a necessary condition for the realisation of the aim, and asserts, truly or falsely, that this condition is (or will be) satisfied. Then our critic is saying that a strong kind of realism is a presupposition of (B*), and that to adopt this aim is, by implication, to *affirm* this presupposition.

It would be irrational for a person X to retain an aim while denying what he takes to be one of its presuppositions; for his denial would imply that his aim is unrealisable. But is it irrational for him to pursue an aim if he cannot vouch for the truth of all its presuppositions? If so, the sceptical results of Part One would mean that it is always irrational to adopt any aim. But even if we combine an affirmative answer to this question with a common-sense view of how much an agent can know, it will still proscribe as irrational the adoption of many aims, especially aims that are rather ambitious, that seem perfectly legitimate. A person who has adopted an ambitious aim may be in no position to spell out all its presuppositions, let alone to assert them all;

and he may be in no position to assert many of those he can spell out. Suppose his aim is to be the next president of the United States; then one presupposition of his aim is that he will stay alive at least until Inauguration Day. But he is in no position to *assert* this presupposition: he may get assassinated. Or consider Columbus setting out in 1492 to discover new lands. This aim had thousands of presuppositions (for instance, that his ship would not be capsized by a whale) that he was in no position to assert.

So my answer to our critic is that a rational man who espouses (B*) as the aim for science is not thereby obliged to *assert* the kind of realism it presupposes. It is enough that he has no positive reason to deny it.

Must a phenomenalist reject (B*)? Let us approach this by considering what implications phenomenalism has for some other human activity that seems to involve continuous interaction with an external physical reality, mountaineering say. A rather ingenuous critic might have addressed Berkeley or Mach along these lines:- A mountaineer is trying to scale a high and forbidding mountain. His aim is not to *dream* that he has reached the summit but physically to conquer this large and solid chunk of physical reality. According to you, his aim has a false presupposition: he believes himself to be negotiating physical rocks, ice, crevasses, etc., but he is really only dreaming all this. There is no physical reality. Then does not your position imply that if he had a true understanding of the matter and if he were a rational man, he would give up his mountaineering ambitions, since he would realise that they rest upon a false presupposition?

I am confident that Berkeley and Mach would have completely repudiated this alleged implication of their position. They would have insisted that their ontology leaves the mountaineer's world, with all its hazards and exertions, completely intact. Phenomenalism, they might say, has no more tendency to undermine the rationality of mountaineering than of cooking. We might say that while phenomenalism turns the physical world into a kind of dream, it is a dream packed with physical detail. Phenomenalism says that it is *exactly as if* we lived in a physical world, though we do not in fact do so.

But if phenomenalism leaves intact the worlds of the mountaineer, the deep-sea diver, the explorer, the astronaut, not to mention the steeplechaser and the coal miner, why should it call for a drastic contraction of the physicist's world? The world of an experimental physicist contains, let us say, a Wilson cloud chamber which enables him, or so he believes, to observe the paths of subatomic particles. It is widely supposed that phenomenalism eliminates the particles but leaves intact the cloud chamber and its vapour trails. But it seems to me that phenomenalism is not entitled to engage in selective elimination. In the case of the mountaineer, phenomenalism does not leave intact his boots and pickaxe but eliminate the mountaintop while it is behind cloud. Phenomenalism dematerialises everything but leaves everything *in situ*. It turns our physicist's world into a kind of dream world in which it is *exactly as if* there is a Wilson cloud chamber which enables him to observe the paths of subatomic particles. When Hylas, the spokesman for common sense and scientific realism, finally capitulates to Philonous's skilful advocacy of immaterialism, he comforts himself with this thought:

> I am by nature lazy; and this would be a mighty abridgement in knowledge. What doubts, what hypotheses, what labyrinths of amusement, what fields of disputation, what an ocean of false learning, may be avoided by that single notion of immaterialism. (*1713*, p. 259)

But Hylas was wrong. Theoretical physics is not rendered otiose, or easier, by immaterialism. If the physicists working on the Manhattan Project had been immaterialists to a man, their work would have been not a whit less arduous. Every theory and equation that was used would still have had to be used. True, propositions about nuclear fission etc. would have been given an *as if* interpretation; but so would statements about, say, the geographical location of Hiroshima or the effects of the explosion there. It seems to me that phenomenalism should no more have adverse implications for the activities of realist-minded physicists than for the activities of realist-minded mountaineers. But I will not press this point. My main answer remains that, for (B*) to be a tenable aim, physical realism must be possibly true; which it is.

A "science" consisting solely of level-0 reports and deductions therefrom would comply with propositions (I) to (III) in §1.1 above, but it would, as Salmon said, be barren. Something else is needed. Some philosophers have held that this something else should be a synthetic principle, others that it should be some kind of nondeductive relation of confirmation. So far as I know it was Popper who first suggested that it should be neither of these things but an *aim* for science. An aim has no truth value and it cannot help to justify a hypothesis. In Popper's system, the hypotheses of science are neither sanctioned from above nor supported from below. Despite the intensive experimental process to which they are subjected, they remain free-floating and unjustified.

However, an aim can justify rational preferences among unjustified theories. As Popper put it: 'We can never justify a theory. But we can sometimes "justify" (in a different sense) our preference for a theory' (1974a, p. 119).

But as I indicated at the beginning of chapter 4, the idea of defeating rationality-scepticism by introducing an aim for science is attended by the dangerous possibility that there may be rival, perhaps mutually incompatible, aims so that a preference among theories that is rational according to one aim is irrational according to another. However, our freedom of choice here is restricted in various important ways. Unlike Bacon and Descartes, who considered themselves to be starting virtually from scratch, we have before us the massive development of science since their day, and any aim that we propose must be in line with that. It would be fatuous to propose an aim that would have required science to develop in some quite different direction. Husserl seemed to be doing that (see § 4.45 above); but I think that he should be understood, not as calling upon existing science to make a radical change of course, but as calling for a quite different kind of science.

Popper's proposal that we take increasing verisimilitude as the aim for science has impressed many people as having just the right kind of objective appropriateness to the *de facto* history of science. Does not that history teach us that we must not require the latest and best in a developing sequence of scientific theories to be true, but that we can and should require each theory in the sequence to be a better approximation to the truth than its predecessor?

But while one desideratum, or material adequacy requirement, for a proposed aim for science is that it should fit this great human enterprise, at once so daring and so disciplined, there are also philosophical desiderata. And for me a main one is that the proposed aim should enable us to defeat rationality-scepticism, by enabling us to make rational choices between competing theories, *without* introducing anything vulnerable to Humean criticism; without, in other words, introducing any inductive assumption. Now if corroboration appraisals, or summaries of how well theories have performed so far, are to guide rational preferences without infringing this requirement, they must not be endowed with any predictive implications. But if one theory is in fact closer to the truth than a rival, then its future performance should be better. This poses the following dilemma: either (i) corroboration appraisals are regarded as providing *some* justification, however slight, for verisimilitude appraisals, in which case some inductive assumption, however attenuated, is introduced and this answer to Hume is vulnerable to Humean criticism; or (ii) corroboration appraisals provide *no* justification for verisimilitude appraisals, in which case this aim leaves us in the dark as to why, or whether, we should prefer well corroborated theories, and we have no answer to rationality-scepticism.

I tried to deal with the problem posed by the possibility of alternative aims for science by introducing a number of adequacy requirements for such an aim and then seeking out the highest aim that still satisfies them. I did not really argue for these adequacy requirements. In every argued case there have to be stopping points somewhere and these requirements are the stopping points in my case for the rationality of science. Although I hardly expect them to be challenged, I must admit that, if they were, that might have serious repercussions.

This optimum aim (B*), as I was recently arguing, is not vulnerable to Humean scepticism despite its strongly realist presuppositions. Perhaps I may briefly recapitulate how, given that perceptual experiences are the only permitted raw materials in our theory of scientific knowledge, and deductions the only permitted inferences, the introduction of (B*) makes possible the rational reconstruction of the soaring edifice of physical science. The reconstruction takes place at two main levels. We found in §7.43 that a physical theory could never be sufficiently augmented so as to become testable against nothing but level-0 reports of perceptual experience. So, at the behest of (B*), level-1 statements are introduced into the empirical basis, these being regarded, not as justified by, but as helping to explain, and as testable against, perceptual experiences. A well-articulated scientific theory has a well-defined corroborable content; and the interplay between this and the empirical basis determines its corroboration ranking. (We saw that if one theory dominates another on one "observational"/"theoretical" cut, then it will normally continue to dominate it on another cut.) And although he who accepts Hume's point may not assert that the best corroborated theory in its field is the one that is closest to the truth, he is entitled to assert that it is the one that best fulfils the optimum aim for science.

Is the answer offered by this book to Einstein's two questions a cheering or a depressing one? Let us take his second question first: To what extent are the general results of science true? Our answer to this is that the theories of science, marvellous creations though they be, are at best possibly true. They are under the negative control of experience and this control is very severe, but it provides no positive support for them, gives them no inductive lift, leaves them floating in the ocean of uncertainty. I recognise that many people are likely to find this a gloomy message. A reader has said that the book involves a total capitulation to the depth-pole of the Bacon-Descartes ideal at the expense of its security pole. And while I can reply that we did retrieve *something* from the latter, namely our component (A*), I agree that for most people this will not be nearly enough. A book that seeks to answer Hume while *accepting* his central negative thesis is bound to contain a pessimistic element and I shall not try to belittle or soft-pedal this. But I will make some mitigating points. The first and most important one is that if we replace questions about the inductive probability, degree of confirmation, or whatever, of scientific hypotheses or common-sense beliefs by questions about the *rationality of accepting* them, then I believe that the theory of knowledge developed above delivers verdicts that are at least as cheerfully positive as those yielded by any inductivist theory. Probability-scepticism was taken on board but rationality-scepticism was repelled.

My next point is that the theory does not call for any exorcism of belief. There is a widely held view, encapsulated in our proposition (IV), to the effect that rational people should apportion their degree of belief in hypotheses to the latters' degree of confirmation; and probability-scepticism in conjunction with *that* proposition does indeed imply that rational people should have no positive beliefs about the external world. But we rejected that proposition in favour of proposition (IV*), to the effect that it is rational to accept a statement that is (or is part of) the best explanation one has found for some explanandum. And probability-scepticism in conjunction with this proposition no longer has the above implication. What matters, on the present view, is that the hypotheses one accepts should be rationally accepted. It says neither that a rationally accepted hypothesis should, nor that it should not, be accompanied by a feeling of belief.

My last point in mitigation is this. If Hume's negative thesis is *true* and probability-scepticism is correct, then any theory of knowledge that purports to avoid the pessimistic element in our answer to Einstein's question must somewhere be relying on something that cannot provide what it is supposed to provide. A more optimistic theory will involve what is, objectively speaking, an element of pretence; and it will be in danger of crumbling when this element is detected. (I am not for one moment saying that the author of such a theory will be consciously aware of any element of pretence in it; for example, I have no doubt that Mill genuinely believed that the fundamental principle of induction is inductively derivable from low-level inductions.) I am inclined to invert the famous quotation from Russell, given in §1.14, to read: 'Humean scepticism is logically impeccable and there is an element of well-intentioned insincerity in any philosophy that pretends to reject it.' Someone with a complaint of whose cause he is ignorant may prefer the doctors to be vaguely reassuring; or he may prefer to know the worst and then try to cope with it as best he can. My objective in this book has been to know the worst and then to show that we can cope with it far better than might have been feared; which brings me to our answer to Einstein's first question.

Actually, we have not answered quite the question he asked, namely 'What goal will and can be reached by the science to which I am dedicating myself?' For the aim of science has been taken to be, not to reach a given goal, but to journey along a road with no known end. Moreover, no prognostications have been offered as to whether science will in fact continue to succeed in fulfilling this aim; we can only say that it has fulfilled it with astonishing success hitherto. So let us recast his question as: 'What is the proper aim of the science to which I am dedicating myself?' And to this question our answer is, of course, to achieve theories that are ever deeper and more unified, and ever more predictively powerful and exact. That answer, especially its first half, is one that Einstein himself and, I believe, all great scientists would have endorsed. Yet during the heroic period, beginning around 1905, when science was penetrating to ever deeper levels, the predominant philosophical schools were, as we saw in chapter 4, sternly admonishing science to remain at the phenomenal surface. We diagnosed this hostility to depth as resulting from the fatal attraction exerted by the security-pole of the Bacon-Descartes ideal.

The pessimistic element in our neo-Popperian theory of knowledge is the recognition that this desire for epistemological security, though deep-seated, is a yearning for a will-o'-the-wisp, for something we cannot have. The pretence that we can have it, if only we will lower our aim sufficiently, had to be buttressed by a series of increasingly severe constraints on scientific theorising. When we throw off that pretence, we free science from all these constraints and allow it to raise its aim high. If it has a pessimistic element, ours is also a liberating theory.

Dates in parentheses in the text are normally those of the original publication of the work in question. Where page references are to a later version, this is given by the last entry in the bibliography against that date.

Agassi, Joseph
1959   'Corroboration versus Induction', *British Journal for the Philosophy of Science*, 9, pp. 311–317.
1966   'Sensationalism', *Mind*, 75, pp. 1–24; reprinted in Agassi, *1975*, pp. 92–126.
1975   *Science in Flux*, Dordrecht: Reidel.

Armstrong, David M.
1973   *Belief, Truth and Knowledge*, Cambridge: University Press.
1975   'Towards a Theory of Properties', *Philosophy*, 50, pp. 145–155.

Austin, John L.
1946   'Other Minds', *Aristotelian Society*, supplementary vol. xx; reprinted in Austin, *1961*, pp. 44–84.
1961   *Philosophical Papers*, Oxford: Clarendon Press.
1962   *Sense and Sensibilia*, reconstructed from manuscript notes by G.J. Warnock, Oxford: Clarendon Press.

Ayer, Alfred J.
1956   *The Problem of Knowledge*, Harmondsworth: Penguin.
1959   (ed.), *Logical Positivism*, London: George Allen & Unwin.
1971   'Conversation with A.J. Ayer', in Bryan Magee (ed.), *1971*, pp. 48–65.
1974   'Truth, Verification and Verisimilitude', in Schilpp (ed.), *1974*, pp. 684–691.

Bacon, Francis
1620   *Novum Organum*, in J. Spedding, R.L. Ellis, and D.D. Heath (eds.), *The Works of Francis Bacon*, London: Longman, 1857–59, vol. 4. References (e.g. II, v) are to book and numbered aphorism.

Bar-Hillel, Yehoshua

1965   (ed.), *Logic, Methodology and Philosophy of Science.*
(Proceedings of the 1964 International Congress for
Logic, Methodology, and Philosophy of Science), Am-
sterdam: North Holland.

1974   'Popper's Theory of Corroboration', in Schilpp (ed.),
1974, i, pp. 332–348.

Barker, Stephen F.

1961   'The Role of Simplicity in Explanation', in Feigl and
Maxwell (eds.), *1961*, pp. 265–274.

Bartley, III, William W.

1962   *The Retreat to Commitment*, New York: Alfred A.
Knopf.

1964   'Rationality versus the Theory of Rationality', in Bunge
(ed.), *1964*, pp. 3–31.

1968   'Theories of Demarcation between Science and Metaphysics', in Lakatos and Musgrave (eds.), *1968*, pp. 40–
64.

Baumrin, Bernard M.

1963   (ed.), *Philosophy of Science: The Delaware Seminar*, vol.
2, New York: John Wiley.

Bentham, Jeremy

1776   *A Fragment on Government*, F.C. Montagne (ed.), Oxford: University Press, 1891.

Berkeley, George

1710   *A Treatise concerning the Principles of Human Knowledge,* in Berkeley, *1948–57*, vol. 2.

1713   *Three Dialogues Between Hylas and Philonous*, in Berkeley, *1948–57*, vol. 2.

1948–57   *The Works of George Berkeley, Bishop of Cloyne*, A.A.
Luce and T.E. Jessop (eds.), 9 vols, London: Nelson.

Black, Max

1950   (ed.), *Philosophical Analysis*, Cornell: University Press.

1954   (ed.), *Problems of Analysis*, London: Routledge & Kegan
Paul.

1954   ' "Pragmatic" Justifications of Induction', in Black
(ed.), *1954*, pp. 157–208.

1958   'Self-Supporting Inductive Arguments', reprinted in
Black, *1962*, pp. 209–218.

1962   *Models and Metaphors*, Cornell: University Press.

1966   'The Raison d'Être of Inductive Argument', *British Journal for the Philosophy of Science*, 17, pp. 177–204.

Boltzmann, Ludwig
1905   Essay 16, in *Populäre Schriften*; tr. by Paul Foulkes, in Brian McGuinness (ed.), *Ludwig Boltzmann: Theoretical Physics and Philosophical Problems*, (Vienna Circle Collection, vol. 5), Dordrecht: Reidel, 1974.

Bolzano, Bernard
1837   *Wissenschaftlehre*; tr. by Rolf A. George, as *Bernard Bolzano, Theory of Science*, Oxford: Blackwell, 1972.

Boole, George
1854   'Solution of a Question in the Theory of Probabilities', *The Philosophical Magazine*, 7, reprinted in Boole, *1952*, pp. 270–273.
1952   *Studies in Logic and Probability*, R. Rhees (ed.), London: Watts.

Born, Max
1936   *The Restless Universe*, tr. by Winifred M. Deans; 2nd ed., New York, Dover, 1951.

Braithwaite, Richard B.
1953   *Scientific Explanation*, Cambridge: University Press.

Bridgman, Percy W.
1927   *The Logic of Modern Physics*, New York: Macmillan.

Bunge, Mario
1964   (ed.), *The Critical Approach to Science and Philosophy*, Glencoe: Free Press.
1967   *Scientific Research*, 2 vols., Berlin: Springer-Verlag.
1968   'The Maturation of Science', in Lakatos and Musgrave (eds.), *1968*, pp. 120–137.

Butler, Joseph
1736   *The Analogy of Religion, Natural and Revealed, to the Constitution and Course of Nature*; reprinted in Bohn's Libraries, London: George Bell, 1889.

Butts, Robert E., and Hintikka, Jaakko
1977a   (eds.), *Foundational Problems in the Special Sciences*, Dordrecht: Reidel.
1977b   (eds.), *Basic Problems in Methodology and Linguistics*, Dordrecht: Reidel.

Carnap, Rudolf

1928   *Der Logische Aufbau der Welt*; English translation in Carnap *1967*, pp. 1–300.

1928   'Scheinprobleme in der Philosophie: Das Fremdpsychische und der Realismusstreit'; English translation in Carnap, *1967*, pp. 301–343.

1942   *Introduction to Semantics*; reprinted in Carnap, *1961*.

1950   *Logical Foundations of Probability*, London: Routledge & Kegan Paul.

1952   *The Continuum of Inductive Methods*, Chicago: University Press.

1961   *Introduction to Semantics and Formalization of Logic*, Cambridge: Harvard University Press.

1963   'Replies and Systematic Expositions', in Schilpp (ed.), *1963*, pp. 859–1013.

1966   *Philosophical Foundations of Modern Physics*, Martin Gardner (ed.), New York: Basic Books.

1967   *The Logical Structure of the World and Pseudoproblems in Philosophy*, English translation of Carnap, *1928* and 1928 by Rolf A. George, London: Routledge & Kegan Paul.

1968a   'The Concept of Constituent-Structure', in Lakatos (ed.), *1968*, pp. 218–220.

1968b   'Inductive Logic and Inductive Intuition', in Lakatos (ed.), *1968*, pp. 258–267.

Caspar, Max

1948   *Johannes Kepler*, Stuttgart; tr. by C. Doris Hellman, as *Kepler*, London and New York: Abelard-Schuman, 1959.

Causey, Robert

1977   *Unity of Science*, Dordrecht: Reidel.

Chase, William G., and Simon, Herbert A.

1973   'Perception in Chess', *Cognitive Psychology*, 4, pp. 55–81.

Chisholm, Roderick

1966   *Theory of Knowledge*; 2nd ed., Englewood Cliffs, N.J.: Prentice Hall, 1977.

Clifford, William K.

1886   *Lectures and Essays*, Leslie Stephen and Frederick Pollock (eds.), London: Macmillan.

Cohen, L. Jonathan, and Hesse, Mary B.
1980   (eds.), *Applications of Inductive Logic*, Oxford: University Press.

Cohen, Robert S., Feyerabend, Paul K. and Wartofsky, Marx W.
1976   (eds.), *Essays in Memory of Imre Lakatos*, Dordrecht: Reidel.

Cohen, Robert S., and Seeger, Raymond J.
1970   (eds.), *Ernst Mach, Physicist and Philosopher*, Dordrecht: Reidel.

Colodny, Robert G.
1970   (ed.), *Pittsburgh Studies in the Philosophy of Science*, vol. 4, Pittsburgh: University Press.

Condorcet, Marie-Jean-Antoine-Nicholas
1795   *L'Esquisse d'un tableau historique des progres de l'esprit humain*; tr. by June Barraclough, as *Sketch for a Historical Picture of the Progress of the Human Mind*, London: Weidenfeld and Nicholson, 1955.

Crossley, John N. and Dummett, Michael A.E.
1965   (eds.), *Formal Systems and Recursive Functions*, Amsterdam: North-Holland.

Dalton, John
1808   *A New System of Chemical Philosophy*, part 1, Manchester: Bickerstaff.

De Groot, A.D.
1965   *Thought and Choice in Chess*, The Hague: Mouton

Descartes, René
1642   *Meditationes de prima philosophia*; tr. as *Meditations on First Philosophy*, in H.S. Haldane and G.R.T. Ross (eds.), *Philosophical Works of Descartes*, 2nd rev. ed., Cambridge: University Press, 1931.

Dixon, N.F.
1966   'The Beginnings of Perception'; reprinted in Brian M. Foss (ed.), *New Horizons in Psychology 1*, Harmondsworth: Penguin, 1972.

Duhem, Pierre
1906   *La Théorie Physique: Son Objet, Sa Structure*; 2nd French ed., tr. by Philip P. Wiener, as *The Aim and Structure of Physical Theory*, Princeton: University Press, 1954.

1908   ΣΩZEIN TA ΦAINOMENA: *Essai sur la notion de*

théorie physique de Platon à Galilée; tr. by Edmund Doland and Chaninah Maschler, as *To Save the Phenomena: An Essay on the Idea of Physical Theory from Plato to Galileo*, Chicago: University Press, 1969.

Earman, John, and Glymour, Clark

1980a   'Relativity and Eclipses: The British Eclipse Expeditions of 1919 and Their Predecessors', *Historical Studies in the Physical Sciences*, 11, pp. 49–85.

1980b   'The Gravitational Redshift as a Test of General Relativity: History and Analysis', *Studies in the History and Philosophy of Science*, 11, pp. 175–214.

Einstein, Albert

1920   *Relativity, the Special and General Theory: A Popular Exposition*, tr. by Robert W. Lawson, London: Methuen; reprinted in Methuen's University Paperback series, 1960.

1949   'Autobiographical Notes', in Schilpp (ed.), *1949*, pp. 2–95.

Evans-Pritchard, Edward E.

1937   *Witchcraft and Magic Among the Azande*, Oxford: Clarendon Press.

Feigl, Herbert

1950   'De Principiis non disputandum ... ?', in Black (ed.), *1950*, pp. 113–147.

Feigl, Herbert, and Maxwell, Grover

1961   (eds.), *Current Issues in the Philosophy of Science*, New York: Holt, Rinehart & Winston.

Feigl, Herbert, Scriven, Michael, and Maxwell, Grover

1958   (eds.), *Minnesota Studies in the Philosophy of Science*, vol. 2, Minneapolis: University of Minnesota Press.

Festinger, L., Reicken, H.W., Jr., and Schachter, S.

1956   *When Prophecy Fails*, Minnesota: University Press.

Feyerabend, Paul K.

1974   'Zahar on Einstein', *British Journal for the Philosophy of Science*, 25, pp. 25–28.

1975   *Against Method*, London: NLB.

Feyerabend, Paul K., and Maxwell, Grover

1966   (eds.), *Mind, Matter and Method: Essays in Philosophy and Science in Honor of Herbert Feigl*, Minneapolis: University of Minnesota Press.

Fisher, Seymour, and Greenberg, Roger D.

1977   *The Scientific Credibility of Freud's Theories and Therapy*, Hassocks: Harvester.

Frege, Gottlob

1892   'Über Sinn und Bedeutung', *Zeitschrift für Philosophie und Philosophische Kritik*, 100, pp. 25–50; tr. as 'On Sense and Reference' in Frege, *1952*, pp. 56–78.

1952   *Translations from the Philosophical Writings of Gottlob Frege*, Peter Geach and Max Black (eds.), Oxford: Basil Blackwell.

Freud, Sigmund

1930   *Civilization and Its Discontents*, in Freud, *1966–74*, vol. 21, pp. 64–145.

1966–74   *The Standard Edition of the Complete Psychological Works of Sigmund Freud*, James Strachey (ed.), 24 vols., London: Hogarth.

Gellner, Ernest A.

1968   'The New Idealism—Cause and Meaning in the Social Sciences', in Lakatos and Musgrave (eds.), *1968*, pp. 377–406.

1974   *Legitimation of Belief*, Cambridge: University Press.

Giedymin, Jerzy

1982   *Science and Convention*, Oxford: Pergamon Press.

Gilbert, William

1600   *De Magnete*; tr. by P. Fleury Mottelay, 1893; reprinted, Dover, 1958.

Gillies, Donald A.

1973   *An Objective Theory of Probability*, London: Methuen.

Glanvill, Joseph

1676   *Essays on Several Important Subjects in Philosophy and Religion*, London: Facsimile reproduction, Friedrich Fromman Verlag, 1970.

Glymour, Clark

1970   'On Some Patterns of Reduction', *Philosophy of Science*, 37, pp. 340–353.

1980   *Theory and Evidence*, Princeton: University Press.

Goodman, Nelson

1947   'The Problem of Counterfactual Conditionals', *Journal of Philosophy*, February 1947; reprinted in Goodman, *1954*, pp. 13–34.

BIBLIOGRAPHY

1954   *Fact, Fiction and Forecast*, London: Athlone Press.

Graves, John C.
1974   'Uniformity and Induction', *British Journal for the Philosophy of Science*, 25, pp. 301–318.

Gregory, Richard L.
1972   *Eye and Brain: the Psychology of Seeing*, 2nd ed. (1st ed. 1966), London: Weidenfeld and Nicholson.

Grünbaum, Adolf
1960   'The Duhemian Argument', *Philosophy of Science*, 27, pp. 75–87.
1971   'Can We Ascertain the Falsity of a Scientific Hypothesis?' in Mandelbaum (ed.), *1971*, pp. 69–129.
1976   'Is Falsifiability the Touchstone of Scientific Rationality? Karl Popper versus Inductivism', in Cohen, Feyerabend, and Wartofsky (eds.), *1976*, pp. 213–250.
1983   'Can Psychoanalytic Theory be Cogently Tested "on the Couch"?', in Laudan (ed.), *1983*, pp. 143–309.

Hahn, Hans, Neurath, Otto, and Carnap, Rudolf
1929   'Wissenschaftliche Weltauffassung: Der Wiener Kreis'; tr. as 'The Scientific Conception of the World: The Vienna Circle', in Neurath, *1973*, pp. 299–318.

Hamlyn, David
1970   *The Theory of Knowledge*, London: Macmillan.

Hare, Richard M.
1952   *The Language of Morals*, Oxford: Clarendon Press.

Harris, John H.
1974   'Popper's Definitions of Verisimilitude', *British Journal for the Philosophy of Science*, 25, pp. 160–166.

Harrison, Jonathan
1962   'Knowing and Promising', *Mind*, 71, pp. 443–457.

Hayek, Friedrich A.
1963   'Rules, Perception and Intelligibility', *Proceedings of the British Academy*, 48; reprinted in Hayek, *1967*, pp. 43–63.
1967   *Studies in Philosophy, Politics and Economics*, London: Routledge & Kegan Paul.

Heath, Peter
1955   'Intentions', *Aristotelian Society*, supplementary vol. 29, pp. 147–164.

BIBLIOGRAPHY

Helmholtz, Hermann von
1977   *Hermann Helmholtz: Epistemological Writings*, Robert
            S. Cohen and Yehuda Elkana (eds.), Dordrecht: Reidel.

Hempel, Carl G.
1945   'Studies in the Logic of Confirmation', *Mind*, 54,
            pp. 1–26 and 97–121; reprinted in Hempel, *1965*, pp.
            3–46.
1965   *Aspects of Scientific Explanation*, New York: Free Press.
1966   *Philosophy of Natural Science*, Englewood Cliffs, N.J.:
            Prentice Hall.

Hesse, Mary
1961   *Forces and Fields*, London: Nelson.
1974   *The Structure of Scientific Inference*, London: Macmil-
            lan.

Hilpinen, Risto
1968   *Rules of Acceptance and Inductive Logic*, Amsterdam:
            North-Holland.

Hintikka, Jaakko
1965a   'Distributive Normal Forms in First Order Logic', in
            Crossley and Dummett (eds.), *1965*, pp. 48–91; re-
            printed in Hintikka, *1973*, pp. 242–286.
1965b   'Towards a Theory of Inductive Generalisation', in Bar-
            Hillel (ed.), *1965*, pp. 274–288.
1968   'Induction by Enumeration and Induction by Elimi-
            nation', in Lakatos (ed.), *1968*, pp. 191–216.
1973   *Logic, Language-Games and Information*, Oxford: Clar-
            endon Press.
1974   *Knowledge and the Known*, Dordrecht: Reidel.

Holton, Gerald J.
1970   'Mach, Einstein, and the Search for Reality', in Cohen
            and Seeger (eds.), *1970*, pp. 165–199.
1973   *Thematic Origins of Scientific Thought: Kepler to Ein-
            stein*, Cambridge: Harvard University Press.

Holton, Gerald, and Roller, Duane H.D.
1958   *Foundations of Modern Physical Science*, London: Ad-
            dison-Wesley.

Horwich, Paul
1978   'A Peculiar Consequence of Nicod's Criterion', *British
            Journal for the Philosophy of Science*, 29, pp. 262–263.

Howson, Colin

1973   'Must the Logical Probability of Laws be Zero?', *British Journal for the Philosophy of Science*, 24, pp. 153–163.

Hübner, Kurt

1978   'Some Critical Comments on Current Popperianism on the Basis of a Theory of System Sets' in Radnitzky and Andersson (eds.), *1978*, pp. 279–289.

Hume, David

1739–40   *A Treatise of Human Nature*, book 1: 'Of the Understanding', book 2: 'Of the Passions', book 3: 'Of Morals', in Selby-Bigge (ed.), *1888*.

1745   *A letter from a Gentleman to His Friend in Edinburgh*; Ernest C. Hassner and John V. Price (eds.), Edinburgh: University Press, 1967.

1748   *Enquiries Concerning the Human Understanding*; in Selby-Bigge (ed.), *1902*.

Husserl, Edmund

1931   *Méditations Cartésiennes*; tr. by Dorion Cavins, as *Cartesian Meditations*, The Hague: Martinus Nijhoff, 1973.

Jaffe, Bernard

1960   *Michelson and the Speed of Light*, New York: Doubleday.

Jeans, James

1947   *The Growth of Physical Science*, Cambridge: University Press.

Jeffrey, Richard C.

1965   *The Logic of Decision*, New York: McGraw Hill.

1968   'Probable Knowledge', in Lakatos (ed.), *1968*, pp. 166–190.

Jeffreys, Harold

1939   *Theory of Probability*, Oxford: Clarendon Press.

1948   *Theory of Probability*, 2nd rev. ed. of Jeffreys, *1939*, Oxford: Clarendon Press.

1957   *Scientific Inference*, Cambridge: University Press.

1961   *Theory of Probability*, 3rd rev. ed. of Jeffreys, *1939*, Oxford: Clarendon Press.

Jeffreys, Harold, and Wrinch, Dorothy

1921   'On Certain Fundamental Principles of Scientific Enquiry', *Philosophical Magazine*, 42, pp. 269–298.

Jungk, Robert
1956   *Heller als tausend Sonnen*; tr. as *Brighter than 1000 Suns*, London: Gollancz and Hart-Davis, 1958.

Kant, Immanuel
1781/87   *Kritik der reinen Vernunft*, 1st ed. 1781, 2nd ed. 1787; tr. by Norman Kemp Smith, as *Immanuel Kant's Critique of Pure Reason*, London: Macmillan, 1953. References (e.g. A66 = B91) are to the pages of the 1st and 2nd German editions.

1783   *Prolegomena zu einer jeden künftigen Metaphysik die als Wissenschaft wird auftreten können*; tr. by Peter G. Lucas, as *Prolegomena to any future Metaphysics that will be able to present itself as a Science*, Manchester: University Press, 1953.

Kekes, John
1976   *A Justification of Rationality*, Albany: SUNY Press.

Kemeny, John G.
1953   'The Use of Simplicity in Induction', *The Philosophical Review*, 62, pp. 391–408.

Kemeny, John G., and Oppenheim, Paul
1956   'On Reduction', *Philosophical Studies*, 7, pp. 6–19.

Kepler, Johannes
1605   Letter to Herwart von Hohenberg, February 10, 1605; quoted in Holton, *1973*, p. 72.

Keynes, John Maynard
1921   *A Treatise on Probability*, London: Macmillan.

Koertge, Noretta
1969   *A Study of Relations between Scientific Theories: A Test of the General Correspondence Principle*, Ph.D. Thesis, University of London.

Krajewski, Wladyslaw
1977   *Correspondence Principle and Growth of Science*, Dordrecht: Reidel.

Kuhn, Thomas S.
1961   'The Function of Measurement in Modern Physical Science', *Isis*, 52, pp. 161–190; reprinted in Kuhn, *1977*, pp. 178–224.

1962   *The Structure of Scientific Revolutions*, Chicago: University Press.

1977   *The Essential Tension*, Chicago: University Press.

Kyburg, Henry E., Jr.
1965   'Probability, Rationality and a Rule of Detachment', in Bar-Hillel (ed.), *1965*, pp. 301–310.
1968   'The Rule of Detachment in Inductive Logic', in Lakatos (ed.), *1968*, pp. 98–119.
Kyburg, Henry E., Jr., and Nagel, Ernest
1963   (eds.), *Induction: Some Current Issues*, Middletown, Conn.: Wesleyan University Press.
Lakatos, Imre
1962   'Infinite Regress and Foundations of Mathematics', *Aristotelian Society Supplementary Volume*, 36, pp. 155–84; reprinted in Lakatos, *1978*, ii, pp. 3–23.
1968   (ed.), *The Problem of Inductive Logic*, Amsterdam: North Holland.
1968   'Changes in the Problem of Inductive Logic' in Lakatos (ed.), *1968*, pp. 315–417; reprinted in Lakatos, *1978*, i, pp. 128–200.
1970   'Falsification and the Methodology of Scientific Research Programmes' in Lakatos and Musgrave (eds.), *1970*, pp. 91–196; reprinted in Lakatos, *1978*, i, pp. 8–101.
1971   'History of Science and its Rational Reconstruction', in Roger C. Buck and Robert S. Cohen (eds.), *Boston Studies in the Philosophy of Science*, 8, pp. 91–135; reprinted in Lakatos, *1978*, i, pp. 102–138.
1974   'Popper on Demarcation and Induction', in Schilpp (ed.), *1974*, pp. 241–273; reprinted in Lakatos, *1978*, i, pp. 139–167.
1978   *Philosophical Papers*, vol. 1: *The Methodology of Scientific Research Programmes*, vol. 2: *Mathematics, Science and Epistemology*, John Worrall and Gregory Currie (eds.), Cambridge: University Press.
Lakatos, Imre, and Musgrave, Alan
1968   (eds.), *Problems in the Philosophy of Science*, Amsterdam: North Holland.
1970   (eds.), *Criticism and the Growth of Knowledge*, Cambridge: University Press.
Laplace, Pierre S.
1798–1825 *Traité de Mécanique Celeste*, 5 vols., Paris: Bachelier.
1812   *Essai Philosophique sur les Probabilités*, tr. by F.W. Trus-

cott and F.L. Emory, as *A Philosophical Essay on Probabilities*, New York: Dover, 1951.

1824   *Exposition du Système du Monde*, Paris: Bachelier, tr. by Henry H. Harte, as *The System of the World*, 2 vols., Dublin: University Press, 1830.

Laudan, Larry

1977   *Progress and its Problems*, London: Routledge & Kegan Paul.

1983   (ed.), *Mind and Medicine: Explanation and Evaluation in Psychiatry and Medicine*, Berkeley: University of California Press.

Leibniz, Gottfried Wilhelm

1678   Letter to Herman Conring, 19 March 1678, in Loemker (ed.), *1956*, pp. 284–293.

1679   Letter to John Frederick, Duke of Brunswick-Hanover, in Loemker (ed.), *1956*, pp. 397–403.

1710   Letter to Des Bosses, in Loemker (ed.), *1956*, pp. 973–974.

1715–16   'The Controversy between Leibniz and Clarke', in Loemker (ed.), *1956*, pp. 1095–1169.

Lenin, V.I.

1908   *Materialism and Empirio-Criticism*, in Lenin, *Collected Works*, vol xiii, London: Lawrence and Wishart, 1938.

Le Roy, Édouard

1899–1900   'Science et Philosophie', *Revue de Métaphysique et de Morale*, 7, pp. 375–425, 503–562, 708–731; 8, pp. 37–72.

Levinson, Paul

1982   (ed.), *In Pursuit of Truth: Essays in Honour of Karl Popper's 80th Birthday*, Hassocks: Harvester.

Lévy-Bruhl, Lucien

1927   *L'Ame Primitive*; tr. by Lilian A. Clare, as *The 'Soul' of the Primitive* (intro. by E.E. Evans-Pritchard), London: Allen & Unwin, 1965.

Lewis, Clarence I.

1946   *An Analysis of Knowledge and Valuation*, La Salle: Open Court.

Locke, John

1690   *An Essay Concerning Human Understanding*; A.S. Pringle-Pattison (ed.), Oxford: Clarendon Press, 1924. References (e.g. II, iv, 6) are to book, chapter, and section.

Loemker, Leroy E.
1956   (ed.), *Gottfried Wilhelm Leibniz: Philosophical Papers and Letters*, 2 vols., Chicago: University Press.

Lorenz, Konrad
1965   *Evolution and Modification of Behavior*, Chicago: University Press.

Lukasiewicz, Jan, and Tarski, Alfred
1930   'Untersuchungen über den Aussagenkalkül'; English translation in Tarski, *1956*, pp. 38–59.

Mach, Ernst
1872   *Die Geschichte und die Wurzel des Satzes von der Erhaltung der Arbeit*; tr. by Philip E.B. Jourdain, as *History and Root of the Principle of the Conservation of Energy*, Chicago: Open Court, 1911.

1883   *Die Mechanik in ihrer Entwickelung*, 9th German ed. 1933; tr. by Thomas J. McCormack, as *Science of Mechanics*, 5th English ed., La Salle: Open Court, 1942.

1886   *Beiträge zur Analyse der Empfindungen*, 5th German ed. 1906; tr. by C.M. Williams and Sydney Waterlow, as *The Analysis of Sensations*, (intro. by S. Szasz), New York: Dover, 1959.

1906   *Erkenntnis und Irrtum*, 2nd ed. (1st ed., 1905); 5th German ed. 1926; tr. by Thomas J. McCormack and Paul Foulkes, as *Knowledge and Error*, (intro. by Erwin N. Hiebert), (Vienna Circle Collection, vol. 3), Dordrecht: Reidel, 1976.

MacIntyre, Alasdair C.
1959   'Hume on "Is" and "Ought" ', *The Philosophical Review*, 68, pp. 451–468.

Magee, Bryan
1971   (ed.), *Modern British Philosophy*, London: Secker and Warburg.

Malcolm, Norman
1942   'Moore and Ordinary Language', in Schilpp (ed.), *1942*, pp. 345–368.

1952   'Knowledge and Belief', *Mind*, 61, reprinted in Malcolm, *1963*, pp. 58–72.

1963 *Knowledge and Certainty*, Englewood Cliffs, N.J.: Prentice Hall.

Mandelbaum, Maurice

1964 *Philosophy, Science and Sense Perception*, Baltimore: Johns Hopkins Press.

1971 (ed.), *Observation and Theory in Science*, Baltimore: Johns Hopkins Press.

Maxwell, James Clerk

1890 *The Scientific Papers of James Clerk Maxwell*, W.D. Niven (ed.), 2 vols., Cambridge: University Press.

Mill, John Stuart

1843 *A System of Logic*, 8th ed., 1872, in J.M. Robson (ed.), *Collected Works of John Stuart Mill*, vols. 7 and 8, Toronto: University Press, 1973–4. References (e.g. III, iii, 1) are to book, chapter, and section.

Miller, David W.

1974a 'Popper's Qualitative Theory of Verisimilitude', *British Journal for the Philosophy of Science*, 25, pp. 166–177.

1974b 'On the Comparison of False Theories by their Bases', *British Journal for the Philosophy of Science*, 25, pp. 178–188.

1975 'The Accuracy of Predictions', *Synthese*, 30, pp. 159–191.

1980 'Science without Induction', in Cohen and Hesse (eds.), *1980*, pp. 111–129.

1982 'Conjectural Knowledge: Popper's Solution of the Problem of Induction', in Levinson (ed.), *1982*, pp. 17–49.

Mises, Richard von

1928 *Wahrscheinlichkeit, Statistik und Wahrheit*; tr. by Hilda Geiringer, as *Probability, Statistics and Truth*, London: Allen & Unwin, 1957.

Moore, George E.

1903 *Principia Ethica*, Cambridge: University Press.

1959 *Philosophical Papers*, London: Allen & Unwin.

Musgrave, Alan E.

1974 'Logical versus Historical Theories of Confirmation', *British Journal for the Philosophy of Science*, 25, pp. 1–23.

1975 'Popper and "Diminishing Returns from Repeated

Tests" ', *Australasian Journal of Philosophy*, 53, pp. 248–253.

1976 'Method or Madness', in Cohen, Feyerabend, and Wartofsky (eds.), *1976*, pp. 457–491.

1978 'Evidential Support, Falsification, Heuristics, and Anarchism', in Radnitzky and Andersson (eds.), *1978*, pp. 181–201.

Nagel, Ernest

1961 *The Structure of Science,* London: Routledge & Kegan Paul.

1963 'Carnap's Theory of Induction', in Schilpp (ed.), *1963*, pp. 785–825.

Neurath, Otto

1933 'Protokollsätze', *Erkenntnis*, 3, pp. 204–214; tr. as 'Protocol Sentences', in Ayer (ed.), *1959*, pp. 199–208.

1973 *Otto Neurath: Empiricism and Sociology*, Robert S. Cohen and Marie Neurath (eds.), (Vienna Circle Collection, vol. 1), Dordrecht: Reidel.

Newton, Isaac

1729 *Mathematical Principles of Natural Philosophy*, tr. by Andrew Motte; revised translation by Florian Cajori; Berkeley: University of California Press, 1947.

1959–77 *The Correspondence of Isaac Newton*, 7 vols.; vols. 1–3, H.W. Turnbull (ed.); vol. 4, J.F. Scott (ed.); vols. 5–7, A. Rupert Hall and Laura Tilling (eds.); Cambridge: University Press.

Newton-Smith, William H.

1981 *The Rationality of Science*, London: Routledge & Kegan Paul.

Niiniluoto, Ilkka

1977 'On the Truthlikeness of Generalizations', in Butts and Hintikka (eds.), *1977b*, pp. 121–147.

Oakeshott, Michael

1962 *Rationalism in Politics*, London: Methuen.

Oddie, Graham James

1979 *The Comparability of Theories by Verisimilitude and Content*, Ph.D. Thesis, University of London.

1981 'Verisimilitude Reviewed', *British Journal for the Philosophy of Science*, 32, pp. 237–265.

O'Hear, Anthony
1975 'Rationality of Action and Theory-Testing in Popper', *Mind*, 84, pp. 273–283.
1980 *Karl Popper*, London: Routledge and Kegan Paul.

Oppenheim, Paul, and Putnam, Hilary
1958 'Unity of Science as a Working Hypothesis', in Feigl, Scriven, and Maxwell (eds.), *1958*, pp. 3–36.

Pannekoek, Anton
1951 *De Groei van ons Werelbeeld*; tr. as *A History of Astronomy*, New York: Interscience Publishers, 1961.

Pascal, Blaise
1670 *Pensées sur la Religion*; tr. by H.F. Stewart, as *Pascal's Pensées*, London: Routledge & Kegan Paul, 1950.

Passmore, John
1978 *Science and its Critics*, London: Duckworth.

Peirce, Charles Sanders
1891 'The Architecture of Theories', *The Monist*, 1, pp. 161–176; in Peirce, *1931–58*, vol. 6, chap. 1.
1892 'The Doctrine of Necessity Examined', *The Monist*, 2, pp. 321–337; in Peirce, *1931–58*, vol. 6, chap. 2.
1893 'The Fixation of Belief' (with revisions), in Peirce, *1931–58*, vol. 5, pp. 223–247.
1901 'The Logic of Drawing History from Ancient Documents', in Peirce, *1931–58*, vol. 7, chap. 3.
1931–58 *Collected Papers of Charles Sanders Peirce*, 8 vols.; vols. 1–6, Charles Hartshorne and Paul Weiss (eds.); vols. 7–8, Arthur W. Burks (ed.), Cambridge: Harvard University Press. A reference, e.g. 6.526, is to vol. 6, paragraph 526.

Planck, Max
1948 *Wissenschaft Selbstbiographie*; tr. by F. Gaynor as *Scientific Autobiography and Other Papers*, London: Greenwood Press, 1949.

Poincaré, Henri
1902 *La Science et l'hypothèse*; tr. by G.B. Halsted, 1913, as *Science and Hypothesis*; reprinted, New York: Dover, 1952.
1905 *La Valeur de la Science*; tr. by G.B. Halsted, 1913, as *The Value of Science*; reprinted, New York: Dover, 1958.

BIBLIOGRAPHY

Polanyi, Michael
1952   'The Stability of Beliefs', *British Journal for the Philosophy of Science*, 3, pp. 217–232.
1958   *Personal Knowledge*; 2nd rev. ed., 1962, London: Routledge & Kegan Paul.
Popkin, Richard H.
1960   *The History of Scepticism from Erasmus to Descartes*, Assen: Van Gorcum.
1979   *The History of Scepticism from Erasmus to Spinoza*, Berkeley: University of California Press. (Revised and expanded edition of Popkin, *1960*.)
Popper, Karl R.
1934   *Logik der Forschung*; English translation in Popper, *1959*.
1940   'What is Dialectic?' *Mind*, 49, pp. 403–426; reprinted in Popper, *1963*, pp. 312–335.
1945   *The Open Society and Its Enemies*, 2 vols.; 4th rev. ed., 1962, London: Routledge & Kegan Paul.
1957   *The Poverty of Historicism*, London: Routledge & Kegan Paul.
1957   'The Aim of Science', *Ratio*, 1; reprinted in Popper, *1972*, pp. 191–205.
1959   *The Logic of Scientific Discovery*, tr. of Popper, *1934*, with new preface, footnotes, and appendices, London: Hutchinson.
1962   'Facts, Standards and Truth', addendum to 4th ed. of Popper *1945*, ii, pp. 369–396.
1963   *Conjectures and Refutations*, London: Routledge & Kegan Paul.
1966   *Of Clouds and Clocks*, St. Louis: Washington University; reprinted in Popper, *1972*, pp. 206–255.
1966   'A theorem on Truth-Content', in Feyerabend and Maxwell (eds.), *1966*, pp. 343–353.
1968   'On Rules of Detachment and So-Called Inductive Logic', in Lakatos (ed.), *1968*, pp. 130–139.
1972   *Objective Knowledge*, Oxford: Clarendon Press.
1974a   'Intellectual Autobiography', in Schilpp (ed.), *1974*, pp. 3–181.
1974b   'Replies to My Critics', in Schilpp (ed.), *1974*, pp. 961–1197.

373

1982a   *Realism and the Aim of Science*, W.W. Bartley, III (ed.), London: Hutchinson.

1982b   *The Open Universe: an Argument for Indeterminism*, W.W. Bartley, III (ed.), London: Hutchinson.

Popper, Karl R., and Eccles, John C.

1977   *The Self and Its Brain*, Berlin: Springer-Verlag.

Porter, P.B.

1954   'The Hidden Man', *American Journal of Psychology*, 69, pp. 550–51.

Post, Heinz

1971   'Correspondence, Invariance and Heuristics', *Studies in the History and Philosophy of Science*, 2, pp. 213–255.

Putnam, Hilary

1963   ' "Degree of Confirmation" and Inductive Logic', in Schilpp (ed.), *1963*, pp. 761–783; reprinted in Putnam, *1979*, i, pp. 270–292.

1974   'The "Corroboration" of Theories', in Schilpp (ed.), *1974*, pp. 221–240; reprinted in Putnam, *1979*, i, pp. 250–269.

1979   *Philosophical Papers*, 2nd. ed., vol. 1: *Mathematics, Matter and Method*; vol. 2: *Mind, Language and Reality*, Cambridge: University Press.

Quine, Willard Van Orman

1951   'Two Dogmas of Empiricism', *The Philosophical Review*, 60, pp. 20–43; reprinted in Quine, *1953*, pp. 20–46.

1953   *From a Logical Point of View*, Cambridge: Harvard University Press.

1960   *Word and Object*, Cambridge: M.I.T. Press.

1966   *The Ways of Paradox and Other Essays*, New York: Random House.

Radnitzky, Gerard, and Andersson, Gunnar

1978   (eds.), *Progress and Rationality in Science*, Dordrecht: Reidel.

Railton, Peter

1978   'A Deductive-Nomological Model of Probabilistic Explanation', *Philosophy of Science*, 45, pp. 206–226.

Ramsey, Frank P.

1931   *The Foundations of Mathematics*, R.B. Braithwaite (ed.), London: Routledge & Kegan Paul.

1931   'Theories' in Ramsey, *1931*, pp. 212–236.

Reichenbach, Hans
  1938   *Experience and Prediction*, Chicago: University Press.
Reid, Thomas
  1763   Letter to David Hume, 18 March 1763, in Reid, *1895*,
                pp. 91–92.
  1895   *The Works of Thomas Reid D.D.*, William Hamilton
                (ed.), 2 vols., 8th ed., London: Longmans, Green.
Rey, Abel
  1907   *La Théorie Physique chez les physiciens contemporains*,
                Paris: F. Alcan.
Robinson, Richard
  1964   *An Atheist's Values*, Oxford: Clarendon Press.
Rose, Hilary, and Rose, Stephen
  1976a   (eds.), *The Radicalisation of Science*, London: Macmil-
                lan.
  1976b   (eds.), *The Political Economy of Science*, London: Mac-
                millan.
Russell, Bertrand
  1914   *Our Knowledge of the External World*; 2nd rev. ed. 1926,
                London: Allen & Unwin.
  1917   *Mysticism and Logic*, London: Allen & Unwin.
  1919   *Introduction to Mathematical Philosophy*; 2nd ed., Lon-
                don: Allen & Unwin, 1920.
  1946   *Hı .ory of Western Philosophy*, London: Allen & Unwin.
  1948   *Human Knowledge: Its Scope and Limits*, London: Allen
                & Unwin.
Ryle, Gilbert
  1949   *The Concept of Mind*, London: Hutchinson.
Sabra, A.I.
  1967   *Theories of Light from Descartes to Newton*, London:
                Oldbourne.
Salmon, Wesley
  1957   'Should we attempt to Justify Induction?', *Philosophical
                Studies*, 8, pp. 45–47.
  1961   'Vindication of Induction', in Feigl and Maxwell (eds.),
                1961, pp. 245–256.
  1963a   'On Vindicating Induction', in Kyburg and Nagel (eds.),
                1963, pp. 27–41.
  1963b   'Inductive Inference', in Baumrin (ed.), *1963*, pp. 341–
                368.

1965   'Consistency, Transitivity, and Inductive Support', *Ratio, 7*, pp. 164–169.

1966   *The Foundations of Scientific Inference*, 2nd ed. 1967, Pittsburgh: University Press.

1968   'The Justification of Inductive Rules of Inference', in Lakatos (ed.), *1968*, pp. 24–43.

1971   (with Richard Jeffrey and James G. Greene), *Statistical Explanation and Statistical Relevance*, Pittsburgh: University Press.

1981   'Rational Prediction', *British Journal for the Philosophy of Science*, 32, pp. 115–125.

Santayana, George

1923   *Scepticism and Animal Faith*; reprinted, New York: Dover, 1955.

Schaffner, Kenneth F.

1967   'Approaches to Reduction', *Philosophy of Science*, 34, pp. 137–147.

Scheffler, Israel

1961   'A Rejoinder on Confirmation', *Philosophical Studies*, 12, pp. 19–20.

1963   *The Anatomy of Inquiry*, New York: Alfred A. Knopf.

Schilpp, Paul Arthur

1942   (ed.), *The Philosophy of G.E. Moore* (The Library of Living Philosophers), Chicago: Northwestern University.

1949   (ed.), *Albert Einstein: Philosopher-Scientist* (The Library of Living Philosophers), Evanston; reprinted New York: Harper.

1963   (ed.), *The Philosophy of Rudolf Carnap* (The Library of Living Philosophers), La Salle: Open Court.

1974   (ed.), *The Philosophy of Karl Popper* 2 vols. (The Library of Living Philosophers), La Salle: Open Court.

Schlick, Moritz

1918   *Allgemeine Erkenntnislehre*; 2nd German ed. 1925; tr. by Albert E. Blumberg, as *General Theory of Knowledge*, (intro. by Albert E. Blumberg and Herbert Feigl), New York: Springer-Verlag, 1974.

1921   Notes and Comments, in Hermann von Helmholtz, *Schriften zur Erkenntnistheorie*; tr. by Malcolm F. Lowe, in Cohen and Elkana (eds.), *1977*.

1926   'Erleben, Erkennen, Metaphysik'; trans. Peter Heath, as 'Experience, Cognition and Metaphysics', in Schlick, *1979*, ii, pp. 99–111.

1931   'Die Kausalität in der gegenwärtigen Physik', *Die Naturwissenschaften*, 19, pp. 145–162; tr. by Peter Heath, as 'Causality in Contemporary Physics', in Schlick, *1979*, ii, pp. 176–209.

1979   *Philosophical Papers*, Henk L. Mulder and Barbara F.B. van de Velde-Schlick (eds.), 2 vols., (Vienna Circle Collection, vol. 11) Dordrecht: Reidel.

Scriven, Michael

1959   'Explanation and Prediction in Evolutionary Theory', *Science*, 130, pp. 477–482.

Selby-Bigge, L.A.

1888   (ed.), *Hume's Treatise*, Oxford: Clarendon Press.

1902   (ed.), *Hume's Enquiries*, 2nd ed., Oxford: Clarendon Press.

Shimony, Abner

1970   'Scientific Inference' in Colodny R.G. (ed.), *1970*, pp. 79–172.

Sobocinski, B.

1955–56   'On Well Constructed Axiom Systems', *Rocznik Polskiego Towarzystwa Naukowego na Obczyznie*, 6, pp. 54–70.

Sommers, Fred

1963   'Types and Ontology', *Philosophical Review*, 72, pp. 327–363.

Stegmüller, Wolfgang

1977   *Collected Papers on Epistemology, Philosophy of Science and History of Philosophy*, 2 vols., Dordrecht: Reidel.

1980   'Two Successor Concepts to the Notion of Statistical Explanation', in Wright (ed.), *1980*, pp. 37–52.

Stove, David C.

1973   *Probability and Hume's Inductive Scepticism*, Oxford: Clarendon Press.

Strawson, Peter R.

1952   *Introduction to Logical Theory*, London: Methuen.

1958   'On Justifying Induction', *Philosophical Studies*, 9, pp. 20–21.

1959   *Individuals*, London: Methuen.

Szumilewicz, Irena
1977   'Incommensurability and the Rationality of the Development of Science', *British Journal for the Philosophy of Science*, 25, pp. 156–160.

Tarski, Alfred
1930   'Fundementale Begriffe der Methodologie der deduktiven Wissenschaften', *Monatschefte für Mathematik und Physik*, 37, pp. 361–404; translated in Tarski, 1956, pp. 60–109.
1956   *Logic, Semantics, Metamathematics*, tr. by J.H. Woodger, Oxford: Clarendon Press.

Tichý, Pavel
1974   'On Popper's Definitions of Verisimilitude', *British Journal for the Philosophy of Science*, 25, pp. 155–160.

Tuomela, Raimo
1973   *Theoretical Concepts*, New York: Springer-Verlag.

Unger, Peter
1975   *Ignorance: A Case for Scepticism*, Oxford: Clarendon Press.

Urbach, Peter
1981   'On the Utility of Repeating the "Same" Experiment', *Australasian Journal of Philosophy*, 59, pp. 151–162.

Van Fraassen, Bas C.
1980   *The Scientific Image*, Oxford: Clarendon Press.

Vernon, M.D.
1952   *A Further Study of Visual Perception*, Cambridge: University Press.

Waismann, Friedrich
1930–31   'Logische Analyse des Wahrscheinlichkeitsbegriffs'; tr. as 'The Logical Analysis of the Concept of Probability', in Waismann, 1977, pp. 4–21.
1977   *Philosophical Papers*, Brian McGuinness (ed.), (Vienna Circle Collection, vol. 8), Dordrecht: Reidel.

Walter, W. Grey
1953   *The Living Brain*; reprinted Harmondsworth: Penguin, 1961.

Watkins, John W. N.
1957   'Farewell to the Paradigm Case Argument', *Analysis*, 18, pp. 25–33.

1958   'Confirmable and Influential Metaphysics', *Mind*, 67, pp. 345–365.

1968a   'Non-Inductive Corroboration' in Lakatos (ed.), *1968*, pp. 61–66.

1968b   'Hume, Carnap and Popper', in Lakatos (ed.), *1968*, pp. 271–282.

1974   'Otto Neurath', *British Journal for the Philosophy of Science*, 25, pp. 343–352.

1975   'Metaphysics and the Advancement of Science', *British Journal for the Philosophy of Science*, 26, pp. 91–121.

1978a   'Minimal Presuppositions and Maximal Metaphysics', *Mind*, 87, pp. 195–209.

1978b   'The Popperian Approach to Scientific Knowledge', in Radnitzky and Andersson (eds.), *1978*, pp. 23–43.

1978c   'Corroboration and the Problem of Content-Comparison', in Radnitzky and Andersson (eds.), *1978*, pp. 339–378.

Westfall, Richard S.

1980   *Never At Rest: A Biography of Isaac Newton*, Cambridge: University Press.

Wheeler, John Archibald

1977   'Genesis and Observership', in Butts and Hintikka (eds.), *1977a*, pp. 3–33.

Whewell, William

1837   *History of the Inductive Sciences, from the Earliest to the Present Times*, 3 vols., London: Parker, 3rd ed. 1857.

1840   *The Philosophy of the Inductive Sciences, Founded upon their History*, 2 vols., London: Parker; new edition 1847.

White, Morton

1956   *Toward Reunion in Philosophy*, Cambridge: Harvard University Press.

Whittaker, Edmund T.

1951   *A History of the Theories of Aether and Electricity*; reprinted New York: Harper Torchbooks, 1960, 2 vols.

Wilkes, Kathleen

1980   'Brain States', *British Journal for the Philosophy of Science*, 31, pp. 111–129.

Wilson, Curtis A.

1969   'From Kepler's Laws, So-Called, to Universal Gravitation: Empirical Factors', *Archive for History of Exact Science*, 6, pp. 86–170.

Wittgenstein, Ludwig

1922 *Tractatus Logico-Philosophicus*, London: Routledge & Kegan Paul.

1953 *Philosophical Investigations*, tr. by G.E.M. Anscombe, Oxford: Blackwell.

1969 *On Certainty*, tr. by Denis Paul and G.E.M. Anscombe, Oxford: Blackwell.

Worrall, John

1978 'The Ways in Which the Methodology of Scientific Research Programmes Improves on Popper's Methodology', in Radnitzky and Andersson (eds.), *1978*, pp. 45–70.

Wright, George Henrik von

1980 (ed.), *Logic and Philosophy*, The Hague: Martinus Nijhoff.

Yoshida, Ron M.

1977 *Reduction in the Physical Sciences*, Halifax: Philosophy in Canada: Monograph Series 4.

Zahar, Elie

1973 'Why did Einstein's Programme supersede Lorentz's?', *British Journal for the Philosophy of Science*, 24, pp. 95–123 and 223–262.

INDEX OF SYMBOLS

. . . .

Library of Congress Cataloging in Publication Data

Watkins, John W. N. (John William Nevill)
    Science and scepticism.

    Bibliography: p.
    Includes indexes.
    1. Science—Philosophy. 2. Skepticism. 3. Rationalism. 4. Knowledge, The-
ory of. I. Title.
Q175.W298 1984   501   84-42555
ISBN 0-691-07294-9
ISBN 0-691-10171-X (pbk.)

John Watkins is Professor of Philosophy at The London School of Economics and
Political Science

Ingram Content Group UK Ltd.
Milton Keynes UK
UKHW020320200523
422043UK00007B/402